T0183956

Phenology of Ecosystem Processes

Asko Noormets

Editor

Phenology
of Ecosystem Processes

Applications in Global Change Research

 Springer

Editor
Asko Noormets
Department of Forestry and Environmental Resources
North Carolina State University
920 Main Campus Drive
Raleigh NC 27695
USA

ISBN 978-1-4899-8108-0   ISBN 978-1-4419-0026-5 (eBook)
DOI 10.1007/978-1-4419-0026-5
Springer Dordrecht Heidelberg London New York

© Springer Science+Business Media, LLC 2009
Softcover re-print of the Hardcover 1st edition 2009

Cover image: The "phenological clocks" of different processes (GEP – gross ecosystem productivity, ET – evapotranspiration, ER – ecosystem respiration) indicate the intensity of a given flux by the width of the colored band (A. Noormets and K. Kramer). The background is a mosaic of photos from Harvard Forest EMS Tower webcam, courtesy of Andrew Richardson.

Printed on acid-free paper

Springer is part of Springer Science+Business Media (www.springer.com)

Foreword

Phenology, the study of the timing of biological organisms and processes like leaf-out and flowering, has a long and rich history. Many of our best and longest records started with amateur scientists at their estates in Europe and on royal grounds in Asia; there our predecessors measured the timing of leaf out, flowering and the arrival of birds. Among the longest phenology on record are those recording the timing of the cherry bloom in Japan (8th Century), the timing of the wine harvest in France, from the 1300s (Chuine *et al.* 2004) and the timing of arrival of spring at the Marsham estate in Britain (1736-1947) (Sparks and Carey 1995).

In recent years phenology has gained resurgence in interest and importance; a web of science search reveals over 8000 citations under the keyword 'phenology' and about 5000 of these papers have been published in the last decade. One reason phenology is gaining importance is due to the fact that it is proving to be an independent record on global warming and global change. It is becoming widely documented that spring is occurring earlier and earlier at many locations across the globe due to global warming (Menzel 2001; Menzel *et al.* 2006). There is much concern about this trend because asynchronies may occur between pollinators and beneficial insects and between food sources for birds, insects and animals, etc. (Parmesan 2006).

A variety of new technologies are helping advance the study of phenology, and are contributing to this renaissance in the field. With the launching of Earth observing satellites we are producing a long record of changes in the greening of the biosphere (Myneni *et al.* 1997). Newer technologies, like web cameras (Richardson *et al.* 2007) and eddy covariance measurements (Gu *et al.* 2003; Baldocchi *et al.* 2005), are providing automated, continuous and areally-averaged measures of phenology. And these new technologies are being complemented by an expansion of phenology gardens and networks across the US, Europe and Asia, which manually study key indicator plants, like lilac (Schwartz 2003).

The current book will add new perspective and information on many of the new areas of phenology. In particular its has material that is not widely included in previous books on phenology (Schwartz 2003). For instance, it contains several chapters examining the connection between ecosystem carbon fluxes and phenology (Barr *et al.*, Billmark and Griffis), a chapter on the relationship between soil respiration and phenology (Davidson and Holbrook), several investigating the combined use of carbon flux measurements and remote sensing on detecting phenology (Xiao *et al.*, Reed *et al.*), another on the role of microclimate (Richardson and O'Keefe), and one on the roles of phenology on process-level forest modeling (Kramer and Hänninen). The information contained in this book will provide background material for understanding the consequence of changing phenology in terms of feedbacks between the biosphere and the climate system.

November 23, 2008   Dennis D. Baldocchi
University of California Berkeley

References

Baldocchi, D.D., Black, T.A., Curtis, P.S., Falge, E., Fuentes, J.D., Granier, A., Gu, L., Knohl, A., Pilegaard, K., Schmid, H.P., Valentini, R., Wilson, K., Wofsy, S., Xu, L. and Yamamoto, S. (2005) Predicting the onset of net carbon uptake by deciduous forests with soil temperature and climate data: a synthesis of FLUXNET data. Int. J. Biometeorol. 49, 377–387.

Chuine, I., Yiou, P., Viovy, N., Seguin, B., Daux, V. and Ladurie, E.L. (2004) Grape ripening as a past climate indicator. Nature 432, 289–290.

Gu, L., Post, W.M., Baldocchi, D.D., Black, T.A., Verma, S., Vesala, T. and Wofsy, S. (2003) Phenology of vegetation photosynthesis. In: Schwartz, M.D. (Ed.) *Phenology: An Integrative Science.* Kluwer Academic Publishers, Dordrecht. pp. 467–485.

Menzel, A. (2001) Trends in phenological phases in Europe between 1951 and 1996. Int. J. Biometeorol. 44, 76–81.

Menzel, A., Sparks, T.H., Estrella, N., Koch, E., Aasa, A., Ahas, R., Alm-Kubler, K., Bissolli, P., Braslavska, O., Briede, A., Chmielewski, F.M., Crepinsek, Z., Curnel, Y., Dahl, A., Defila, C., Donnelly, A., Filella, Y., Jatcza, K., Mage, F., Mestre, A., Nordli, O., Penuelas, J., Pirinen, P., Remisova, V., Scheifinger, H., Striz, M., Susnik, A., van Vliet, A.J.H., Wielgolaski, F.E., Zach, S. and Zust, A. (2006) European phenological response to climate change matches the warming pattern. Global Change Biol. 12, 1969–1976.

Myneni, R.B., Keeling, C.D., Tucker, C.J., Asrar, G. and Nemani, R.R. (1997) Increased plant growth in the northern high latitudes from 1981 to 1991. Nature 386, 698–702.

Parmesan, C. (2006) Ecological and evolutionary responses to recent climate change. Annu. Revi. Ecol. Evol. Syst. 37, 637–669.

Richardson, A.D., Jenkins, J.P., Braswell, B.H., Hollinger, D.Y., Ollinger, S.V. and Smith, M.L. (2007) Use of digital webcam images to track spring green-up in a deciduous broadleaf forest. Oecologia 152, 323–334.

Schwartz, M.D. (2003) *Phenology: An Integrative Enviornmental Science.* Kluwer Academic Publishers, Dordrecht, pp. 592.

Sparks, T. H. and Carey, P.D. (1995) The responses of species to climate over 2 centuries - an analysis of the Marsham phenological record, 1736-1947. J. Ecol. 83, 321–329.

Preface

The effect of warming temperatures on biological processes has been well documented (Badeck *et al.* 2004; Parmesan and Yohe 2003), and is evidenced by changes in the timing of discernible life cycle events, like leaf-out and flowering of plants, and migration and reproduction of animals. It is implicit that these life cycle events are representative indicators of a change in some underlying process. Ever more sophisticated general circulation and ecosystem productivity models have narrowed the boundaries of uncertainty sufficiently to bring attention to the effect of the seasonal timing of ecosystem processes, notably carbon and water exchange. It is becoming increasingly evident that both interannual and regional variation have a strong phenological component (Baldocchi 2008). The associated changes in surface energy balance and partitioning (Wilson and Baldocchi 2000) both affect and are driven by vegetation phenology (Alessandri *et al.* 2007; Baldocchi 2008; Morisette *et al.* 2008). Quantifying the seasonality of these processes is required for constraining ecosystem productivity models (Kramer *et al.* 2002), refining remote sensing (RS) estimates of ecosystem properties (Morisette *et al.* 2008) and narrowing the uncertainty bounds on global biogeochemical models (Olesen *et al.* 2007). While the vegetation-index-based assessments (e.g. Goetz *et al.* 2005) broadly corroborate ground-based observations of long-term trends of lengthening growing season (Menzel 2000, 2003; Menzel *et al.* 2005), the patterns of interannual variation in land surface reflectance and vegetation processes do not always coincide (Badeck *et al.* 2004; Fisher *et al.* 2007). We hypothesize that the power of RS monitoring of vegetation processes would be improved if the calibration of the reflectance data was done against the process of interest (as opposed to validating a RS gross productivity product against a degree-day model of bud-break, for example). This is all the more important when considering that even ground-based observations may yield conflicting results when data collected with different methods is compared, because they may entail different (and sometimes implicit) assumptions (Parmesan 2007). Furthermore, process-based approach is required because even closely related processes do not have the same environmental drivers and same sensitivities to them. For example, the onset of ecosystem respiration is generally delayed in relation to gross productivity in temperate deciduous and boreal conifer forests (Falge *et al.* 2002). While continuous in nature, the driving factors of these processes vary seasonally (Davidson and Holbrook, current volume; Carbone and

Vargas 2008). Thus, the changes in ecosystem processes, including biogeochemical fluxes, exhibit *phenological* change, as per the definition of phenology by Lieth (1974): *"Phenology is the study of the timing of recurrent biological events, the causes of their timing with regard to biotic and abiotic forces, and the interrelation among phases of the same or different species"*.

The recent increased interest in the seasonality of ecosystem processes has already revealed several novel aspects, some of which force us to reconsider earlier paradigms and assumptions. For example, the observed seasonality of tropical rainforest carbon balance has been found to be opposite to all earlier model predictions (Saleska *et al.* 2003) and strongly influenced by the degree of anthropogenic disturbance (Huete *et al.* 2008). The long-held view of urban heat island effect on the timing of bud-break is challenged by the latest global analysis (Gazal *et al.* 2008). And a common picture has emerged from previously divergent pieces of evidence about the effect of delayed autumn senescence on forest carbon balance (Piao *et al.* 2008). Several novel findings also emerge from the syntheses presented in the current volume. Notably, some phenological patterns seem reflected in diurnal cycles, potentially providing a novel insight into continuities across temporal scales. Billmark and Griffis (Chapter 6) report that the rate of morning increase in isotopic discrimination changes seasonally, whereas Davidson and Holbrook (Chapter 8) discuss how the diurnal hysteresis in the relationship between soil respiration and temperature indicates seasonal changes in the primary driving factor. In all, the chapters in the current volume present examples of how phenology is measured and considered in various analyses of ecosystem biogeochemical processes, give a brief overview of the background of each question, and propose new approaches for quantifying phenological patterns. The recognition of the urgency of climate change related issues (Gore 2006), the potential implications of disparate responses in ecologically related organisms (Fussmann *et al.* 2007; Parmesan 2007), and calls for more realistic representation of seasonal changes in regional climate models (Morisette *et al.* 2008), have brought much attention to phenology. We hope that the current collection of studies helps those new to the field get an overview of its scope, provides a reference to people active in the field, and serves as an educational aid for courses on climate change and ecosystem ecology. The current volume is not intended to present a comprehensive overview of the field of land surface phenology. The two chapters on this (10 and 11) only highlight the most common contact points with ecosystem ecology and provide an example of how these two approaches have been applied together. Upon completing this book, we hope the reader will develop his or her own vision of the seasonality of ecosystem processes, detectable as distinctly as the purple of an opening bud of a lilac.

Acknowledgements This book grew out from a session "Phenology and ecosystem processes" at the 91st Annual Meeting of the Ecological Society of America (ESA), held in Memphis, Tennessee (USA), in August, 2006. The session itself was inspired by a stimulating discussion with Mark Schwartz at the 8th Annual Meeting of the Chequamegon Ecosystem-Atmosphere Study (ChEAS) in 2005. The current volume includes four chapters based on the presentations made at the ESA meeting, and seven new contributions, significantly broadening the scope covered in 2006. Sincere thanks to Janet Slobodien of Springer for the invitation to develop the material presented at that meeting into this book. It has been a rewarding experience. I am grateful to the Southern Global Change Program of US Forest Service for support and accommodation throughout the preparation of this volume.

December 11, 2008   Asko Noormets
Raleigh

References

Alessandri, A., Gualdi, S., Polcher, J. and Navarra, A. (2007) Effects of land surface-vegetation on the boreal summer surface climate of a GCM. J. Clim. 20, 255–278.

Badeck, F.W., Bondeau, A., Bottcher, K., Doktor, D., Lucht, W., Schaber, J. and Sitch, S. (2004) Responses of spring phenology to climate change. New Phytol. 162, 295–309.

Baldocchi, D.D. (2008) 'Breathing' of the terrestrial biosphere: lessons learned from a global network of carbon dioxide flux measurement systems. Aust. J. Bot. 56, 1–26.

Carbone, M.S. and Vargas, R. (2008) Automated soil respiration measurements: new information, opportunities and challenges. New Phytol. 177, 295–297.

Falge, E., Baldocchi, D.D., Tenhunen, J., Aubinet, M., Bakwin, P.S., Berbigier, P., Bernhofer, C., Burba, G., Clement, R., Davis, K.J., Elbers, J.A., Goldstein, A.H., Grelle, A., Granier, A., Guddmundsson, J., Hollinger, D., Kowalski, A.S., Katul, G., Law, B.E., Malhi, Y., Meyers, T., Monson, R.K., Munger, J.W., Oechel, W., Paw U, K.T., Pilegaard, K., Rannik, Ü., Rebmann, C., Suyker, A., Valentini, R., Wilson, K. and Wofsy, S. (2002) Seasonality of ecosystem respiration and gross primary production as derived from FLUXNET measurements. Agric. For. Meteorol. 113, 53–74.

Fisher, J.I., Richardson, A.D. and Mustard, J.F. (2007) Phenology model from surface meteorology does not capture satellite-based greenup estimations. Global Change Biol. 13, 707–721.

Fussmann, G.F., Loreau, M. and Abrams, P.A. (2007) Eco-evolutionary dynamics of communities and ecosystems. Funct. Ecol. 21, 465–477.

Gazal, R., White, M.A., Gillies, R., Rodemaker, E., Sparrow, E. and Gordon, L. (2008) GLOBE students, teachers, and scientists demonstrate variable differences between urban and rural leaf phenology. Global Change Biol. 14, 1568–1580.

Goetz, S.J., Bunn, A.G., Fiske, G.J. and Houghton, R.A. (2005) Satellite-observed photosynthetic trends across boreal North America associated with climate and fire disturbance. Proc. Natl. Acad. Sci. 102, 13521–13525.

Gore, A. (2006) An Inconvenient Truth: The Planetary Emergency of Global Warming and What We Can Do About It. Rodale Books, New York, pp. 328.

Huete, A.R., Restrepo-Coupe, N., Ratana, P., Didan, K., Saleska, S.R., Ichii, K., Panuthai, S. and Gamo, M. (2008) Multiple site tower flux and remote sensing comparisons of tropical forest dynamics in Monsoon Asia. Agric. For. Meteorol. 148, 748–760.

Kramer, K., Leinonen, I., Bartelink, H.H., Berbigier, P., Borghetti, M., Bernhofer, C., Cienciala, E., Dolman, A.J., Froer, O., Gracia, C.A., Granier, A., Grünwald, T., Hari, P., Jans, W., Kellomäki, S., Loustau, D., Magnani, F., Markkanen, T., Matteucci, G., Mohren, G.M.J., Moors, E., Nissinen, A., Peltola, H., Sabate, S., Sanchez, A., Sontag, M., Valentini, R. and Vesala, T. (2002) Evaluation of six process-based forest growth models using eddy-covariance measurements of CO_2 and H_2O fluxes at six forest sites in Europe. Global Change Biol. 8, 213–230.

Lieth, H. (Ed.) (1974) Phenology and seasonality modeling. Springer, New York, pp. 444.

Menzel, A. (2000) Trends in phenological phases in Europe between 1951 and 1996. Int. J. Biometeorol. 44, 76–81.

Menzel, A. (2003) Plant phenological anomalies in Germany and their relation to air temperature and NAO. Clim. Change 57, 243–263.

Menzel, A., Sparks, T.H., Estrella, N. and Eckhardt, S. (2005) 'SSW to NNE' - North Atlantic Oscillation affects the progress of seasons across Europe. Global Change Biol. 11, 909–918.

Morisette, J.T., Richardson, A.D., Knapp, A.K., Fisher, J.I., Graham, E.A., Abatzoglou, J., Wilson, B.E., Breshears, D.D., Henebry, G.M., Hanes, J.M. and Liang, L. (2008) Tracking the rhythm of the seasons in the face of global change: phenological research in the 21st century. Front. Ecol. Environ. 6, doi:10.1890/070217.

Olesen, J.E., Carter, T.R., Diaz-Ambrona, C.H., Fronzek, S., Heidmann, T., Hickler, T., Holt, T., Quemada, M., Ruiz-Ramos, M., Rubaek, G.H., Sau, F., Smith, B. and Sykes, M.T. (2007) Uncertainties in projected impacts of climate change on European agriculture and terrestrial ecosystems based on scenarios from regional climate models. Clim. Change 81, 123–143.

Parmesan, C. (2007) Influences of species, latitudes and methodologies on estimates of phenological response to global warming. Global Change Biol. 13, 1860–1872.

Parmesan, C. and Yohe, G. (2003) A globally coherent fingerprint of climate change impacts across natural systems. Nature 421, 37–42.

Piao, S.L., Ciais, P., Friedlingstein, P., Peylin, P., Reichstein, M., Luyssaert, S., Margolis, H., Fang, J.Y., Barr, A., Chen, A.P., Grelle, A., Hollinger, D.Y., Laurila, T., Lindroth, A., Richardson, A.D. and Vesala, T. (2008) Net carbon dioxide losses of northern ecosystems in response to autumn warming. Nature 451, 49–53.

Saleska, S.R., Miller, S.D., Matross, D.M., Goulden, M.L., Wofsy, S.C., da Rocha, H.R., de Camargo, P.B., Crill, P., Daube, B.C., de Freitas, H.C., Hutyra, L., Keller, M., Kirchhoff, V., Menton, M., Munger, J.W., Pyle, E.H., Rice, A.H. and Silva, H. (2003) Carbon in amazon forests: Unexpected seasonal fluxes and disturbance-induced losses. Science 302, 1554–1557.

Wilson, K. and Baldocchi, D. (2000) Seasonal and interannual variability of energy fluxes over a broadleaved temperate deciduous forest in North America. Agric. For. Meteorol. 100, 1–18.

Contents

Contributors

Dennis Baldocchi Department of Environmental Science, Policy and Management, University of California Berkeley 137 Mulford Hall #3114, CA 94720, USA

Alan Barr Climate Research Division, Environment Canada, 11 Innovation Boulevard, Saskatoon, SK S7N 3H5, Canada

Kaycie A. Billmark Department of Soil, Water, and Climate, University of Minnesota-Twin Cities, 1991 Upper Buford Circle, St. Paul, MN 55108, USA

Chandrashekhar Biradar Department of Botany and Microbiology, University of Oklahoma, 101 David L. Boren Boulevard, Norman, OK 73019, USA

T. Andrew Black Faculty of Land and Food Systems, University of British Columbia, 135-2357 Main Mall, Vancouver, BC V6T 1Z4, Canada

Jiquan Chen Department of Environmental Sciences, University of Toledo, 2801 West Bancroft Street, Toledo, OH 43606, USA

Eric A. Davidson The Woods Hole Research Center, 149 Woods Hole Road, Falmouth, MA 02540-1644, USA

Ankur Desai Department of Atmospheric and Oceanic Sciences, University of Wisconsin - Madison, 1225 West Dayton Street, Madison, WI 53706, USA

Lawrence B. Flanagan Department of Biological Sciences, University of Lethbridge, 4401 University Drive, Lethbridge, Alberta, T1K 3M4, Canada

Timothy J. Griffis Department of Soil, Water, and Climate, University of Minnesota-Twin Cities, 1991 Upper Buford Circle, St. Paul, MN 55108, USA

Lianhong Gu Environmental Sciences Division, Oak Ridge National Laboratory, P.O. Box 2008, Building 1509, Oak Ridge, TN 37831, USA

Julian L. Hadley Harvard Forest, Harvard University, 324 North Main Street, Petersham, MA 01366, USA

Heikki Hänninen Department of Biological and Environmental Sciences, University of Helsinki, 1 Viikinkaaari, Helsinki, FIN-00014, Finland

N. Michele Holbrook Organismic and Evolutionary Biology, Harvard University, 16 Divinity Avenue, Cambridge, MA 02138, USA

David Y. Hollinger Northeast Research Station, USDA Forest Service, 271 Mast Road, Durham, NH 03824, USA

Koen Kramer Alterra, Centre of Ecosystem Studies, Wageningen University and Research Centre, P.O. Box 47, 6700 AA, Wageningen, Netherlands

Harry McCaughey Department of Geography, Queen's University, 99 University Avenue, Kingston, Ontario K7L 3N6, Canada

J. William Munger Department of Earth and Planetary Sciences, Harvard University, 20 Oxford Street, Cambridge, MA 01238, USA

Asko Noormets Department of Forestry and Environmental Resources, North Carolina State University, 920 Main Campus Drive, Raleigh, NC 27695, USA

John O'Keefe Harvard Forest, Harvard University, 374 North Main Street, Petersham, MA 01366, USA

Wilfred M. Post Environmental Sciences Division, Oak Ridge National Laboratory, P.O. Box 2008, Building 1509, Oak Ridge, TN 37831, USA

Bradley C. Reed Geographic Analysis and Monitoring, U.S. Geological Survey, 12201 Sunrise Valley Drive, Reston, VA 20192, USA

Andrew D. Richardson Complex Systems Research Center, University of New Hampshire, 8 College Road, Durham, NH 03824, USA

Mark D. Schwartz Department of Geography, University of Wisconsin-Milwaukee, 2200 East Kenwood Boulevard, Milwaukee, WI 53201, USA

Andrew E. Suyker School of Natural Resources, University of Nebraska, 3310 Holdrege Street, Lincoln, NE 68583, USA

Shashi B. Verma School of Natural Resources, University of Nebraska, 3310 Holdrege Street, Lincoln, NE 68583, USA

Timo Vesala Department of Physical Sciences, University of Helsinki, Gustaf Hällströmin katu 2a, Helsinki, FIN-00014, Finland

Steve C. Wofsy Department of Earth and Planetary Sciences, Harvard University, 20 Oxford Street, Cambridge, MA 01238, USA

Weixing Wu Institute of Geographical and Natural Resource Research, Chinese Academy of Sciences, 11a Datun Road, Beijing 100101, China

Xiangming Xiao Department of Botany and Microbiology, University of Oklahoma, 101 David L. Boren Boulevard, Norman, OK 73019, USA

Huimin Yan Institute of Geographical and Natural Resource Research, Chinese Academy of Sciences, 11a Datun Road, Beijing 100101, China

Junhui Zhang Institute of Applied Ecology, Chinese Academy of Sciences, 72 Wenhua Road, Shenhe District, Shenyang 110016, China

Part I
Phenological Phenomena

Climatic and Phenological Controls of the Carbon and Energy Balances of Three Contrasting Boreal Forest Ecosystems in Western Canada

Alan Barr, T. Andrew Black, and Harry McCaughey

Abstract Seasonal and interannual variability in the carbon and energy cycles of boreal forests are controlled by the interaction of climate, ecophysiology and plant phenology. This study analyses eddy-covariance data from mature trembling aspen, black spruce and jack pine stands in western Canada. The seasonal cycles of the surface carbon and energy balances were tightly coupled to the seasonal cycle of soil temperature. The contiguous carbon-uptake period was ~50 days longer for the black spruce and jack pine stands than the trembling aspen stand, with 30 days difference in spring and 20 days difference in autumn. The black spruce and jack pine carbon-uptake period spanned the warm season, with gross ecosystem photosynthesis beginning during spring thaw and continuing until air temperature dropped to below freezing in autumn. In contrast, the trembling aspen carbon-uptake period was determined by the timing of leaf emergence and senescence, which occurred well after spring thaw and before autumn freeze. Regression analysis identified spring temperature as the primary factor controlling annual net ecosystem production at all three sites, through its influence on the onset of the growing season. Precipitation and soil water content had significant but secondary influences on the annual carbon fluxes. The impact of spring warming on annual net ecosystem production was 2–3 times greater at the deciduous-broadleaf than the evergreen-coniferous sites, confirming the high sensitivity of boreal deciduous-broadleaf forests to spring warming. The analysis confirmed the pivotal role of phenology in the response of northern ecosystems to climate variability and change.

A. Barr (✉)
Environment, Canada
e-mail: alan.barr@ec.gc.ca

T. A. Black
Land and Food Systems, University of British Columbia, Vancouver, BC, Canada
e-mail: andrew.black@ubc.ca

H. McCaughey
Department of Geography, Queen's University, Kingston, ON, Canada
e-mail: mccaughe@queensu.ca

A. Noormets (ed.), *Phenology of Ecosystem Processes*,
DOI 10.1007/978-1-4419-0026-5_1, © Springer Science+Business Media, LLC 2009

3

Abbreviations

APAR   absorbed photosynthetically-active radiation (mol m^{-2} y^{-1}) (Eqn. 8a, 8b)
CUP   carbon-uptake period (days)
DOY   day of year
EGS   end date of growing season (Table 2)
EF   evaporative fraction (Eqn. 3)
F_N net ecosystem production (μmol m^{-2} s^{-1}) or (g C m^{-2} y^{-1})
F_P gross ecosystem photosynthesis (μmol m^{-2} s^{-1}) or (g C m^{-2} y^{-1})
F_R ecosystem respiration (μmol m^{-2} s^{-1}) or (g C m^{-2} y^{-1})
GS   growing season
H sensible heat flux density (W m^{-2})
LAI   leaf area index
LGS   length of growing season (days, Table 2)
NDVI   broadband estimate of the normalized difference vegetation index (Eqn. 7)
OGS   onset date of growing season (Table 2)
P annual total precipitation (mm)
P_{2y} total precipitation from current and previous years (mm)
PAR   photosynthetically-active radiation
Q sum of surface storage energy flux densities (W m^{-2})
R_n net radiation flux density (W m^{-2})
R_{sd} global incoming shortwave flux density (W m^{-2})
SWC   soil volumetric water content
T_a air temperature above the forest canopy (°C)
T_s soil temperature at 5, 10 or 20-cm depth from the top of the surface organic horizon (°C)
β Bowen ratio (Eqn. 4)
ΣD cumulative degree days (°C days, Eqn. 5)
λE latent heat flux density (W m^{-2})

1   Introduction

The exchanges of carbon, water and energy between the land surface and the atmosphere are controlled by land cover characteristics, climate and plant phenology. Inter-annual climatic variability exerts a strong influence on the terrestrial carbon and energy balances, although its influence is often indirect, mediated through plant phenology (Goulden et al. 1996; Myneni et al. 1997; Barr et al. 2004). In boreal ecosystems, phenology controls the seasonal onset and ending of the carbon-uptake period (CUP), thereby affecting net ecosystem production (F_N) (Goulden et al. 1996; Black et al. 2000; White and Nemani 2003; Barr et al. 2004; Churkina et al. 2005; Baldocchi et al. 2008). It also influences seasonal changes in the partitioning of net radiation into sensible, latent and surface storage heat fluxes (Wilson and Baldocchi 2000; Kljun et al. 44). These in turn affect the global carbon cycle and climate system (Sellers et al. 1997).

Continuous time series of eddy-covariance fluxes are just now becoming long enough to enable analysis of the interactions of climatic variability and plant phenology. They show promise to elucidate the key phenological controls on the seasonal cycles of canopy conductance and photosynthetic capacity (Grelle et al. 1999; Gu et al. 2003; Barr et al. 2007). They also have potential to shed new light on the processes that control the spring startup and autumn shutdown of photosynthesis, identifying triggers and thresholds that are consistent across sites and among years (Baldocchi et al.2008).

1.1 Boreal Forests

The boreal forest is one of the earth's most extensive ecosystems. It occupies 12 million km^2 or 8% of the earth's land surface, spanning North America and Eurasia at latitudes of 50–70°N (Hogg 2002). It comprises a mosaic of contrasting land covers, resulting from the interplay of topography, soil drainage, soil water holding capacity, disturbance history and ecological succession. The boreal forests of western Canada have peatlands in the poorly-drained deeper depressions, black spruce treed wetlands in the more subtle lowland depressions, jack pine forests on the well-drained sandy uplands, and mixed-wood and trembling aspen forests on the well-drained loamy uplands (Hall et al. 1997). Superimposed is the spatial pattern of disturbance history. Fire is the dominant disturbance regime (Stocks et al. 2002), with a return period of 50–200 years (Bonan and Shugart 1989). Harvesting plays a less important role because of the boreal forest's low net primary productivity (Landsberg and Gower 1997).

The climate of the boreal forest is characterized by long cold winters and extreme seasonal variations in temperature (Black et al. 2005). Average annual air temperature ranges from −10 to +3°C (Hogg 2002). The ground remains snow covered for 4–8 months of the year. Summers are short, but the growing season is sufficiently long, warm and moist to sustain tree growth. Snowmelt and rainfall are usually sufficient to minimize soil water stress during the growing season (Hogg 1997; Black et al. 2005).

1.2 Phenology

In this study, we define plant phenology broadly to include seasonal changes in plant development and metabolism. Plants are usually well adapted to the seasonality of their environment, and shifts in their seasonal phenology provide compelling evidence that ecosystems are being influenced by climate change (Cleland et al. 2007). In cold regions, plant phenology is an integral component of the adaptation to extreme seasonal variations in temperature (Leith 1974).

Boreal evergreen-coniferous and deciduous-broadleaf forests have contrasting phenological strategies. They share winter dormancy whereby growth is inhibited at low temperatures (Savitch et al. 2002; Espinosa-Ruiz et al. 2004). They differ in the seasonal cycles of leaf area index (LAI) and photosynthetic capacity. The CUP of deciduous forests is determined by the phenology of leaf emergence and senescence (Goulden et al. 1996; Barr et al. 2004). In contrast, the CUP of boreal evergreen conifers, which maintain green foliage throughout the cold season, is determined primarily by the timing of spring thaw and autumn freeze (Goulden et al. 1998; Monson et al. 2005). Boreal conifers have phenological and physiological adaptations to enable their foliage to withstand extreme cold in winter (Öquist et al. 2001; Savitch et al. 2002; Slaney 2006) while efficiently and rapidly up- and down-regulating their photosynthetic capacity in spring and autumn (Ensminger et al. 2004). The cold-hardening process is triggered in autumn by cool temperatures and diminishing daylength (Huner et al. 1993; Lindgren and Hällgren 1993). An almost complete suppression of photosynthesis during the cold season is followed by a rapid recovery of photosynthesis in spring, within days of exposure to non-freezing temperatures (Ottander et al. 1995; Ensminger et al. 2008). The recovery rate depends on air temperature (Lundmark et al. 1988; Kramer et al. 2000) and the antecedent soil temperature (Ensminger et al. 2008). The foliage of boreal conifers is well adapted to minimize frost damage during these transitional periods (Häkkinen and Hari 1988; Gaumont-Guay et al. 2003; Ensminger et al. 2008).

1.3 Inter-Annual Variability

The exchanges of carbon, water and energy from boreal forest ecosystems to the atmosphere are sensitive to inter-annual climatic variability. Vapor pressure deficit is a primary determinant of evapotranspiration from upland forests (Mackay et al. 2007). Spring temperature has a dominant, positive influence on F_N in both deciduous-broadleaf (Goulden et al. 1996; White et al. 1999; Black et al. 2000; Baldocchi et al. 2001, 2005; Carrara et al. 2003; White and Nemani 2003; Barr et al. 2002, 2004, 2007) and evergreen-coniferous forests (Goulden et al. 1998; Hollinger et al. 2004; Krishnan et al. 2008; MacMillan et al. 2008), with higher sensitivity among deciduous-broadleaf than evergreen-coniferous forests (Black et al. 2005; Churkina et al. 2005; Welp et al. 2007; Baldocchi et al. 2008). At the landscape scale, the major source of inter-annual variability in F_N from boreal black spruce stands of different ages is the impact of spring temperature on the gross ecosystem photosynthesis (F_P) of the older stands (McMillan et al. 2008). At regional scales in northern latitudes, the seasonal amplitude of atmospheric CO_2 concentration, and hence the magnitude of the carbon sink, is positively correlated with the timing of spring thaw (MacDonald et al. 2004). Other climatic controls on F_N include summer drought (Griffis et al. 2004; Hollinger et al. 2004; Angert et al. 2005; Ciais et al. 2005; Krishnan et al. 2006; Barr et al. 2007; Kljun et al. 2007; Welp et al. 2007), growing-season temperature (Ciais et al. 2005), autumn temperature (Hollinger et al. 2004), and annual temperature (Morgenstern et al. 2004; Dunn et al. 2007). Piao et al. (2008)

showed a large-scale negative impact of autumn warming on F_N in northern ecosystems, related to an increase in autumn ecosystem respiration.

The impact of temperature on annual F_N is often indirect, mediated through plant phenology. Barr et al. (2004) showed a remarkable synchrony between the seasonal cycles of F_N, F_P and LAI in a boreal trembling aspen forest. The phenological controls on F_P are related to the rate of change in photosynthesis with expanding leaf area (Brooks et al. 1996; Gratani and Ghia 2002). The positive response of F_N to spring temperature is related to an associated increase in the CUP, particularly to the high-productivity days that are added in spring (Black et al. 2000; Baldocchi et al. 2001; Carrara et al. 2003; White and Nemani 2003; Barr et al. 2004; Churkina et al. 2005). The effects of climate and phenology are often expressed in consort. Richardson et al. (2007) demonstrated that 40% of the variance in annual F_N in a temperate red spruce forest was due to environmental drivers and 55% to the biotic response.

The primary objectives of this chapter are: to characterize the seasonal cycles of the surface energy and carbon balances of three representative boreal forest stands; to identify the climatic and phenological factors that control the seasonality of the energy and carbon cycles, with particular attention to the transitional seasons of spring and autumn; to relate interannual variability in carbon and water fluxes to climatic and phenological controls; and to compare and contrast deciduous-broadleaf versus evergreen-coniferous plant functional types. A secondary objective is to compare different approaches for delineating the growing season.

2 Study Sites and Measurements

2.1 Site Description

The forest-atmosphere exchanges of carbon, water and energy were measured above three mature boreal forest stands in the southern boreal forest of central Saskatchewan, Canada. The sites were established in 1994 as part of the Boreal Ecosystem-Atmosphere Study (BOREAS, Sellers et al. 1997) and have continued since 1997 as part of the Boreal Ecosystem Research and Monitoring Sites (BERMS) and Fluxnet-Canada (Margolis et al. 2006) programs. Table 1 summarizes the salient site characteristics. Mean annual, January and July air temperatures at Waskesiu Lake (53.92°N, 106.07°W), 1971–2000, were 0.4, −17.9 and 16.2°C, respectively. Mean annual precipitation at Waskesiu Lake was 467 mm, 30% of which fell as snow.

The Old Aspen site regenerated after a wildfire in 1919 and has a trembling aspen overstory (*Populus tremuloides* Mich. with scattered *Populus balsamifera* L.) and hazelnut understory (*Corylus cornuta* Marsh.). The subsurface geology consists of clay-rich glacial till, below a 10-cm organic forest-floor and a 30-cm silt loam mineral soil horizon.

Table 1 Physical characteristics of the three boreal forest stands

	Old Aspen	Old Black Spruce	Old Jack Pine
Dominant tree species	*Populus tremuloides* Mich.	*Picea mariana (Mill.)* B.S.P.	*Pinus banksiana* Lamb.
Year of last disturbance	1919	~1914	~1879
Measurement years	1997–2006	2000–2006	2000–2006
Latitude (°N)	53.629	53.987	53.916
Longitude (°W)	106.200	105.117	104.690
Elevation (m)	601	629	579
Canopy height (m)	21	10	14
Stem density[a] (ha^{-1})	980	5,900	1,190
Leaf area index	3.7–5.2[b]	3.8[c]	2.6[c]
Organic soil depth (cm)	8–10	20–30	0–5

[a]Gower et al. (1997).
[b]Aspen overstory and hazelnut understory (Barr et al. 2004).
[c]Overstory only (Chen et al. 2006).

The Old Black Spruce site regenerated after a wildfire in 1879, resulting in a stand dominated by black spruce (*Picea mariana* (Mill.) B.S.P.), with sparsely distributed tamarack (*Larix laricina* Du Roi). The forest floor consists of a nearly continuous feather moss community (e.g., *Hylocomium splendens*, *Pleurozium schreberi*), with irregular patches of hummocky peat (*Sphagnum* spp.) in wetter areas. The soil has a 20–30 cm peat layer overlying waterlogged sand, with imperfect to poor drainage.

The Old Jack Pine site, which originated after wildfire in 1914, is located in the flat portion of a glacial outwash plain. The overstory is a pure stand of jack pine (*Pinus banksiana* Lamb.). The understory is dominated by reindeer lichen (*Cladonia spp.*) and infrequent, scattered clumps of green alder (*Alnus viridis spp. crispa* (Ait.) Turrill). The sandy soil is extremely well drained and nutrient poor.

2.2 Carbon, Water and Energy Fluxes

The carbon balance of a vegetated surface,

$$F_N = F_P - F_R \tag{1}$$

shows the partitioning of F_N, the net exchange of carbon between an ecosystem and the atmosphere, into components for carbon uptake F_P and carbon release by ecosystem respiration (F_R). F_N is positive when the ecosystem is gaining carbon from the atmosphere. Both F_P and F_R are given positive signs. All terms in Eqn. (1) have units of (μmol m^{-2} s^{-1}), (g C m^{-2} d^{-1}) or (g C m^{-2} y^{-1}).

The energy balance of a vegetated surface may be written as:

$$R_n = H + \lambda E + Q \tag{2}$$

where R_n is the net radiation flux density, H is the sensible heat flux density, λE is the latent heat flux density, and Q is the surface storage flux density (the sum of the soil heat flux density, the biomass heat storage flux density, the photosynthetic energy flux density, and terms for freeze-thaw events and heat storage in the snow pack). Our measurements of Q pertain to the warm season only and do not include freeze-thaw or snowpack heat storage, neither of which was measured in this study. All terms in Eqn. (2) have units of W m^{-2}. Two derived parameters characterize the partitioning of surface available energy $(R_n - Q)$ between H and λE: the evaporative fraction (EF):

$$EF = \lambda E/(R_n - Q) = \lambda E/(H + \lambda E) \tag{3}$$

and the Bowen ratio (β):

$$\beta = H/\lambda E \tag{4}$$

The estimation of EF and β was limited to periods when the ratios were well defined, i.e. the numerator was well above zero. To circumvent the issue of energy-balance nonclosure by the eddy-covariance method (Barr et al. 2006), $(R_n - Q)$ can be replaced by $(H + \lambda E)$ in Eqn. (3). The energy balance closure fraction was ~85% at all three sites (Barr et al. 2006).

F_N, H and λE were measured using the eddy-covariance technique at twice the height of the forest canopy. At all three sites, the micrometeorological conditions were nearly ideal. Measurement details are given in Griffis et al. (2003) and Barr et al. (2006). To ensure that the EC measurements were comparable among sites, similar instrumentation, data acquisition hardware and post-processing software were used at all sites. Gaps in flux data were filled and F_N was partitioned into F_P and F_R using the standard methods of Fluxnet-Canada (Barr et al. 2004).

2.3 Derived Biophysical Variables

2.3.1 Degree Days

Cumulative, growing-season, degree days (ΣD) were calculated on a daily basis as:

$$\Sigma D = \sum_{d=d_i}^{d_f} \max(T_a - T_a^*, 0^\circ C) \tag{5}$$

where d is the day of year (DOY), d_i and d_f are the initial and final values of d for growing-season degree-day accumulation, T_a is mean daily air temperature, and T_a^* is the base temperature (5°C for trembling aspen, 0°C for black spruce and jack pine). We identified d_i for each year as the first day in spring when daily mean soil temperature at 5-cm depth exceeded -0.2°C, coinciding with the onset of snowmelt.

2.3.2 Growing-Season Length, Onset and End

Several approaches were used to delineate the onset DOY (OGS), end DOY (EGS) and length (LGS) of the growing season (GS) (Table 2). This study uses the term growing season in a general sense to encompass a number of diverse approaches to define the warm-season, photosynthetic period in climatic regions with a distinct cold season. The approaches fall into three broad categories: (1) those based on the annual temperature cycle, including thermal indices and the delineation of frozen, thawing/freezing and thawed periods; (2) those based on seasonal changes in the normalized difference vegetation index (NDVI); and (3) those derived directly from measured

Table 2 Criteria to delineate the onset OGS, end EGS and length LGS of the growing season

Onset	End	Duration	Delineation
OGS^{Fp}	EGS^{Fp}	$LGS^{Fp}(CUP)$	Core, contiguous photosynthetic period: first and last DOY of detectable photosynthesis during the contiguous CUP, assigned by visual inspection of the F_N time series
$OGS^{\Sigma Fp}$	$EGS^{\Sigma Fp}$	$LGS^{\Sigma Fp}$	A robust, automated estimate of the CUP: DOY when cumulative F_P first exceeded 1% and 99% of the annual total
OGS^{Sink}	EGS^{Sink}	LGS^{Sink}	Net carbon-uptake period: DOY of spring minimum and autumn maximum in cumulative F_N
$OGS^{\lambda E}$	$EGS^{\lambda E}$	$LGS^{\lambda E}$	Transpiration period: first and last DOY when daily mean λE exceeded a threshold (25% of the annual 90th percentile) for at least five consecutive days
OGS^{Thaw}_{Onset}	$EGS^{Freeze}_{Contiguous}$	$LGS^{Thaw}_{Extended}$	Extended warm season: first DOY of contiguous thaw period in spring[a] and contiguous frozen period in autumn[b]
OGS^{Thaw}_{End}	$EGS^{Freeze}_{Earliest}$	LGS^{Thaw}_{Core}	Core warm season: last DOY of contiguous spring thaw period[a], first autumn freeze[b]
$OGS^{\Sigma D}$	n/a	n/a	DOY when cumulative degree days (Eqn. 5) first exceeded a preset threshold (trembling aspen: 120°C days using a base temperature of 5°C; black spruce and jack pine: 40°C days using a base temperature of 0°C)
$OGS^{\Sigma T_a}$	$EGS^{\Sigma T_a}$	$LGS^{\Sigma T_a}$	Warm season: DOY when the sum of positive values of daily mean T_a (in °C) first exceeded 5% and 95% of the annual total
$OGS^{\Sigma T_s}$	$EGS^{\Sigma T_s}$	$LGS^{\Sigma T_s}$	Warm season: DOY when the sum of positive values of daily mean soil temperature T_s at 5-cm depth (in °C) first exceeded 2% and 98% of the annual total
OGS^{NDVI}	EGS^{NDVI}	LGS^{NDVI}	Green period: first and last DOY when daily NDVI exceeded a preset threshold (trembling aspen, 0.6; black spruce and jack pine, 0.5) for at least five consecutive days

[a]Spring thaw is defined as a multi-day period in March to May with daily maximum T_a above 0°C, daily mean 5-cm and 20-cm T_s near 0°C, and shallow soil volumetric water content above the frozen winter value.
[b]The frozen period includes all days when daily maximum T_a or T_s remain below 0°C.

fluxes, based on the onset and cessation of photosynthesis or transpiration in the annual F_N and λE time series. Category 3 includes the CUP. Details are given in Table 2.

2.3.3 Annual Centroids

The annual centroid DOY (d_c), a robust, weighted measure of the timing of the annual peak of flux and climatic variables, was calculated as:

$$d_c = \left(\sum_{i=1}^{365} d_i m_i \right) / \left(\sum_{i=1}^{365} m_i \right)$$ (6)

where d_i is the DOY and m_i is the daily mean. For variables such as T_s, T_a, R_n, H and F_N which became negative in the cold season, the annual centroid was calculated from positive m_i values only.

2.3.4 Normalized Difference vegetation Index

The NDVI is a spectral index used to characterize temporal and spatial changes in the greenness of vegetated surfaces. A broadband estimate of NDVI (Huemmrich et al. 1999) was calculated from daily radiation fluxes as:

$$\text{NDVI} = 1 - (P_u / P_d) / (R_{su} / R_{sd})$$ (7)

where P_u and P_d are up- and down-welling flux densities of photosynthetically-active radiation (PAR) and R_{su} and R_{sd} are up- and down-welling flux densities of shortwave radiation.

2.3.5 Canopy PAR Absorption

PAR absorption by the canopy foliage (APAR) was estimated during the photosynthetically-active period based on LAI and PAR extinction in the canopy. For the two-canopy trembling aspen stand, APAR was estimated following Chen et al. (1999) and Barr et al. (2007) as:

$$\text{APAR} = (P_d - P_u)[\underbrace{1 - e^{-k_a \Omega_{ae} L_a}}_{\text{aspen}} + \underbrace{e^{-k_a (\Omega_{ae}(L_a + S) + \delta B_{ae})} (1 - e^{-k_h \Omega_h L_h})}_{\text{hazelnut}}]$$ (8a)

where L_a and L_h are trembling aspen and hazelnut LAI (which vary seasonally and inter-annually, Barr et al. 2004), k_a and k_h are PAR extinction coefficients for the trembling aspen and hazelnut canopies (0.540 and 0.756, respectively, Blanken et al. 1997), Ω_{ae} is the trembling aspen leaf clumping index (0.69), S is the (hemi-surface) trembling aspen stem area index (0.40), B_{ae} is the effective (hemi-surface) trembling aspen branch area index (0.44), δ is the fraction of B_{ae} that is exposed to gaps in the canopy (Kucharik et al. 1998) and Ω_h is the hazelnut leaf clumping index (0.98). For the black spruce and jack pine canopies, APAR was estimated following Griffis et al. (2003) as

$$\text{APAR} = (P_d - P_u)(1 - e^{-kL_e}) \tag{8b}$$

where k is the PAR extinction coefficient (0.430 black spruce and 0.523 jack pine, Griffis et al. 2003) and L_e is the effective LAI (2.72 black spruce and 1.76 jack pine, Chen et al. 2006). Annual APAR was computed by integrating Eqns. (8a) and (8b) over the photosynthetically-active period, determined from the seasonal cycles of L_a and L_h at the trembling aspen site (Barr et al. 2004) and by periods of non-freezing T_a and T_s at the conifer sites. Annual APAR effectively combines LGS with daylength, weighting each day in the GS by daily APAR.

2.3.6 Regression Analysis

The dependence of annual H, λE, F_N, F_R and F_P on climatic and GS variables was evaluated by ordinary least-squares linear regression and multiple linear regression. Multiple linear regression was done for all possible combinations of independent variables. The independent variables included annual R_n and PAR, APAR (Eqns. 8a and 8b), OGS, EGS and LGS (Table 2), T_a and T_s (mean annual, April–May, June–August and September–October), mean warm-season soil volumetric water content (SWC), annual precipitation (P), and 2-year cumulative precipitation (P_{2y}, including the previous year). The independent variables were ranked according to the regression's goodness of fit (p, the probability of a type-one error); also reported is the coefficient of determination (r^2).

3 Seasonality of the Carbon, Water and Energy Cycles

3.1 Climate

Figure 1 shows mean annual cycles of precipitation, snow depth, air temperature, soil temperature and NDVI (Eqn. 7) for the three forest stands, averaged from 1997 to 2006. 59% of the annual precipitation fell during the four summer months of June to September. 28% fell as snow, evenly distributed from October to April. Snow depth peaked in March at 0.3–0.6 m. The snowpack duration varied from 5 to 6 months, with remarkable cross-site consistency in its timing and depth. Unlike the annual cycle of T_a, which was virtually identical among sites, the annual cycle of T_s differed among sites in both phase and amplitude. The annual amplitude of mean daily T_s at 5-cm depth, measured from the top of the surface organic horizon, was highest at the jack pine site (21°C), intermediate at the trembling aspen site (16°C) and lowest at the black spruce site (11°C), reflecting site differences in the depth of the forest-floor organic soil horizon (Table 1) and the presence of a moss understory at the black spruce site. The T_s annual cycle lagged the T_a annual cycle by 16 days at the trembling aspen and jack pine sites compared with 26 days at the black spruce site, reflecting the extended soil thaw period at the black spruce site.

3.2 Leaf Phenology and Greening

The broadleaf-deciduous and evergreen-coniferous forest stands had contrasting annual NDVI cycles (Fig. 1, lower panel), with distinct signatures of winter snow cover, spring snowmelt, spring greening and autumn senescence. During the snow-covered period, the leafless trembling aspen NDVI (0.18 ± 0.11) was 50% lower than evergreen black spruce (0.36 ± 0.14) and jack pine (0.34 ± 0.14) NDVI. The high variability of NDVI during the snow-covered period resulted from the intermittent loading and unloading of intercepted snow on the forest canopy. During April snowmelt, NDVI rose sharply to ~0.40 (trembling aspen), ~0.53 (black spruce) and ~0.58 (jack pine). During the warm season, NDVI at the deciduous-broadleaf

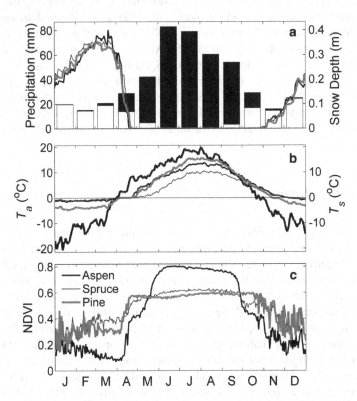

Fig. 1 Seasonal climatology of the three forest sites, 1997–2006, including: (**a**) *upper panel –* mean monthly total precipitation (*bars*, snow in *white*, rain in *black*, averaged among sites) and snow depth (lines, by site); (**b**) *middle panel –* 7-day mean air temperature T_a (*black thick line*, averaged among sites) and 10-cm soil temperature T_s (*lines*, by site); and (**c**) *lower panel -* daily mean NDVI (Eqn. 7).

trembling aspen site underwent large changes, in contrast with the much subtler warm-season changes at the evergreen black spruce and jack pine sites. Greening at the trembling aspen site, as evidenced by spring NDVI first exceeding 0.42, was delayed until 19 ± 6 days after thaw and began when the 5-cm T_s reached ~5°C and ΣD (Table 2) reached 70–80°C days. Aspen NDVI rose throughout May from 0.40 to 0.80, associated with leaf emergence and expansion (Barr et al. 2004). At the jack pine site, the post-snowmelt NDVI value was similar to the broad warm-season NDVI plateau, with an initial decline from 0.58 to 0.55 associated with needle fall in May followed by an increase to 0.60 following bud burst and needle expansion in June and July. At the black spruce site, a subtle May rise in NDVI, from 0.55 to 0.60, was associated with tamarack leafout, which accounted for ~10% of the summer LAI. At the trembling aspen and black spruce stands, NDVI dropped during aspen, hazelnut and tamarack leaf senescence in late September and early October, followed by an additional drop in NDVI with the onset of the winter snowpack at all three sites.

The seasonal NDVI transitions have different phenological significance for evergreen-coniferous and deciduous-broadleaf forests. NDVI undergoes two major transitions each year, one associated with the disappearance and reappearance of the winter snowpack, the other associated with changes in LAI during the snow-free period. For deciduous-broadleaf forests, the OGS and EGS are determined by spring leafout and autumn leaf senescence. NDVI provides a robust indicator of both because of its strong dependence on LAI. For evergreen-coniferous forests, NDVI provides a robust indicator of the OGS but not the EGS. The OGS occurs during spring thaw, which coincides with snowmelt and the associated rise in NDVI. However, the EGS occurs during autumn freeze, which is decoupled from snow accumulation and the associated drop in NDVI in some years, depending on the timing of snowfall in relation to the onset of persistent, freezing temperatures.

3.3 Surface Energy Balance

The surface energy balances (Eqn. 2) of the deciduous-broadleaf and evergreen-coniferous stands had characteristic differences in the seasonality of H and λE (Fig. 2). The annual centroids (Eqn. 6, Table 3) showed the annual cycles of R_{sd} and R_n to be nearly symmetrical around the summer solstice. The T_a and λE annual cycles lagged the R_{sd} annual cycle by ~4 weeks, peaking in mid July. In contrast, H peaked 6–8 weeks earlier than λE, ~4 weeks, 2 weeks and 10 days before the summer solstice for trembling aspen, black spruce, and jack pine, respectively. The three stands had similar energy balances during the cold season and spring thaw (October to April) but diverged after trembling aspen leafout in May, with mean June–September β (Eqn. 4) of 0.93 for black spruce and 1.26 for jack pine versus 0.25 for trembling aspen. Inter-annual variability in monthly H and λE was high throughout the warm season, from May to September.

The differences in evaporative fraction (EF, Eqn. 3) between plant functional types were striking (Fig. 3). At the coniferous sites, EF rose sharply to ~0.2 with

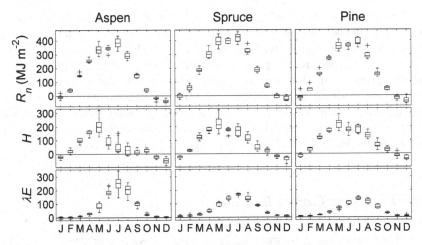

Fig. 2 Box plots showing inter-annual variation in monthly total net radiation (R_n), sensible heat (H) and latent heat (λE) flux densities above three boreal forest stands, 1997–2006 for the trembling aspen and 2000–2006 for the black spruce and jack pine sites. The boxes show the median (*centre line*) and inter-quartile range. The whiskers show either the maxima and minima or 1.5 times the inter-quartile range from the median (with outliers shown as + symbols).

Table 3 Timing (mean date ± s.d.) of the annual centroid (Eqn. 6)

	Aspen	Spruce	Pine
T_a	07/16 ± 3	07/16 ± 3	07/16 ± 3
5-cm T_s	08/01 ± 3	08/02 ± 3	07/30 ± 3
R_{sd}	06/20 ± 2	06/20 ± 2	06/20 ± 2
R_n	06/18 ± 2	06/18 ± 2	06/19 ± 2
H	05/22 ± 9	06/05 ± 6	06/09 ± 6
λE	07/15 ± 4	07/09 ± 5	07/12 ± 4
F_N	07/13 ± 7	06/22 ± 10	06/09 ± 6
F_R	07/19 ± 4	07/23 ± 4	07/23 ± 4
F_P	07/17 ± 5	07/16 ± 5	07/19 ± 4

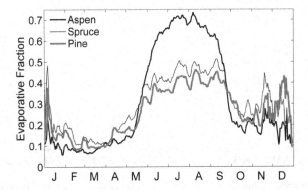

Fig. 3 Mean annual cycle of the daytime evaporative fraction, averaged over 1997–2006 for the trembling aspen site and 2000–2006 for the black spruce and jack pine sites. The values are 7-day running means.

the onset of thaw in the second week of April, followed by a second distinct rise in May and a gradual increase with increasing T_a during late May to mid July, to a seasonal maximum of ~0.47 for black spruce and ~0.41 for jack pine. It began to drop in late September, reaching ~0.2 by the end of October. At the trembling aspen site, the spring rise in EF was delayed until leafout in May, during which EF rose dramatically, reaching its annual peak of ~0.7 at the beginning of August. It dropped sharply with leaf senescence in September. During June to September, when the canopy was foliated, trembling aspen EF was ~50% higher than conifer EF. During the leafless period (October to April), it was ~30% lower.

3.4 Ecosystem Carbon Balance

As with the surface energy balance (Fig. 2), the annual ecosystem carbon cycle (Eqn. 1) had characteristic deciduous–coniferous differences (Fig. 4). In particular, the F_P annual cycle at the trembling aspen site was shorter, more intense and had higher inter-annual variability than at the black spruce or jack pine sites. The F_P and F_R annual cycles peaked in mid July, slightly after T_a and before T_s (Table 3). However, F_R peaked a little later than F_P, with a difference in the annual centroids of 2 days for trembling aspen, 7 days for black spruce and 4 days for jack pine. The F_R - F_P offset shifted the F_N annual cycle towards spring and contributed to a mid-summer depression in F_N at the conifer sites, limiting the period of positive F_N to May and June (Griffis et al. 2003; Black et al. 2005; Bergeron et al. 2006). Unlike H and λE, where

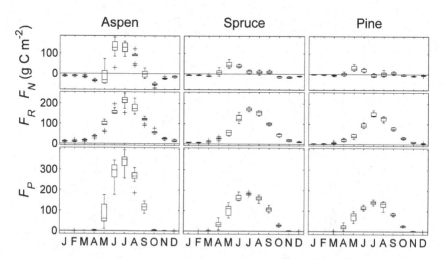

Fig. 4 Box plot showing inter-annual variation in monthly total net ecosystem production (F_N), ecosystem respiration (F_R) and gross ecosystem photosynthesis (F_P) from three boreal forest stands, 1997–2006 for the trembling aspen and 2000–2006 for the black spruce and jack pine sites.

inter-annual variability remained high from May to September, monthly F_N, F_R and F_P were most variable in spring, with extreme variability in F_N and F_P during May.

4 Delineating the Spring and Autumn Transitions

4.1 Snowmelt, Soil Thaw and Spring Onset

Spring thaw comprises multiple, coordinated processes as shown in Fig. 5. The transitional thaw period was characterized by simultaneous changes in temperature and moisture: T_a rose to above freezing; the snow pack and soil began to thaw, a multi-day transition during which the snowpack melted and the thawing snow and soil layers remained isothermal at ~0°C (Monson et al. 2005); and SWC increased abruptly as the soil thawed and snowmelt infiltrated the soil. The contiguous thaw period lasted for 17 ± 5 days (mean ± s.d.) for trembling aspen, 18 ± 13 days for black spruce and 17 ± 8 days for jack pine. At the evergreen black spruce and jack pine sites, the startup of photosynthesis and transpiration occurred during thaw, as seen in the escalating positive excursions of F_N and λE (Fig. 5, lower panels). The recovery of photosynthesis was rapid and reversible, with repeated startups in some years (Monson et al. 2005), associated with the ability of boreal conifers to respond opportunistically to short periods above freezing through the rapid up- and down-regulation of photosynthesis (Ensminger et al. 2004). Ensminger et al. (2008) demonstrated that the speed of photosynthetic recovery of Scots pine in spring depends on the severity of the antecedent soil freezing; intermittent frosts during or after the thaw period delayed but did not severely inhibit photosynthetic recovery, whereas refreezing of the soil severely retarded but did not completely inhibit the recovery process. At the deciduous trembling aspen site, the startup of photosynthesis and transpiration was delayed by the thermal requirements of leafout; photosynthesis was first detected 26 ± 7 days after thaw. At all sites, the startup of photosynthesis was more distinct than the startup of transpiration because transpiration was obscured by comparable, concurrent changes in surface evaporation.

4.2 Autumn Senescence and Freeze

As with spring startup, autumn shutdown was controlled by different processes at the deciduous and evergreen stands (Fig. 6). Aspen photosynthesis and transpiration were shut down by leaf senescence in late September in all years, as seen in the simultaneous decline in NDVI, λE and F_N in Fig. 6. The consistent timing of leaf senescence among years, well before the onset of freezing, suggests control by daylength (Morgenstern 1996). The timing of leaf senescence was largely independent of T_a, although in 1 of the 10 study years it was hastened by an early frost, i.e. when

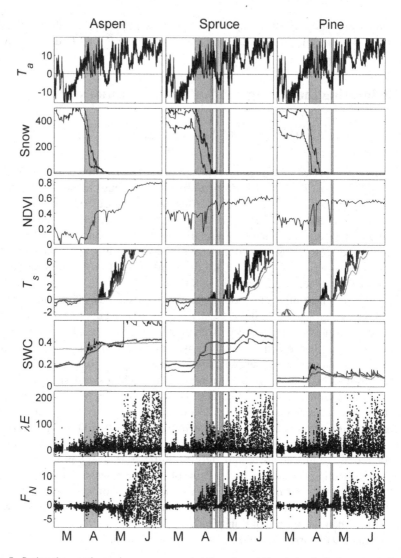

Fig. 5 Spring thaw and growing-season onset at three boreal forest stands during a typical year (2005), showing 30-min mean: air temperature (T_a, °C); snow depth (m, 2–3 locations); NDVI (Eqn. 7); soil temperature (T_s, °C, *black* 5-cm, *gray* 20-cm, *light gray* 50-cm), soil volumetric water content (SWC, *black* 0–15 cm, *gray* 15–30 cm, *light gray* 30–60 cm); latent heat flux density (λE, W m^{-2}); and net ecosystem production (F_N, μmol m^{-2} s^{-1}). The *shaded area* shows the thaw period, identified as days with positive daily maximum T_a and near-zero daily mean T_s, following a post-cold-season increase in soil liquid SWC.

daily minimum T_a dropped below 0°C. In contrast, conifer photosynthesis and transpiration were insensitive to mild frost events and were shut down only when daily maximum T_a dropped below 0°C, which we will refer to as freezing. The freezing of the soil did not factor into autumn shutdown because T_s remained positive well after subzero T_a had initiated the cessation of photosynthesis.

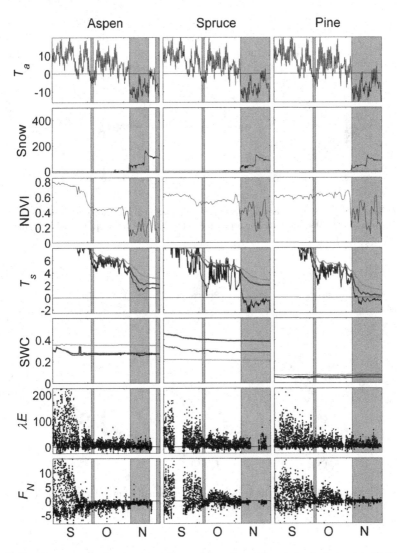

Fig. 6 Growing-season end and autumn freeze at three boreal forest stands during a typical year (2000). The panels are the same as in Fig. 5. The *shaded area* shows the frozen period, identified as days with daily maximum T_a or daily mean T_s of below 0°C.

4.3 Seasonal Cycles of Net Ecosystem Production and Evapotranspiration in Relation to Soil Temperature

The seasonal cycles of F_N and λE showed a remarkable degree of coupling to the seasonal T_s cycle (Figs. 7 and 8). At the evergreen black spruce and jack pine sites, F_N and λE rose sharply during April thaw, reaching more than 50% of annual

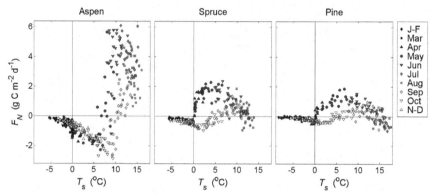

Fig. 7 Ten-day mean net ecosystem production (F_N) in relation to 5-cm soil temperature (T_s), shown for each year (1997–2006 for trembling aspen and 2000–2006 for black spruce and jack pine).

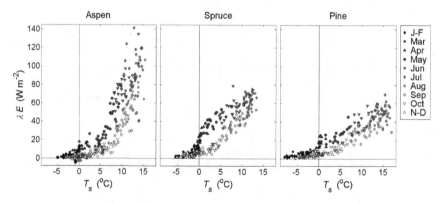

Fig. 8 Ten-day mean latent heat flux density (λE) in relation to 5-cm soil temperature (T_s), shown for each year (1997–2006 for trembling aspen and 2000–2006 for black spruce and jack pine).

maximum F_N and ~40% of maximum λE at T_s of just above 0°C. As the soil continued to warm, λE and F_N responded differently. λE continued to rise over the entire T_s range whereas F_N peaked during May and June at T_s of ~5°C and then dropped as the soil warmed in July and August, becoming negative at T_s above 13°C and 11°C for black spruce and jack pine, respectively. The decline in F_N at high T_s resulted from the differential responses of F_R and F_P to temperature. In contrast to the evergreen conifers, the deciduous trembling aspen's spring rise in F_N and λE was delayed by the thermal requirements of leafout until about four weeks after spring thaw, coincident with T_s of 5–7°C. In autumn, the trembling aspen decline in F_N and λE occurred during leaf senescence in late September, earlier and at higher T_s than the conifer decline. The hysteresis in the F_N versus T_s and λE versus T_s relationships was caused primarily by the 6-week lag between R_{sd} and T_s and secondarily by the seasonal drawdown of soil water.

4.4 Growing-Season Indices

Growing-season length (LGS) and onset and end dates (OGS and EGS respectively), were estimated for all site-years using several independent approaches, based on T_a, T_s, ΣD, spring thaw, autumn freeze, F_N, F_P, λE and NDVI (Table 4). Because the onset and end of photosynthesis in the F_N time series gave the most distinct and meaningful demarcation of the GS, we used LGS^{F_P}, OGS^{F_P} and EGS^{F_P} (Table 2) as reference standards to evaluate the other indices. LGS^{F_P} was ~50 days shorter for trembling aspen (133 ± 11 days, mean ± s.d.) than black spruce (185± 15 days) or jack pine (183 ± 13 days), with ~30 days difference in OGS^{F_P} and ~20 days difference in EGS^{F_P}. The GS indices that best matched the F_P-based indices differed between the broadleaf-deciduous and evergreen-coniferous stands, related to differences in phenology. At the evergreen black spruce and jack pine sites, the GS was best characterized by the timing of spring thaw and autumn freeze. The similarity in the values of the GS indices between the black spruce and jack pine stands was remarkable. At the deciduous trembling aspen site, the best GS indices were those that captured leaf emergence and senescence, i.e. degree-day accumulation in spring and spring and autumn NDVI. The GS indices derived from the λE time series had poor agreement with those derived from the F_N time series, particularly in autumn. Surprisingly, other indices based on seasonal changes in energy partitioning between H and λE, derived from the Bowen ratio, EF and surface conductance (not shown), also performed poorly, because the derived variables were inherently noisy, even after wet-canopy periods had been excluded from the analysis. In general, OGS indices outperformed EGS indices (Table 4); OGS was more distinct and thus more easily and accurately identified. The best indices showed OGS to be more slightly variable among years than EGS, particularly at the trembling aspen site.

Like spring startup, autumn shutdown at the conifer sites was reversible, with repeated shutdowns in some years (Fig. 6). This led to ambiguity in the definition of conifer OGS, EGS and LGS. This ambiguity was resolved by differentiating between the core and extended GS, where the extended GS also includes periods of carbon uptake outside the core, contiguous CUP. This distinction was not needed at the trembling aspen site, where the much shorter GS did not include any carbon-uptake periods outside the core. Of the seven years of measurements at the evergreen-coniferous sites, five years for black spruce and four years for jack pine had a temporary shutdown in spring following the initial startup of photosynthesis, and two years for both sites had a secondary period of carbon uptake in autumn following the initial shutdown of photosynthesis.

Table 4 Onset (OGS) and end (EGS) dates of the growing season (mean ± s.d.), as delineated by multiple approaches (Table 2) for the period 1997–2006 for the trembling aspen site and 2000–2006 for the black spruce and jack pine sites. Also shown is a comparison of these approaches vis-à-vis the OGS^{Fp} and EGS^{Fp} standard methods, showing bias (mean ± r.m.s error) and linear-regression r^2 and statistical significance (* 5%, ** 1%)

	Old Aspen			Old Black Spruce			Old Jack Pine		
	Date	Bias	r^2	Date	Bias	r^2	Date	Bias	r^2
OGS^{Fp}	5/19 ± 9			4/19 ± 12			4/20 ± 11		
OGS$^{\Sigma Fp}$	5/15 ± 7	−4 ± 6	0.59**	4/18 ± 11	−1 ± 11	0.33	4/23 ± 9	2 ± 8	0.47
OGS^{Sink}	5/25 ± 10	6 ± 3	0.91**	4/19 ± 11	0 ± 11	0.32	4/21 ± 11	1 ± 10	0.36
OGS$^{\lambda E}$	5/22 ± 11	4 ± 5	0.81**	4/21 ± 7	2 ± 9	0.40	4/19 ± 8	−1 ± 13	0.00
OGS$^{Thaw}_{Onset}$	4/05 ± 7	−44 ± 10	0.05	4/03 ± 7	−16 ± 10	0.33	4/06 ± 6	−14 ± 10	0.22
OGS$^{Thaw}_{End}$	4/23 ± 6	−26 ± 7	0.46*	4/29 ± 9	10 ± 9	0.47	4/26 ± 7	6 ± 5	0.85**
OGS^{SD}	5/19 ± 10	0 ± 2	0.95**	4/17 ± 12	−2 ± 11	0.31	4/20 ± 11	0 ± 9	0.40
OGS$^{\Sigma Ta}$	4/26 ± 7	−23 ± 8	0.30	4/27 ± 8	8 ± 9	0.41	4/27 ± 9	7 ± 9	0.41
OGS$^{\Sigma Ts}$	5/08 ± 7	−12 ± 7	0.50*	5/17 ± 7	28 ± 9	0.46	5/12 ± 7	22 ± 9	0.38
OGS^{NDVI}	5/20 ± 9	1 ± 2	0.97**	4/24 ± 6	5 ± 10	0.22	4/18 ± 6	−2 ± 10	0.23
EGS^{Fp}	9/29 ± 4			10/21 ± 10			10/21 ± 9		
EGS$^{\Sigma Fp}$	9/24 ± 3	−5 ± 3	0.62**	10/17 ± 7	−4 ± 10	0.15	10/18 ± 6	−3 ± 8	0.33
EGS^{Sink}	9/17 ± 5	−12 ± 4	0.32	9/29 ± 8	−22 ± 12	0.04	10/10 ± 9	10 ± 13	0.03
EGS$^{\lambda E}$	9/22 ± 5	−8 ± 7	0.07	10/03 ± 6	−18 ± 12	0.00	10/04 ± 11	−16 ± 10	0.01
EGS$^{Freeze}_{First}$	10/19 ± 11	20 ± 13	0.28	10/19 ± 11	−2 ± 2	0.95**	10/19 ± 6	−2 ± 3	0.95**
EGS$^{Freeze}_{Contiguous}$	11/07 ± 11	39 ± 13	0.12	11/07 ± 15	17 ± 15	0.09	11/10 ± 11	21 ± 12	0.06
EGS$^{\Sigma Ta}$	10/04 ± 8	5 ± 11	0.35	10/02 ± 9	−19 ± 13	0.04	10/02 ± 11	−19 ± 12	0.01
EGS$^{\Sigma Ts}$	11/03 ± 8	35 ± 11	0.44*	10/22 ± 9	1 ± 12	0.03	10/21 ± 9	1 ± 11	0.04
EGS^{NDVI}	9/29 ± 6	−1 ± 3	0.78**	10/18 ± 23	−3 ± 25	0.00	11/04 ± 7	15 ± 14	0.09

5   Climatic and Phenological Controls

Linear-regression analysis of the dependence of annual fluxes on climatic and GS variables (Sect. 2.4) produced many statistically significant relationships, despite the small sample size (10 years for the trembling aspen site and 7 years for the black spruce and jack pine sites). Table 5 shows selected variables that represent the most consistent and significant results. Of the three sites, the regression relationships were strongest at the trembling aspen site, which had the highest inter-annual variability and the largest sample size. Among fluxes, the relationships were strongest for F_P ($r^2 = 0.38$, mean across all sites and independent variables) and F_N ($r^2 = 0.34$), intermediate for F_R ($r^2 = 0.27$) and weakest for H ($r^2 = 0.23$) and λE ($r^2 = 0.18$). Among GS delineators (Table 2), the OGS group was most closely related to the fluxes ($r^2 = 0.33$, mean across all sites and fluxes), followed by LGS ($r^2 = 0.23$) then EGS ($r^2 = 0.14$).

Although the best independent variables, i.e. those that produced the most significant regression fits, varied by flux and site, many commonalities were observed across fluxes and sites. For all fluxes except λE, the best independent variables were consistent among sites. Annual F_N, F_R and F_P were most closely related to April–May T_s, OGS$^{\Sigma T_s}$, APAR, April–May T_a, annual T_s and annual T_a,

Table 5 Results of least-squares linear regression between annual fluxes (dependent variable) and selected independent variables, showing the slope (when $p \leq 20\%$) and statistical significance (* $p \leq 5\%$, ** $p \leq 1\%$)

Dependent variables	Site	Independent variables						
		OGS$^{\Sigma F_P}$ (days)	OGS$^{\Sigma T_s}$ (days)	APAR (mol m^{-2})	AM-T_s[1] (°C)	T_s[2] (°C)	SWC[3] (%)	P (mm)
H (MJ m^{-2})	Aspen	17.0*	14.5		−94	−220**	−12.3	−0.58
H (MJ m^{-2})	Spruce	11.5				−231	−38*	−0.96*
H (MJ m^{-2})	Pine	6.8					−74	−0.31
λE (MJ m^{-2})	Aspen	−17.7*	−12.4	0.18	90	193**	10.0	
λE (MJ m^{-2})	Spruce		−5.9*	0.15*	69*			
λE (MJ m^{-2})	Pine						44	0.22*
F_N (g C m^{-2})	Aspen	−8.9	−8.2	0.16**	59*	101*		
F_N (g C m^{-2})	Spruce		−2.6**	0.06**	27**	20		
F_N (g C m^{-2})	Pine		−2.9*	0.11	19*	30		
F_R (g C m^{-2})	Aspen	−9.5**	−7.4		48	106**	6.3	0.25
F_R (g C m^{-2})	Spruce	−1.6	−4.2*	0.10	42	44		
F_R (g C m^{-2})	Pine		−2.2	0.11	17	38*		
F_P (g C m^{-2})	Aspen	−18.5**	−15.5*	0.23**	106**	206**		
F_P (g C m^{-2})	Spruce	−3.1	−6.8**	0.16*	69*	64		
F_P (g C m^{-2})	Pine	−2.9	−5.1**	0.22**	36**	68**		

[1]Mean April–May soil temperature at 5-cm depth.
[2]Mean annual soil temperature at 5-cm depth.
[3]Mean warm-season soil volumetric water content at 0–30-cm depth.

with fits that were often significant at the 5% level. April-May T_s and $OGS^{\Sigma T_s}$ gave particularly strong fits, highlighting the importance of spring temperature for the carbon cycle of boreal forests (Baldocchi et al. 2001; 2005; Barr et al. 2004; 2007; Black et al. 2005). The goodness of fit was intermediate for other OGS and LGS variables and inconsistent or poor for EGS, SWC and P. In contrast, annual H was most closely related to water availability (SWC, P and P_{2y}), $OGS^{\Sigma T_s}$, $EGS^{\Sigma T_s}$ and $OGS^{\Sigma T_P}$. The response of λE varied among sites but not by plant functional type. Aspen and black spruce λE, like F_N, F_R and F_P, were closely related to April-May T_s, APAR, $OGS^{\Sigma T_s}$, April-May T_a, annual T_s, and annual T_a. In contrast, jack pine λE, like H, was most closely related to SWC and P, and showed little relation to temperature or OGS indices. In all cases, H responded negatively when λE, F_N, F_R and F_P responded positively, and vice versa. Warmer springs, earlier OGS, higher APAR and increased water availability all promoted F_N, F_R, F_P and λE but suppressed H. The magnitude of the responses varied by plant functional type. The impact of $OGS^{\Sigma T_s}$ and April–May T_s on F_N, F_R and F_P was 2–3 times greater at the trembling aspen than the black spruce and jack pine sites. The positive F_N response at all sites resulted from the differential sensitivity of F_P and F_R to spring warming.

Multiple linear regression with all possible combinations of independent variables highlighted the strong dependence of F_N, F_R and F_P on $OGS^{\Sigma F_{Ts}}$, APAR, and April–May T_s. Simple linear regression with these variables alone performed as well as any multiple combinations. The most significant two-variable regression models used various combinations of (a) APAR; (b) $OGS^{\Sigma T_s}$ or $OGS^{\Sigma T_{Fp}}$; (c) April–May T_s or annual T_s; and (d) SWC, P or P_{2y}. They often combined indicators that were sensitive to spring temperature (i.e. $OGS^{\Sigma T_s}$, April–May T_s and APAR) with indicators of growing-season water availability (i.e. SWC, P or P_{2y}). Although λE, F_N, F_R and F_P are known to depend on water availability (Barr et al. 2007), simple linear regression with SWC or P as the independent variable did not reveal this dependence. The dependence became evident only through multiple linear regression, when SWC or P was paired with APAR, $OGS^{\Sigma T_s}$, T_s or April–May T_s. We conclude that water availability is an important but secondary influence on annual carbon, water and energy fluxes, which are dominated by spring temperature and phenological controls (Law et al. 2002; Barr et al. 2007). The trembling aspen fluxes, in particular, have been shown to be sensitive to drought (Krishnan et al. 2006; Barr et al. 2007). For the black spruce site, Krishnan et al. (2008) found that low soil moisture in the extremely dry year of 2003 caused canopy conductance to decline by 50% and F_P to decline by 20%; however, λE was relatively unaffected since the vapour pressure deficit almost doubled. However, the net impact of drought on F_N at these sites has been shown to be neutral or even positive in some years because both F_P and F_R are suppressed by drought (Griffis et al. 2004; Barr et al. 2007; Krishnan et al. 2008).

The observed response of F_N to spring temperature and OGS is similar to reports from other sites, although not all studies have found a significant response. At the Harvard hardwood forest, White and Nemani (2003) found a strong positive relationship between annual F_N and the CUP (+5.4 g C m^{-2} d^{-1}) but, surprisingly, the relationship between F_N and canopy duration (the number of days in leaf) was not significant. They concluded that canopy duration was not a good determinant of the CUP (but contrast Barr et al.2004, who concluded the opposite using a different approach to estimate the canopy duration, based on high-LAI days only). Table 4 confirms that not all approaches to estimate OGS and LGS are equally robust. Other observations of a significant positive response of F_N to CUP include: +3.5 g C m^{-2} d^{-1} (temperate mixed broadleaf and coniferous forest, Carrara et al. 2003) and +3.2 g C m^{-2} d^{-1} (boreal black spruce, Krishnan et al. 2008). Further analysis of the Harvard forest data set (Urbanski et al. 2007) showed that the climatic controls on annual F_N were partially masked by long-term changes in woody biomass, vegetation succession and disturbance events. In contrast with the mixedwood Harvard forest, the sites in this study are dominated by one tree species. The relative stability of their biomass and species composition over the measurement period undoubtedly contributes to the robust and consistent responses observed among sites (Table 5). However, other studies have found F_N and F_P to be independent of CUP and LGS (Suni et al. 2003b). In a northern boreal black spruce forest, Dunn et al. (2007) reported a significant positive response of annual F_N to mean annual T_a but not LGS or OGS, which they attributed to a similar response of F_P and F_R to OGS and LGS. It may also be related to a long-term shift in site conditions (Urbanski et al. 2007), in the case of Dunn et al. (2007), a gradual rise in the water table that masked the effect of LGS and OGS on F_N.

Two cross-site synthesis studies of forest flux towers have shown that the cross-site response of F_N to CUP is similar to the inter-annual response at one site. Based on data from 11 temperate and boreal broadleaf-deciduous forests, Baldocchi et al. (2008) reported a tight fit between F_N and CUP, with a gain of 5.6 g C m^{-2} for each additional day of carbon uptake. Churkina et al. (2005) reported a similar result for 16 deciduous- broadleaf and six evergreen-coniferous forests, with greater sensitivity of F_N to CUP at deciduous-broadleaf (+5.8 g C m^{-2} d^{-1}) than evergreen-coniferous forests (+3.4 g C m^{-2} d^{-1}). The similarity of these cross-site sensitivities to inter-annual responses at one site is remarkable. It points to fundamental similarities in ecosystem functioning across a broad geographic range of temperate and boreal forests, with different sensitivities for evergreen-coniferous and broadleaf-deciduous forests (Churkina et al. 2005).

We were surprised that the complex analysis in this study produced such simple results– the dominant importance of spring temperature for annual F_N, F_P and F_R (Table 4). We were also surprised that, at all three sites in this study, simple metrics based on temperature alone such as April–May T_s, April–May T_a, OGS^{2Ts}, even annual T_s and annual T_a, were more closely related to annual F_N, F_R and F_P than the more complex OGS or LGS indices derived from the F_N, F_P and NDVI time series. We interpret this as another confirmation of the dominant influence of spring temperature on carbon uptake by boreal forests. In contrast with the simplicity of our results, Luyssaert et al. (2007) found that temperature alone was insufficient to explain seasonal and inter-annual variability in F_N from three temperate and boreal pine forests. They found it necessary to invoke a complex of environmental factors, including temperature, moisture and PAR and their seasonal variations.

The strong, positive response of F_N to T_s and T_a *at the annual time scale* was also surprising. Given the contrasting spring (positive) and autumn (negative) responses of F_N to temperature (Fig. 7; contrast Barr et al. 2004, 2007 with Piao et al. 2008), one would expect the annual F_N–T_s and F_N–T_a relationships to be weak. However, we found positive and statistically significant annual relationships for all three sites. Apparently, the strong F_P response to spring temperature, which promotes annual F_N, dominated the weaker F_R response to autumn temperature, which suppresses annual F_N. Two related observations substantiate this conclusion: monthly F_N was more variable among years in spring than in autumn (Fig. 4); and the linear-regression fits between annual C fluxes (F_N, F_R or F_P) and September-October temperature or EGS were weak.

Two objections have been raised to annual regression analyses such as reported in Table 5. First, cross correlations and interactions are common among environmental variables, making it difficult to isolate the primary controlling factors. This difficulty is partially resolved by identifying relationships that are consistent among multiple sites and years (Lindroth et al. 2008). Second, the results of annual analyses may depend on the dates used to delineate the annual cycle, e.g. January–December vs. July–June (Luyssaert et al. 2007). While the latter argument has merit if the start and end dates occur in periods of high inter-annual variability, it is not valid if the start and end dates occur during an extended quiescent period with low interannual variability, such as the dormant cold-season in this study (Figs. 2 and 4).

6   Temperature Controls on the Seasonal Phenology of Deciduous-Broadleaf and Evergreen-Coniferous Forests

Our results highlight important differences in the interplay of climate, phenology and the carbon cycle between evergreen-coniferous and broadleaf-deciduous boreal forests, associated with differences in their seasonal LAI and F_P cycles. In this study, LGS was 50 days longer for black spruce and jack pine forests than for a trembling aspen forest. The evergreen-coniferous GS spanned the warm season, with F_P beginning during spring thaw and ending during autumn freeze. The deciduous trembling aspen GS, in contrast, was delimited by leaf emergence and senescence. The timing of leaf emergence and senescence, well after spring thaw and before autumn freeze, reduces the risk of frost damage to the non-hardy trembling aspen foliage. Late spring frosts have been linked to abnormal leaf flushing and dieback of trembling aspen (Zalasky 1976). In contrast, the evergreen-conifer foliage, which is cold tolerant (Ensminger et al. 2004, 2008), is able to exploit the transitional seasons of spring and autumn, as evidenced by the rapid rise in F_N and λE during thaw and the temporary up-regulation of F_P during short, warm periods before and after the contiguous, core CUP. Boreal conifers thus maximize carbon uptake during the spring and autumn transitions while being carbon neutral during the core summer months, whereas the more productive trembling aspen forests maximize carbon uptake during the core summer months while losing carbon during the shoulder seasons (Fig. 4).

The seasonal phenology of boreal forests is coupled to the annual temperature cycle. The relative role of T_a and T_s as phenological controls is still debated (Suni et al. 2003a). Simple degree-day models perform well in predicting the timing of bud burst (Häkkinen et al. 1998; Linkosalo 2000) and leaf unfolding and expansion (Lechowicz 1984; Hunter and Lechowicz 1992; Barr et al. 2004). Our results at the trembling aspen site show remarkable agreement among three independent OGS indices (OGS$^{\Sigma D}$, OGS^{Fp} and OGS^{NDVI}), adding support to the utility of degree-day models, albeit using a local ΣD threshold (Table 2). The relationship of deciduous forest OGS to T_s is less well documented. In a synthesis of 11 temperate deciduous forests, Baldocchi et al. (2005) reported close agreement between OGS^{Sink} and the date in spring when mean daily T_s equaled mean annual T_a. However, their relationship was limited to temperate sites where mean annual T_a was well above 0°C; it does not apply to boreal regions. At the boreal trembling aspen site in this study, OGS^{Fp} corresponded to 5-cm T_s of 6.5 ± 1.3°C, which is similar to the coldest of the temperate forest sites in Baldocchi et al. (2005). This may represent a lower T_s limit for the onset of the CUP in boreal and temperate deciduous forests.

In boreal coniferous ecosystems, the spring recovery of photosynthesis is also strongly linked to T_a (Lundmark et al. 1988; Suni et al. 2003a), whereas the role of T_s is less certain. Troeng and Linder (1982) observed that the spring recovery of photosynthesis in Scots pine always began after OGS$^{Thaw}_{Onset}$. Bergh and Linder (1999) concluded that the timing of photosynthetic recovery in Norway spruce was primarily controlled by mean daily T_a and the frequency of severe frosts, but showed a small but significant effect of artificial soil warming. However, Suni et al. (2003a) argued that T_s and soil thaw were not useful criteria for identifying conifer OGS^{Fp} because OGS^{Fp} always preceded OGS$^{Thaw}_{End}$ and was therefore not limited by it. While we agree that OGS^{Fp} precedes OGS$^{Thaw}_{End}$, we do not agree that OGS^{Fp} is independent of T_s. Our observations show that conifer OGS^{Fp} consistently occurred within the thaw period, a 2–3 week period during which the melting snowpack and thawing soil become isothermal at 0°C (Fig. 5). Similar findings were reported by Monson et al. (2005). We conclude that the initiation of spring snowmelt and soil thaw is a necessary antecedent to OGS^{Fp} in boreal coniferous forests.

A related issue is the role of plant water relations in the spring recovery of photosynthesis. Jarvis and Linder (2000) postulated that the startup of photosynthesis in boreal coniferous forests depends on spring thaw and the resulting availability of soil water for transpiration. Suni et al. (2003a) disagreed and argued that, even at freezing temperatures, there is sufficient liquid water storage in the tree bole to satisfy the transpirational demands during spring thaw, even if soil water remains unavailable. Both may be true in part. Although liquid water exists in frozen xylem tissue and soil (Suni et al. 2003a), it is held at such low matric potentials that it is unavailable for transpiration. The freezing liquid water retention curve for soil (Spaans and Baker 1996), which we assume to be similar to the relationship for wood, shows that matric potential drops sharply as the temperature drops below 0°C, e.g., to 2.4 MPa by −2 C. However, when the tree bole is thawed or thawing, water from the bole may be able to supply the small transpirational demand that exists before the soil thaws and soil water becomes available, thus enabling the stomata to open in spring. Mellander et al. (2004) observed severe soil water supply limitations to transpiration in Scots pine for 1 week after the onset of spring thaw.

The small transpirational fluxes were apparently satisfied through stored water from the tree boles. Only after the soil had warmed to well above 0°C was the soil water supply limitation completely removed.

The net impact of climate warming on the carbon, water and energy cycles of northern forests remains uncertain. Because the climate of the boreal forest is characteristically cold, warming might be expected to increase forest productivity (Hogg 2002). This and other studies (Goulden et al. 1996; Black et al. 2000; Hollinger et al. 2004; Barr et al. 2007) show a positive impact of warming on F_P, F_N and λE, particularly if significant warming occurs in spring. The impact of warming is manifest through a longer CUP and an increase in APAR (Barr et al. 2007). The effect is particularly strong for high productivity deciduous-broadleaf forests. However, the direct, positive impact of spring warming on F_P and F_N may be offset by increased drought and high-temperature stress in summer (Ciais et al. 2005) and increased F_R in autumn (Piao et al. 2008). It may also be offset by indirect, climate-induced changes in forest structure, expressed through e.g. a delayed, drought-induced reduction in LAI (Barr et al. 2004; Krishnan et al. 2006), drought-induced forest dieback as recently observed in the transitional aspen parklands of western Canada (Hogg et al. 2008), a northward movement of the boreal treeline (Rupp et al. 2001), increased frequency of disturbance by fire (Kurz et al. 2007) and insects (Kurz et al. 2008), and changes in vegetation succession following disturbance (Chapin et al. 2004).

7   Summary and Conclusions

Climatic impacts on the carbon, water and energy cycles of boreal forest ecosystems are often mediated through plant phenology, with contrasting responses of deciduous-broadleaf and evergreen-coniferous forests. The primary determinant of inter-annual variability in net ecosystem production F_N is spring temperature, via its impact on the timing of leaf emergence and expansion (deciduous-broadleaf forests) and spring thaw (evergreen-coniferous forests). In this study, the core, contiguous carbon-uptake period CUP, as delineated by F_N, was ~50 days longer for black spruce and jack pine stands than a trembling aspen stand, with 30 days difference in spring and 20 days difference in autumn. The seasonal cycles of F_N and evapotranspiration were tightly coupled to the seasonal cycle of soil temperature. The black spruce and jack pine CUP spanned the warm season, beginning during spring thaw and ending when air temperature dropped consistently below freezing in autumn. Positive air temperature and near-zero soil temperature, which indicated soil thaw, were mutual requirements for the startup of photosynthesis in spring. In many years, the evergreen-coniferous sites had secondary periods of carbon uptake outside the contiguous CUP. In contrast, the trembling aspen CUP was determined by the timing of leaf emergence and senescence, which occurred well after spring thaw and before autumn freeze.

Not all approaches to delineate the growing season GS were equally effective, as compared to the CUP. The F_N time series provided distinct GS boundaries whereas the surface energy balance did not. The best climatic delineators of the GS varied by plant functional type. For the deciduous-broadleaf trembling aspen site, the GS was determined by leaf emergence and senescence, which in turn were closely related to degree-day accumulation in spring and NDVI changes in spring and autumn. For the evergreen-coniferous black spruce and jack pine sites, the GS was best defined by spring thaw and autumn freeze.

Regression analysis confirmed the strong dependence of inter-annual variability in carbon, water and energy fluxes on climatic and GS factors. Among sites, the linear regression relationships between annual fluxes and climatic factors were strongest for the trembling aspen site. Among fluxes, the relationships were strongest for gross ecosystem photosynthesis and F_N, and weakest for the sensible and latent heat fluxes. Among the GS delineators, the fluxes were most strongly related to the onset date and most weakly related to the end date. At all sites, the strongest climatic determinants of the annual C fluxes, including photosynthesis and respiration, were based on temperature: April–May mean soil and air temperature, annual mean soil and air temperature, and the GS onset date. The sensitivity of F_N to temperature was 2–3 times greater for the trembling aspen than the black spruce and jack pine ecosystems. At all three sites, the positive response of F_N to spring temperature resulted from the differential sensitivity of photosynthesis and respiration to spring warming. Water availability (soil water content and precipitation) had a significant but secondary influence on the annual C fluxes.

Acknowledgements We gratefully acknowledge the contributions of Charmaine Hrynkiw, Dell Bayne, Erin Thompson, Joe Eley, Alison Theede, Bruce Cole, Craig Smith and Steve Enns, who oversaw the meteorological measurements and data management; Zoran Nesic, Andrew Sauter, Rick Ketler, Dominic Lessard, Dan Finch and Sheila McQueen, who provided laboratory, field and data management support for the flux measurements; and Barry Goodison and Bob Stewart, who championed the BERMS program. Financial support was provided by the Climate Research Division of Environment Canada, the Canadian Forest Service, Parks Canada, the Action Plan 2000 on Climate Change, the Program of Energy Research and Development, the Climate Change Action Fund, the Natural Sciences and Engineering Research Council of Canada, the Canadian Foundation for Climate and Atmospheric Sciences, the BIOCAP Canada Foundation, and the National Aeronautic and Space Administration.

References

Angert, A., Biraud, S., Bonfils, C., Henning, C. C., Buermann, W., Pinzon, J., Tucker, C. J. and Fung, I. (2005) Drier summers cancel out the CO2 uptake enhancement induced by warmer springs. Proc. Natl. Acad. Sci. USA 102, 10823–10827.

Baldocchi, D. D., Falge, E., Gu, L. H., Olson, R., Hollinger, D., Running, S., Anthoni, P., Bernhofer, C., Davis, K., Evans, R., Fuentes, J., Goldstein, A., Katul, G., Law, B., Lee, X. H., Malhi, Y., Meyers, T., Munger, W., Oechel, W., Paw, U. K. T., Pilegaard, K., Schmid, H. P., Valentini, R., Verma, S., Vesala, T., Wilson, K. and Wofsy, S. (2001) FLUXNET: A new tool to study the temporal and spatial variability of ecosystem-scale carbon dioxide, water vapor, and energy flux densities. Bull. Am. Meteorol. Soc. 82, 2415–2434.

Baldocchi, D. D., Black, T. A., Curtis, P. S., Falge, E., Fuentes, J.D., Granier, A., Gu, L., Knohl, A., Pilegaard, K., Schmid, H. P., Valentini, R., Wilson, K., Wofsy, S., Xu, L. and Yamamoto,

S. (2005) Predicting the onset of net carbon uptake by deciduous forests with soil temperature and climate data: a synthesis of FLUXNET data. Int. J. Biomet. 49, 377–387.

Baldocchi, D. (2008) Breathing of the terrestrial biosphere: lessons learned from a global network of carbon dioxide flux measurement systems. Aust. J. Bot. 56, 1–26.

Barr, A. G., Griffis, T.J., Black, T.A., Lee, X., Staebler, R. M., Fuentes, J. D., Chen Z. and Morgenstern, K. (2002) Comparing the carbon balances of boreal and temperate deciduous forest stands. Can. J. For. Res. 32, 813–822.

Barr, A. G., Black, T. A., Hogg, E. H., Kljun, N., Morgenstern, K., and Nesic, Z. (2004) Inter-annual variability in the leaf area index of a boreal aspen-hazelnut forest in relation to net ecosystem production. Agric. For. Meteorol. 126, 237–255.

Barr, A. G., Morgenstern, K., Black, T. A., McCaughey, J. H., and Nesic Z. (2006) Surface energy balance closure by the eddy-covariance method above three boreal forest stands and implications for the measurement of the CO2 flux. Agric. For. Meteorol. 140, 322–337.

Barr, A. G., Black, T. A., Hogg, E. H., Griffis, T. J., Morgenstern, K., Kljun, N., Theede, A., and Nesic, Z. (2007) Climatic controls on the carbon and water balances of a boreal aspen forest, 1994–2003. Global Change Biol. 13, 561–576.

Bergh, J. and Linder, S. (1999) Effects of soil warming on photosynthetic recovery in boreal Norway spruce stands. Global Change Biol. 5, 245–253.

Black, T. A., Chen, W. J., Barr, A. G., Arain, M. A., Chen, Z., Nesic, Z., Hogg, E. H., Neumann, H. H., Yang, P. C. (2000) Increased carbon sequestration by a boreal deciduous forest in years with a warm spring. Geophys. Res. Lett. 27, 1271–1274.

Black, T. A., Gaumont-Guay, D., Jassal, R. S., Amiro, B., Jarvis, P. J., Gower, T., Kelliher, F., Dunn, A. and Wofsy, S. (2005) Measurement of CO2 exchange between boreal forest and the atmosphere. In The Carbon Balance of Terrestrial Biomes, eds. H. Griffiths and P.J. Jarvis, pp. 120–141. Garland Science/BIOS Scientific Publishers, Oxford.

Blanken, P. D., Black, T. A., Yang, P. C., Neumann, H. H., Nesic, Z., Staebler, R., den Hartog, G., Novak, M. D. and Lee, X. (1997) Energy balance and canopy conductance of a boreal aspen forest: partitioning overstory and understory components. J. Geophys. Res. 102, 28915–28928.

Bonan, G. B. and Shugart, H. H. (1989) Environmental factors and ecological processes in boreal forests. Annu. Rev. Ecol. Evol. Syst. 20, 1–28.

Brooks, J. R., Sprugel, D. G. and Hinckley, T. M. (1996) The effects of light acclimation during and after foliage expansion on photosynthesis of Abies amabilis foliage within the canopy. Oecologia 107, 21–32.

Carrara, A., Kowalski, A. S., Neirynck, J., Janssens, I. A., Yuste, J. C. and Ceulemans, R. (2003) Net ecosystem CO2 exchange of mixed forest in Belgium over 5 years. Agric. For. Meteorol. 119, 209–227.

Chen, J. M., Govind, A., Sonnentag, O., Zhang, Y., Barr, A. and Amiro, B. (2006) Leaf area index measurements at Fluxnet-Canada forest sites. Agric. For. Meteorol. 140, 257–268.

Chen,W., Black, T. A., Yang, P., Barr, A. G., Neumann, H.H., Nesic, Z., Novak, M. D., Eley, J., Ketler, R. and Cuenca, C. (1999) Effects of climatic variability on the annual carbon sequestration by a boreal aspen forest. Global Change Biol. 5, 41–53.

Churkina, G., Schimel, D., Braswell, B. H. and Xiao, X. M. (2005) Spatial analysis of growing season length control over net ecosystem exchange. Global Change Biol. 11, 1777–1787.

Ciais, P., Reichstein, M., Viovy, N., Granier, A., Ogée, J., Allard, V., Aubinet, M., Buchmann, N., Bernhofer, Chr., Carrara, A., Chevallier, F., De Noblet, N., Friend, A. D., Friedlingstein, P., Grünwald, T., Heinesch, B., Keronen, P., Knohl, A., Krinner, G., Loustau, D., Manca, G., Matteucci, G., Miglietta, F., Ourcival, J. M., Papale, D., Pilegaard, K., Rambal, S., Seufert, G., Soussana, J. F., Sanz, M. J., Schulze, E. D., Vesala, T. and Valentini, R. (2005) Europe-wide reduction in primary productivity caused by the heat and drought in 2003. Nature 437, 529–533.

Cleland, E. E., Chuine, I., Menzel, A., Mooney, H. A. and Schwarz, M. D. (2007) Shifting seasonal phenology in response to global change. Trends Ecol. Evol. 22, 357–365.

Chapin, F. S. III, Callaghan, T. V., Bergeron, Y., Fukada, M., Johnstone, J. F., Juday, G. and Zimov, S. A. (2004) Global change and the boreal forest: thresholds, shifting states or gradual change? Ambio 33, 361–365.

Dunn, A. L., Barford, C. C., Wofsy, S. C., Goulden, M. L. and Daube, B. C. (2007) A long-term record of carbon exchange in a boreal black spruce forest: means, responses to interannual variability, and decadal trends. Global Change Biol. 13, 577–590.

Ensminger, I., Sveshnikov, D., Campbell, D. A., Funk, C., Jansson, S., Lloyd, J., Shibistova, O. and Öquist, G. (2004) Intermittent low temperatures constrain spring recovery of photosynthesis in boreal Scots pine forests. Global Change Biol. 10, 995–1008.

Ensminger, I., Schmidt, L. and Lloyd J. (2008) Soil temperature and intermittent frost modulate the rate of recovery of photosynthesis in Scots pine under simulated spring conditions. New Phytol. 177, 428–442.

Espinosa-Ruiz, A., Saxena, S., Schmidt, J., Mellerowicz, E., Miskolczi, P., Bako, L. and Bhalerao, R. (2004) Differential stage-specific regulation of cyclin-dependent kinases during cambial dormancy in hybrid aspen. Plant J. 38, 603–615.

Gaumont-Guay, D., Margolis, H. A., Bigras, F. J. and Raulier, F. (2003) Characterizing the frost sensitivity of black spruce photosynthesis during cold acclimation. Tree Physiol. 5, 301–311.

Gower, S. T., Vogel, J. G., Norman, J. M., Kucharik, C. J., Steele, S. J. and Stow, T. K. (1997) Carbon distribution and aboveground net primary production in aspen, jack pine, and black spruce stands in Saskatchewan and Manitoba. Canada. J. Geophys. Res. 102, 29029–29041.

Goulden, M. L., Munger, J. W., Fan, S. -M., Daube, B. C. and Wofsy, S. C. (1996) CO2 exchange by a deciduous forest: response to interannual climate variability. Science 271, 1576–1578.

Goulden, M. L., Wofsy, S. C., Harden, J. W., Trumbore, S. E., Crill, P. M., Gower, S. T., Fries, T., Daube, B. C., Fan, S. M., Sutton, D. J., Bazzaz, A. Munger, J. W. (1998) Sensitivity of boreal forest carbon balance to soil thaw. Science 279, 214–217.

Gratani, L. and Ghia, E. (2002) Changes in morphological and physiological traits during leaf expansion of *Arbutus unedo*. Env. Exp. Bot. 48, 51–60.

Grelle, A., Lindroth, A. and Mölder, M. (1999) Seasonal variation of boreal forest surface conductance and evaporation. Agric. For. Meteorol. 98–99, 563–578.

Griffis, T. J., Black, T. A., Morgenstern, K., Barr, A. G., Nesic, Z., Drewitt, G., Gaumont-Guay, D. and McCaughey, J. H. (2003) Ecophysiological controls on the carbon balances of three southern boreal forests. Agric. For. Met. 117, 53–71.

Griffis, T. J., Black, T. A., Gaumont-Guay, D., Drewitt, G. B., Nesic, Z., Barr, A. G., Morgenstern, K. and Kljun, N. (2004) Seasonal variation and partitioning of ecosystem respiration in a southern boreal aspen forest. Agric. Forest Meteorol. 125, 207–223.

Gu, L., Post, W. M., Baldocchi, D. D., Black, A., Verma, S., Vesala, T. and Wofsy, S. (2003) Phenology of vegetation photosynthesis. In Phenology: an Integrative Science, ed. M.D. Schwartz, pp. 467–488. Dordrecht: Kluwer.

Hall, F. G., Knapp, D. E. and Huemmrich, K. F. (1997) Physically based classification and satellite mapping of biophysical characteristics in the southern boreal forest. J. Geophys. Res. 102, 29567–29580.

Häkkinen, R. and Hari, P. (1988) The efficiency of time and temperature driven regulation principles in plants at the beginning of the active period. Silva Fenn 22, 163–170.

Hogg, E. H. (1997) Temporal scaling of moisture and the forest-grassland boundary in western Canada. Agric. For. Meteorol. 84, 115–122.

Hogg, E. H. (2002) Boreal forest. In: Encyclopedia of Global Environmental Change. Volume 2, The Earth System: Biological and Ecological Dimensions of Global Environmental Change, ed. H. A. Mooney and J. G. Canadell, pp. 179–184. Chichester: Wiley.

Hollinger, D. Y., Aber, J., Dail, B., Davidson, E. A., Goltz, S. M., Hughes, H., LeClerc, M. Y., Lee, J. T., Richardson, A. D., Rodrigues, C., Scott, N. A., Achuatavarier, D. and Walsh, J. (2004) Spatial and temporal variability in forest-atmosphere CO2 exchange. Global Change Biol. 10, 1689–1706.

Huemmrich, K. F., Black, T. A., Jarvis, P. G., McCaughey, J. H. and Hall, F. G., (1999) High temporal resolution NDVI from micrometeorological radiation sensors. J. Geophys. Res. 104, 27935–27944.

Huner, N. P. A., Öquist, G., Hurry, V. M., Krol, M., Falk, S. and Griffith M. (1993) Photosynthesis, photoinhibition and low temperature acclimation in cold tolerant plants. Photosyn. Res. 37, 19–139.

Hunter, A. F. and Lechowicz, M. J. (1992) Predicting the timing of budburst in temperate trees. J. Appl. Ecol. 29, 597–604.

Jarvis, P. and Linder, S. (2000) Constraints to growth of boreal forests. Nature 405, 904–905.

Kljun N., Black, T.A., Griffis, T.J., Barr, A. G., Gaumont-Guay, D., Morgenstern, K., McCaughey, J. H. and Nesic, Z. (2007) Response of net ecosystem productivity of three boreal forest stands to drought. Ecosystems, 10, 1039–1055.

Krishnan, P., Black, T. A., Grant, N. J., Barr, A. G., Hogg, E. H., Jassal, R. S. and Morgenstern, K. (2006) Impact of changing soil moisture distribution on net ecosystem productivity of a boreal aspen forest during and following drought. Agric. Forest Meteorol. 139, 208–223.

Krishnan, P., Black, T. A., Barr, A. G., Grant, N. J., Gaumont-Guay, D., and Nesic, Z. (2008) Factors controlling the interannual variability in the carbon balance of a southern boreal black spruce forest. J. Geophys. Res. 113, D09109, doi:10.1029/2007JD008965.

Kucharik, C. J., Norman, J. M. and Gower, S.T. (1998) Measurements of branch area and adjusting leaf area index indirect measurements. Agric. For. Meteorol. 91, 69–88.

Kurz, W. A., Stinson, G. and Rampley, G. (2007) Could increased boreal forest ecosystem productivity offset carbon losses from increased disturbances? Philos. Trans. R. Soc. Lond. B 363, 2261–2269.

Kurz, W. A., Dymond, C. C., Stinson, G., Rampley, G. J., Neilson, E. T., Carroll, A. L., Ebata, T. and Safranyik, L. (2008) Mountain pine beetle and forest carbon feedback to climate change. Nature 452, 987–990

Landsberg, J.J. and Gower, S.T. (1997) Applications of Physiological Ecology to Forest Management. San Diego, CA: Academic.

Law, B. E., Falge, E., Gu, L., Baldocchi, D. D., Bakwin, P., Berbigier, P., Davis, K., Dolman, A. J., Falk, M., Fuentes, J. D. (2002) Environmental controls over carbon dioxide and water vapor exchange of terrestrial vegetation. Agric. For. Meteorol. 113, 97–120.

Lechowicz, M. J. (1984) Why do temperate deciduous trees leaf out at different times? Adaptation and ecology of forest communities. Amer. Nat. 124, 821–842.

Leith, H. (1974) Phenology and Seasonality Modeling. Springer: Berlin.

Lindgren, K. and Hällgren, J.-E. (1993) Cold acclimation of Pinus contorta and Pinus sylvestris assessed by chlorophyll fluorescence. Tree Physiol. 13, 97–106.

Lindroth, A., Lagergren, F., Aurela, M., Bjarnadottir, B., Christensen, T., Dellwik, E., Grelle, A., Ibrom, A., Johansson, T., Lankreijer, H., Launiainen, S., Laurila, T., Mölder, M., Nikinmaa, E., Pilegaard, K., Sigurdsson, B. D. and Vesala, T. (2008) Leaf area index is the principal scaling parameter for both gross photosynthesis and ecosystem respiration of Northern deciduous and coniferous forests. Tellus B 60, 129–142.

Lundmark, T., Hedén, J. and Hällgren, J.-E. (1988) Recovery from winter depression of photosynthesis in pine and spruce. Trees 2, 110–114.

Luyssaert, S., Janssens, I. A., Sulkava, M., Papale, D., Dolman, A. J., Reichstein, M., Hollmén, J., Martin, J. G., Suni, T., Vesala, T., Loustau, D., Law, B. E. and Moors, E. J. (2007) Photosynthesis drives anomalies in net carbon-exchange of pine forests at different latitudes. Global Change Biol. 13, 2110–2127.

MacDonald, K., Kimball, J. S., Njoku, E., Zimmerman, R. and Zhao, M. (2004) Variability in springtime thaw in the terrestrial high latitudes: monitoring a major control on the biospheric assimilation of atmospheric CO_2 with spaceborne microwave remote sensing. Earth Interact. 8, 1–23.

Mackay, D. S., Ewers, B. E., Cook, B. D. and Davis, K. J. (2007) Environmental drivers of evapotranspiration in a shrub wetland and an upland forest in northern Wisconsin. Water Resour. Res. 43, W03442, doi:10.1029/2006WR005149.

McMillan, A. M. S., Winston, G. C. and Goulden, M. L. (2008) Age-dependent response of boreal forest to temperature and rainfall variability. Global Change Biol. 14, 1–13.

Margolis, H. A., Flanagan, L. B., and Amiro, B.D. (2006) The Fluxnet-Canada Research Network: Influence of climate and disturbance on carbon cycling in forests and peatlands. Agric. For. Meteorol. 140, 1–5.

Mellander, P.-E., Bishop, K. and Lundmark, T. (2004) The influence of soil temperature on transpiration: a plot scale manipulation in a young Scots pine stand. Forest Ecol. Manag. 195, 15–28.

Monson, R. K., Sparks, J. P., Rosenstiel, T. N., Scott-Denton, L. E., Huxman, T. E., Harley, P. C., Turnipseed,A. A., Burns, S. P, Backlund, B. and Jia, H. (2005) Climatic influences on net ecosystem CO2 exchange during the transition from wintertime carbon source to springtime carbon sink in a high-elevation, subalpine forest. Oecologia 146, 130–147.

Morgenstern, E. K. (1996) Geographic Variation in Forest Trees. Vancouver: University of British Columbia Press.

Morgenstern, K., Black, T. A., Humphreys, E. R., Griffis, T. J., Drewitt, G. B., Cai, T., Nesic, Z., Spittlehouse, D. L. and Livingston, N. J. (2004) Sensitivity and uncertainty of the carbon balance of a Pacific Northwest Douglas-fir forest during an El Nino La Nina cycle. Agric. For. Meteorol. 123, 201–219.

Myneni, R. B., Keeling, C. D., Tucker, C. J., Asrar, G. and Nemani, R. R. (1997) Increased plant growth in the northern high latitudes from 1981 to 1991. Nature 386, 698–702.

Ottander, C., Campbell, D. and Öquist, G. (1995) Seasonal changes in photosystem II organisation and pigment composition in Pinus *sylvestris*. Planta 197, 176–183.

Öquist G., Gardestrom, P. and Huner, N.P.A. (2001) Metabolic changes during cold acclimation and subsequent freezing and thawing. In Conifer Cold Hardiness, Vol. 1, ed. S.J. Colombo, pp. 137–163. Dordrecht: Kluwer.

Piao, S., Ciais, P., Friedlingstein, P., Peylin, P., Reichstein, M., Luyssaert, S., Margolis, H., Fang, J., Barr, A., Chen, A., Grelle, A., Hollinger, D. Y., Laurila., T., Lindroth, A., Richardson, A. D. and Vesala, T. (2008) Net carbon dioxide losses of northern ecosystems in response to autumn warming. Nature 451, 49–52.

Richardson, A. D., Hollinger, D. Y., Aber, J., Ollinger, S. V., Braswell, B. (2007) Environmental variation is directly responsible for short- but not longterm variation in forest-atmosphere carbon exchange. Global Change Biol. 13, 788–803.

Rupp, T. S., Chapin. F. S. III and Starfield, A. M. (2001) Modeling the influence of topographic barriers on treeline advance at the forest-tundra ecotone in northwestern Alaska. Climatic Change 48, 399–416.

Savitch, L. V., Leonardos, E. D., Krol, M., Jansson, S., Grodzinski, B., Huner, N. P. A. and Öquist,G. (2002) Two different strategies for light utilization in photosynthesis in relation to growth and cold acclimation. Plant Cell Environ. 25, 761–771.

Sellers, P. J., Hall, F. G., Kelly, R. D., Black, A., Baldocchi, D., Berry, J., Ryan, M., Ranson, J. K., Crill, P. M., Lettenmaier, D. P., Margolis, H., Cihlar, J., Newcomer, J., Fitzjarrald, D., Jarvis, P. G., Gower, S. T., Halliwell, D., Williams, D., Goodison, B., Wickland, D. E. and Guertin, F. E. (1997) BOREAS in 1997: Experiment overview, scientific results, and future directions. J. Geophys. Res. 102, 28731–28769.

Slaney, M. (2006) Impact of Elevated Temperature and [CO₂] on Spring Phenology and Photosynthetic Recovery of Boreal Norway Spruce. Doctoral thesis, Swedish University of Agricultural Sciences, Alnarp, Sweden, 47 pp.

Spaans, E. J. A. and Baker, J. M. (1996) The soil freezing characteristic: its measurement and similarity to the soil moisture characteristic. Soil Sci. Soc. Am. J. 60, 13–19.

Stocks, B. J., Mason, J. A., Todd, J. B., Bosch, E. M., Wotton, B. M., Amiro, B. D., Flannigan, M. D., Hirsch, K. G., Logan, K. A., Martel, D. L. and Skinner, W. R. (2002) Large forest fires in Canada, 1959–1997. J. Geophys. Res. 108, FFR5.1–FFR5.12.

Suni, T., Berninger, F., Vesala, T., Markkanen, T., Hari, P., Mäkelä, A., Ilvesniemi, H., Hänninen, H., Nikinmaa, E., Huttula, T., Laurila, T., Aurela, M., Grelle, A., Lindroth, A., Arneth, A., Shibistova, O. and Lloyd, J. (2003a) Air temperature triggers the recovery of evergreen boreal forest photosynthesis in spring. Global Change Biol. 9, 1410–1426.

Suni, T., Berninger, F., Markkanen, T., Keronen, P., Rannik, Ü. and T. Vesala. (2003b) Interannual variability and timing of growing-season CO₂ exchange in a boreal forest. J. Geophys. Res. 108, 4265, doi:10.1029/2002JD002381.

Troeng, E. and Linder, S. (1982) Gas exchange in a 20-year-old stand of Scots pine. I. Net photosynthesis of current and 1-year-old shoots within and between seasons. Physiol. Plantarum 54, 7–14.

Urbanski, S., Barford, C., Wofsy, S., Kucharik, C., Pyle, E., Budney, J., McKain, K., Fitzjarrald, D., Czikowsky, M. and Munger, J. W. (2007) Factors controlling CO2 exchange on timescales from hourly to decadal at Harvard Forest. J. Geophys. Res. 112, G02020, doi:10.1029/2006JG000293.

Welp, L. R., Randerson, J. T. and Liu, H.P. (2007) The sensitivity of carbon fluxes to spring warming and summer drought depends on plant functional type in boreal forest ecosystems. Agric. For. Meteorol. 147, 172–185.

White, M. A., Running, S. W. and Thornton, P. E. (1999) The impact of growing-season length variability on carbon assimilation and evapotranspiration over 88 years in the eastern US deciduous forest. Int. J. Biometeorol. 42, 139–145.

White, M. A. and Nemani, R. R. (2003) Canopy duration has little influence on annual carbon storage in the deciduous broad leaf forest. Global Change Biol. 9, 967–972.

Wilson, K. B. and Baldocchi, D. D. (2000) Seasonal and interannual variability of energy fluxes over a broadleaved temperate deciduous forest in North America. Agric. For. Meteorol. 100, 1–18.

Zalasky, H., (1976) Frost damage in poplar on the prairies. For. Chron. 52, 61–64.

Characterizing the Seasonal Dynamics of Plant Community Photosynthesis Across a Range of Vegetation Types

Lianhong Gu, Wilfred M. Post, Dennis D. Baldocchi, T. Andrew Black, Andrew E. Suyker, Shashi B. Verma, Timo Vesala, and Steve C. Wofsy

Abstract The seasonal cycle of plant community photosynthesis is one of the most important biotic oscillations to mankind. This study built upon previous efforts to develop a comprehensive framework to studying this cycle systematically with eddy covariance flux measurements. We proposed a new function to represent the cycle and generalized a set of phenological indices to quantify its dynamic characteristics. We suggest that the seasonal variation of plant community photosynthesis generally consists of five distinctive phases in sequence each of which results from the interaction between the inherent biological and ecological processes and the progression of climatic conditions and reflects the unique functioning of plant community at different stages of the growing season. We applied the improved methodology to seven vegetation sites ranging from evergreen and deciduous forests to crop to grasslands and covering both cool-season (vegetation active during cool months,

L. Gu (✉) and W.M. Post
Environmental Sciences Division, Oak Ridge National Laboratory, Oak Ridge, TN, USA
e-mail: lianhong-gu@ornl.gov and wmp@ornl.gov

D.D. Baldocchi
Department of Environmental Science, Policy and Management, University of California
Berkeley, CA, USA
e-mail: baldocchi@nature.berkeley.edu

T.A. Black
Land and Food Systems, University of British Columbia, Vancouver, BC, USA
e-mail: ablack@interchange.ubc.ca

A.E. Suyker and S.B. Verma
School of Natural Resources, University of Nebraska Lincoln, Lincoln, NE, USA
e-mail: asuykerl@unl.edu and sverma1@unl.edu

T. Vesala
Department of Physics, University of Helsinki, Helsinki, Finland
e-mail: timo.vesala@helsinki.fi

S.C. Wofsy
Atmospheric and Environmental Sciences, Harvard University, Cambridge, MA, USA
e-mail: scw@io.harvard.edu

A. Noormets (ed.), *Phenology of Ecosystem Processes*,
DOI 10.1007/978-1-4419-0026-5_2, © Springer Science+Business Media, LLC 2009

e.g. Mediterranean climate grasslands) and warm-season (vegetation active during warm months, e.g. temperate and boreal forests) vegetation types. Our application revealed interesting phenomena that had not been reported before and pointed to new research directions. We found that for the warm-season vegetation type, the recovery of plant community photosynthesis at the beginning of the growing season was faster than the senescence at the end of the growing season while for the cool-season vegetation type, the opposite was true. Furthermore, for the warm-season vegetation type, the recovery was closely correlated with the senescence such that a faster photosynthetic recovery implied a speedier photosynthetic senescence and vice versa. There was evidence that a similar close correlation could also exist for the cool-season vegetation type, and furthermore, the recovery-senescence relationship may be invariant between the warm-season and cool-season vegetation types up to an offset in the intercept. We also found that while the growing season length affected how much carbon dioxide could be potentially assimilated by a plant community over the course of a growing season, other factors that affect canopy photosynthetic capacity (e.g. nutrients, water) could be more important at this time scale. These results and insights demonstrate that the proposed method of analysis and system of terminology can serve as a foundation for studying the dynamics of plant community photosynthesis and such studies can be fruitful.

1 Introduction

The dynamics of plant community photosynthesis consists of diurnal and seasonal cycles. These two cycles are the most important biotic oscillations to mankind. The diurnal photosynthetic cycle is primarily driven by changes in light availability associated with the rotation of the Earth and is thus relatively predictable. The seasonal cycle, however, is more complex. It is a process orchestrated by internal biological mechanisms and driven by systematic changes in a suite of interdependent environmental factors such as temperature, photoperiod, radiation, moisture, and nutrient availability. The study of the plant community photosynthetic cycle at the seasonal time scale can be considered as an extension of plant phenology (Gu et al. 2003a; also see the Preface of current volume). This extension, or "vegetation photosynthetic phenology", represents the functional aspect of plant phenology while traditional plant phenological studies focus on the structural aspect such as budbreak, flowering, leaf coloring and leaf fall. Research on vegetation photosynthetic phenology can enrich the ancient but revived discipline of phenology so that it can become a truly integrative environmental science (Schwartz 2003).

The advance of the eddy covariance technique (Baldocchi et al. 1988; Baldocchi 2003) provides a tool amenable for studying the dynamics of plant community photosynthesis (Falge et al. 2002; Gu et al. 2002, 2003a, b). Global and regional networks of eddy covariance flux tower sites covering a wide range of vegetation types have been formed (Baldocchi et al. 2001; Gu and Baldocchi 2002). Although an eddy covariance system measures only the difference between the gross photosynthesis of the plant community and ecosystem respiration (the net ecosystem exchange,

NEE, of CO_2), NEE can be partitioned using approaches such as response functions (Gu et al. 2002), isotopic analysis (Bowling et al. 2001) and simultaneous measurements of carbonyl sulfide flux (Campbell et al. 2008). Thus there exists a great potential of using flux networks to investigate the dynamics of plant community photosynthesis at multiple time scales. When such investigation is conducted in conjunction with examination of variations in plant community structures, underlying biochemical and physiological processes, and climatic conditions, mechanisms controlling the biological oscillations most important to mankind can be revealed. These efforts could not only enhance the theoretical bases of global change biology and ecology and but also lead to effective tools for terrestrial ecosystem management.

This chapter has two objectives. The first is to describe the improvement we have made to the analytical framework of plant community photosynthesis developed in Gu et al. (2003a). The second is to present the application of the improved framework to an expanded set of vegetation sites. Our effort to improve the methodology was guided by three requirements: easy implementation, general applicability, and straightforward link to ecophysiological processes. The application was conducted to examine how the concepts and method of analysis developed in Gu et al. (2003a) and refined in current study could be used to reveal the dynamics and control of the vegetation photosynthetic cycle. To this end, we analyzed the factors affecting the potential of gross primary production at the seasonal time scale. We were particularly interested in the photosynthetic recovery at the beginning and the photosynthetic senescence at the end of the growing season and how recovery and senescence might be related to each other. Instead of presenting site-specific findings, we focused on emergent, community-level photosynthetic properties across vegetation types.

2   Sites and Data Used in the Present Study

We used data from seven eddy covariance flux sites for analyses conducted in the present paper, including five warm-season (vegetation active during warm months) and two cool-season (active during cool months) vegetation sites. The five warm-season sites were a Scots pine forest in Hyytiälä, Finland (61°51′N, 24°17′E; data from 1997; Rannik et al. 2000), an aspen forest in Prince Albert National Park, Saskatchewan, Canada (53°63′N, 106°20′W; data from 1996; Black et al. 2000), a mixed deciduous forest in Walker Branch Watershed in Tennessee, USA (35°58′N, 84°17′W; data from 1996; Wilson et al. 2000), a mixed hardwood forest in Massachusetts, USA (Harvard Forest, 42°32′N, 72°10′W; data from1992; Goulden et al. 1996), and a native tallgrass prairie in Oklahoma, USA (36°56′N, 96°41′W; data from 1997; Suyker and Verma 2001). These five sites were also used in Gu et al. (2003a). The two cool-season sites were a winter wheat site in Oklahoma, USA (36°45′N, 97°05′W; data from 1997; Burba and Verma 2005) and a grassland site in northern California, USA (38°25′N, 120°57′W; data from 2001; Baldocchi et al. 2004). For details of these sites, please see the citations listed.

Our analysis was based on canopy photosynthetic rates which were derived from NEE in the same way as described in detail in Gu et al. (2002) and Gu et al.

(2003a, b) except for the Harvard Forest site. Harvard Forest Data Archive (http://harvardforest.fas.harvard.edu/data/archive.html) provided values of the gross ecosystem exchange (i.e. the canopy photosynthetic rate as termed here). The Harvard Forest group calculated the gross ecosystem exchange in a similar fashion (i.e. response functions based on temperature and photosynthetically active radiation, see the documentation in the Harvard Forest website). Therefore data from all sites were processed consistently and are thus comparable. The partitioning of NEE avoided some processes important at short time scales including, for example, the influence of soil moisture and newly assimilated photosynthate on soil efflux (e.g. Liu et al. 2006). These omissions, however, do not affect the objectives of this paper which are interested in patterns occurring at the seasonal time scale.

3   Representation of the Seasonal Dynamics of Plant Community Photosynthesis

The seasonal cycle of plant community photosynthesis is described by the temporal variation of the canopy photosynthetic capacity (CPC). The CPC is defined as the maximal gross photosynthetic rate at the canopy level when the environmental conditions (e.g. light, moisture, and temperature) are non-limiting for the time of a year under consideration. This definition takes into account the seasonal variation in climate and thus is different from the definition of the leaf-level photosynthetic capacity, which generally assumes that the light intensity is at a saturating level (i.e. $> 1,000$ μmol photons m^{-2} s^{-1}) and temperature is about 25°C regardless of the season under consideration. In contrast, the environmental conditions under which a particular value of the CPC is realized depend on the time of the year.

The CPC forms the boundary line in the scatter plot of the instantaneous canopy photosynthetic rate against time, assuming data from the whole year are used. In practice, the instantaneous canopy photosynthetic rate is derived from NEE measurements which are generally at an hourly or half-hourly resolution. The boundary line can be adequately represented by the following composite function:

$$A(t) = y_0 + \frac{a_1}{\left[1 + exp\left(-\frac{t - t_{01}}{b_1}\right)\right]^{c_1}} - \frac{a_2}{\left[1 + exp\left(-\frac{t - t_{02}}{b_2}\right)\right]^{c_1}} \quad (1)$$

where $A(t)$ is the CPC in day t; y_0, a_1, a_2, b_1, b_2, c_1, c_1, t_{01}, and t_{02} are empirical parameters to be estimated. As shown later, the new function (Eqn. 1) is extremely flexible and can fit well diverse seasonal cycles of plant community photosynthesis. It is capable of simultaneously representing both the recovery and senescence parts of the growing season. In contrast, the Weibull function of Gu et al. (2003a) treats the two parts separately, creating a discontinuity in the middle of the growing season. The new function also eliminates the IF-THEN condition in the Weibull function, and thus its empirical parameters can be estimated with optimization algorithms that require derivatives (the IF-THEN condition leads to discontinuity in derivatives).

To estimate the parameters in Eqn. (1) from NEE measurements, we use the following iterative procedures:

a. Compute hourly or half-hourly (depending on observational time steps) canopy photosynthetic rates from NEE measurements

b. Select the largest value from each day to form a time series of the daily maximal canopy photosynthetic rate. The time series shall cover the complete seasonal cycle.

c. Fit Eqn. (1) to the obtained time series.

d. For each point in the time series, compute the ratio of the daily maximal canopy photosynthetic rate to the value predicted by Eqn. (1) for the corresponding day with the fitted parameters.

e. Conduct the Grubb's test (NIST/SEMATECH 2006) to detect if there is an outlier in the obtained ratios.

f. If an outlier is detected, remove this outlier and go to Step c.

g. If no outlier is found, remove the data points whose ratios are at least one standard deviation (1σ) less than the mean ratio. The remaining dataset is considered to consist of the canopy photosynthetic capacity at various times of the growing season.

h. Fit Eqn. (1) to the time series of the CPC. Eqn. (1) with the obtained parameters depict the seasonal cycle of plant community photosynthesis and is then used for further analyses (see the next section).

The automated, rigorous statistical procedures outlined above improves on the subjective, visual approach of Gu et al. (2003a). In the new approach, the outlier test and identification of data representing the seasonal cycle of plant community photosynthesis are done through the ratio of the daily maximal canopy photosynthetic rate to the value of canopy photosynthetic capacity predicted by Eqn. (1) in each iteration. This normalization process prevents potential bias in the fitting by eliminating the influence of the systematic temporal variation in the canopy photosynthetic capacity on the testing statistics. We conduct the outlier test out of a concern that a few outliers may greatly distort the fitted seasonal pattern of plant community photosynthesis. The Grubb's test has to be done one point at a time, that is, each time an outlier is found, Eqn. (1) must be readjusted (refit) to remove the effect of this identified outlier. This requirement leads to the iteration between Steps c and f. The outliers detected through this iteration are either of unusually low values which may occur in days with severe, photosynthesis-inhibiting weather conditions, or unreasonably large values which may result from noise in the original NEE measurements. Overall, outliers are few (Figs. 1 and 2).

The dataset free of outliers may still contain fairly low values of daily maximal canopy photosynthetic rates that result from short-term, suppressive weather conditions such as heavy cloud cover which are not part of the climate forcing that drives the seasonal photosynthetic cycle. Therefore, after the outliers are detected and removed, we further examine the deviation of the ratio of the daily maximal canopy

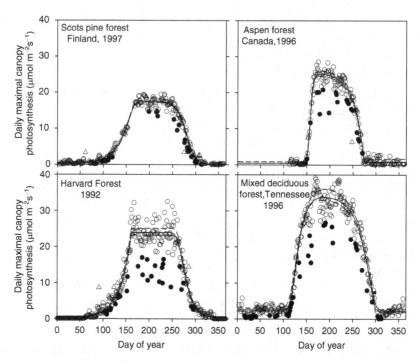

Fig. 1 Seasonal variations of the daily maximal canopy photosynthetic rate for the four forest sites used in this study. *Triangles* denote outliers identified with the Grubb's test. Dots are data points whose corresponding ratios are at least one-standard deviation (1σ) less than the mean ratio. The ratio here refers to the daily maximal canopy photosynthetic rate divided by a value predicted by Eqn. (1) for a given day. The prediction uses parameters obtained through fitting Eqn. (1) to the data that have survived the Grubb's test. The data that have passed both the Grubb's test and the 1σ screening process are considered as the canopy photosynthetic capacity (CPC) and denoted by open circles. The solid line is the regression line of the final fitting of Eqn. (1) to the values of canopy photosynthetic capacity. For comparison, the final regression lines with two different standard deviation criteria (0σ and 2σ) are also shown. These lines are close to each other, indicating that the fitting is insensitive to the choice of filtering criteria.

photosynthetic rate to the predicted CPC. We remove the points with the ratio at least 1σ less than the mean ratio (Step *g*). The choice of the 1σ criterion is based on experiments with varying criteria to screening data affected by short-term weather conditions. Figures 1 and 2 show the fitted curves with different criteria (0σ, 1σ, and 2σ). Overall, these different criteria have only minor influence on the fitted curves. The curves with the 1σ and 2σ criteria are very close to each other. However, the fitted curves with the 0σ criterion deviate relatively more from those with the 1σ or 2σ criteria, indicating that the 0σ criterion may result in too few data to be used to represent the seasonal cycle of plant community photosynthesis and the fitted seasonal patterns with this more restrictive criterion may not be reliable. Therefore, we consider the 1σ criterion as a balanced trade-off between the opposing

Fig. 2 Same as Fig. 1, except for the tallgrass prairie site and the two cool-season vegetation sites (Californian grassland and winter wheat).

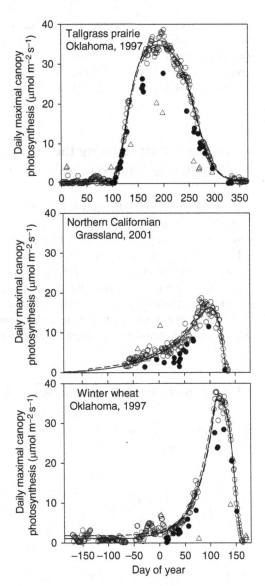

requirements of minimizing the influence of short-term weather variations vs. having a dataset with a sufficient number of samples for a robust fitting of the seasonal pattern.

The fitting for the parameters in Eqn. (1) is done with an optimization package developed as part of the AmeriFlux Data Assimilation Project at the Oak Ridge National Laboratory. Although describing this optimization package is beyond the scope of current study, the quality of fitting shown in Figs. 1 and 2 attests to not only the suitability of Eqn. (1) for quantifying the seasonal dynamics of plant community photosynthesis but also the effectiveness of the optimization package. We have automated the procedures outlined above and the calculations of indices that characterize the seasonal dynamics of plant community photosynthesis.

4   Indices Characterizing the Seasonal Dynamics of Plant Community Photosynthesis

Indices that characterize the seasonal dynamics of plant community photosynthesis facilitate the comparison of different vegetation types across climate zones and the same vegetation in different years for functional disparities and similarities. These indices can also be related directly to environmental variables to reveal how changes in climate conditions affect the carbon assimilation of plant community. Using Eqn. (1), we have revised the set of indices proposed in Gu et al. (2003a) to provide a comprehensive terminological system for quantifying various features in the seasonal dynamics of plant community photosynthesis. A collection of these indices and their definitions is given in the appendix.

4.1   Characterizing the Dynamics in CPC

The growth rate (k) of the CPC is the derivative of the canopy photosynthetic capacity with respect to the day (t) of year:

$$k(t) = \frac{dA(t)}{dt} = \frac{a_1 c_1}{b_1} \frac{exp\left(-\dfrac{t-t_{01}}{b_{01}}\right)}{\left[1+exp\left(-\dfrac{t-t_{01}}{b_{01}}\right)\right]^{1+C_1}} - \frac{a_2 c_2}{b_2} \frac{exp\left(-\dfrac{t-t_{02}}{b_{02}}\right)}{\left[1+exp\left(-\dfrac{t-t_{02}}{b_{02}}\right)\right]^{1+C_2}} \quad (2)$$

The temporal dynamics in the growth rate $k(t)$ of canopy photosynthetic capacity is interesting. Figures 3 and 4 show $k(t)$ for the seven eddy covariance flux sites under this study. All seven sites, which include both warm-season and cool-season types, have one maximum and one minimum in $k(t)$ over the growing season; the maximum occurs early and the minimum late in the growing season. The maximal growth rate of canopy photosynthetic capacity is termed "Peak Recovery Rate" and denoted by k_{PRR}; the day on which this rate occurs is termed "Peak Recovery Day" and denoted by t_{PRD}:

$$k_{PRR} = k\left(t_{PRD}\right) \quad (3)$$

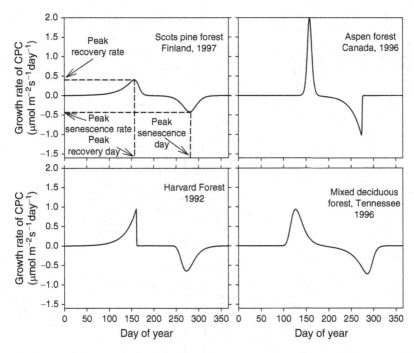

Fig. 3 Temporal variations of the growth rate of canopy photosynthetic canopy (CPC) at the four forest sites.

We further define "Recovery Line" (RL) as the line that passes through the maximum with a slope of k_{PRR}. Its equation can be written as follows:

$$A_{RL}(t) = k_{PRR}t + A(t_{PRD}) - k_{PRR}t_{PRD} \tag{4}$$

where A_{RL} is the canopy photosynthetic capacity predicted by the Recovery Line. Similarly, we term the most negative growth rate of canopy photosynthetic capacity "Peak Senescence Rate" and denote it by k_{PSR} and the day on which k_{PSR} occurs "Peak Senescence Day" and denote it by t_{PSD}:

$$k_{PSR} = k(t_{PSD}) \tag{5}$$

Accordingly, we define "Senescence Line" (SL) as the line that passes through the minimum (the most negative) with a slope of k_{PSR} and describe it by the following equation:

$$A_{SL}(t) = k_{PSR}t + A(t_{PSD}) - k_{PSR}t_{PSD} \tag{6}$$

where A_{SL} is the canopy photosynthetic capacity predicted by the Senescence Line.

Fig. 4 Temporal variations of the growth rate of canopy photosynthetic canopy (CPC) at the tallgrass prairie site and the two cool-season sites (Californian grassland and winter wheat).

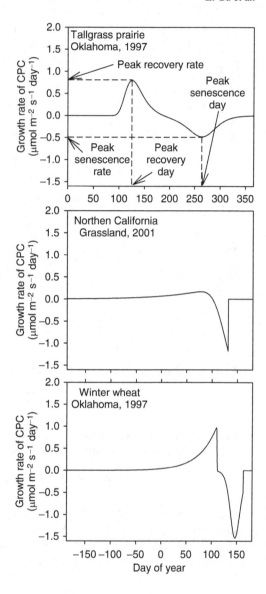

It is very difficult to determine t_{PRD} and t_{PSD} analytically from Eqn. (1). However, they can be approximated by:

$$t_{PRD} \approx t_{01} + b_1 ln(c_1) \tag{7}$$

and

$$t_{PSD} \approx t_{02} + b_2 ln(c_2) \tag{8}$$

Equation (7) is obtained by setting the derivative of the first term in Eqn. (2) with respect to t to zero and solve for t where the first term is at maximum; Eqn. (8) is obtained by setting the derivative of the second term in Eqn. (2) with respect to t to zero and solve for t where the second term is at maximum. Equations (7) and (8) hold because when t is small, the second term in Eqn. (2) is close to zero and when t is large, the first term is close to zero. Alternatively, one could simply compute the value of k for each day of the year and pick up the maximum and the minimum as we did in this study.

The RL and SL defined through the maximum and minimum in the growth rate of canopy photosynthetic capacity capture the two linear features in the temporal variation of canopy photosynthetic capacity very well (Figs. 5 and 6). These two linear features occupy two crucial periods of time in the growing season and dominate the overall shape of the seasonal cycle and thus are important for studying plant community photosynthesis. In Gu et al. (2003a), these linear features are fit with the lines determined by the minima in the radius of curvature. While the minimum in the radius of curvature is a clear mathematical concept, it has no ecological

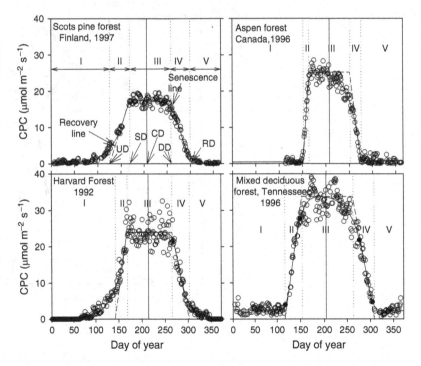

Fig. 5 Temporal variations of canopy photosynthetic capacity (CPC) for the four forest sites. Marked are the five phases of photosynthetic dynamics, upturn day (UD), stabilization day (SD), center day (CD), downturn day (DD), recession day (RD), the recovery line, and the senescence line. The line that parallels the x-axis and links the recovery and senescence lines indicates peak canopy photosynthetic capacity.

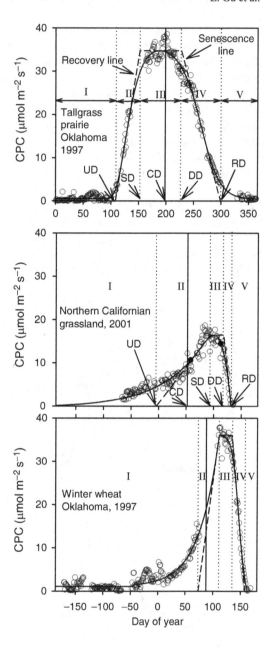

Fig. 6 Same as Fig. 5, except for the tallgrass prairie site and the two cool-season vegetation sites (Californian grassland and winter wheat).

correspondence and thus it is difficult to relate it with any underlying biological or environmental processes. In contrast, it is easy to interpret the ecological meaning of the maximum (minimum) in the growth rate of canopy photosynthetic capacity. Thus the new way of defining the recovery and senescence linear features in $A(t)$ is more desirable.

4.2 Characterizing Canopy Photosynthetic Potential

The area under the curve of $A(t)$ is an indicator of how much carbon dioxide can be potentially assimilated by a plant community over a complete cycle of photosynthesis in a year. Although the actual amount of carbon dioxide assimilated is also influenced by the diurnal cycle and variation in short-term weather conditions, a plant community that maximizes this area can fully realize the potential of carbon dioxide assimilation allowable by variation in climatic conditions in a year. As in Gu et al. (2003a), we term this area "Carbon Assimilation Potential" (u):

$$u = \int_{t_{start}}^{t_{end}} A(t)\,dt \qquad (9)$$

In theory, the above integration could start from the beginning to the end of the growing season, e.g., the start day t_{start} and the end day t_{end} could be set as the first and last day, respectively, for the period of time when the canopy photosynthetic capacity $A > 0$. In practice, it is very difficult to determine these two dates exactly from data as A changes very gradually at the beginning and end of the growing season. However, for the purpose of calculating the carbon assimilation potential u, it is not necessary to determine t_{start} and t_{end} exactly as long as one whole seasonal cycle of photosynthesis is included between t_{start} and t_{end}. This is because the two tails of A contribute little to u. Therefore we conveniently set $t_{start} = 1$ and $t_{end} = 365$ for warm-season vegetation sites (Figs. 5 and 6a) and $t_{start} = -185$ and $t_{end} = 180$ for cool-season vegetation sites (June–June, Fig. 6b, c). Clearly, here we don't intend to use t_{start} (t_{end}) to denote the start (end) of the growing season.

The peak canopy photosynthetic capacity over a complete seasonal cycle of plant community photosynthesis and the day on which this peak occurs should contain useful information about the function of the vegetation and its interaction with the climate. We use A_p to denote the peak canopy photosynthetic capacity:

$$A_p = max\left\{ \left| A(t), t_{start} < t < t_{end} \right| \right\} \qquad (10)$$

We use t_p to denote the day on which the peak canopy photosynthetic capacity occurs. t_p is called "Peak Canopy Photosynthetic Capacity Day" or simply "Peak Capacity Day."

4.3 The Five Phases of the Seasonal Cycle of Plant Community Photosynthesis

As shown in Figs. 5 and 6, the seasonal cycle of plant community photosynthesis can be divided into five consecutive phases:

Phase I.   Pre-phase, a slowly crawling-up stage at the beginning of the growing season.

Phase II. Recovery phase, a rapid recovery and expansion period.

Phase III. Stable phase, a relatively steady stage in the middle of the growing season.
Phase IV. Senescence phase, a rapidly declining stage after the stable phase.
Phase V. Termination phase, a fading stage towards the end of the growing season.

Although different vegetation types may show different characteristics in their seasonal cycles of photosynthesis, the similarity among them is also striking. Throughout a year, plant canopies undergo systematic changes in anatomy, biochemistry and physiology; understanding how these systematic changes interact with seasonal marches in climatic conditions to determine canopy carbon fixation is vital to understanding the functioning of plant communities and the terrestrial carbon cycle. For example, in deciduous canopies, both leaf area index (LAI) and leaf photosynthetic capacities increase in spring and remain relatively stable in the middle of the growing season and then decrease in fall (Wilson et al. 2000; Hikosaka 2003; Niinemets et al. 2004). Many understory plant species take advantage of the high light period prior to canopy closure in early spring or after leaf fall in autumn to fix carbon dioxide and accumulate carbohydrates to prepare for new growth (Routhier and Lapointe 2002; Richardson and O'Keefe, current volume). These biological and ecological processes produce both transient and steady features in the seasonal dynamics of plant community photosynthesis. Understanding processes operating in and factors controlling the transition between different phases of plant community photosynthesis should be an interesting research task for plant ecologists.

4.4   Transitions Between Phases

We name the transitions between the consecutive phases identified above "Upturn Day" (t_U), "Stabilization Day" (t_S), "Downturn Day" (t_D), and "Recession Day" (t_R), respectively. We set the upturn day at the intersection between the recovery line and the *x*-axis and the recession day at the intersection between the senescence line and the *x*-axis. The upturn day and recession day are calculated from Eqns. (4) and (6), respectively, as follows:

$$t_U = t_{PRD} - A \frac{t_{PRD}}{k_{PRR}} \tag{11}$$

$$t_R = t_{PSD} - A \frac{t_{PSD}}{k_{PSR}} \tag{12}$$

The stabilization day and downturn day are set at the days on which the peak canopy photosynthetic capacity A_P is predicted to occur based on the RL equation (4) and the SL equation (6), respectively. These two dates are given by:

$$t_S = t_{PRD} + \frac{\left(A_P - At_{PRD}\right)}{k_{PRR}} \tag{13}$$

$$t_D = t_{PSD} + \frac{\left(A_P - At_{PSD}\right)}{k_{PSR}} \tag{14}$$

Note that in the present paper the terms of upturn day, stabilization day, downturn day, and recession day are used somewhat differently from those in Gu et al. (2003a). They were the names for the four minima in the radius of curvature of $A(t)$ in that previous paper. The four turning points defined by RL and SL have similar meanings as intended in Gu et al. (2003a). Therefore, we continue to use the same terms in the present paper.

4.5 Effective Growing Season Length

Although it is very difficult to determine unequivocally dates for the start and end of the growing season, the upturn day and recession day come close, particularly for the five sites where plants grow in the summer (Figs. 5 and 6a). For these warm-season vegetation sites, the area under the curve of $A(t)$ between t_U and t_R accounts for more than 90% of the corresponding canopy carbon assimilation potential u (97% in tallgrass prairie, 94% in Scots pine and aspen forests, 92% in Harvard Forest and the mixed forest in Tennessee, 83% in California grassland and 75% in winter wheat). Therefore, t_U and t_R may be used to approximate the start and end, respectively, of the growing season for the warm-season vegetation type. However, there is still substantial photosynthesis (~20%) outside the period between t_U and t_R for the cool-season vegetation type. Consequently, for these sites, t_U and t_R are not good markers for the growing season; nevertheless, they can be used to indicate the "active period" of the growing season. Similar functions can be played by the peak recovery and senescence days which can be used to mark the period of the growing season during which the photosynthetic activity of the plant community is strong.

We can also use the standard deviation of the "growing days" to measure the length of the growing season. To do so, we first define the mean or Center Day (t_C) of the growing season as follows:

$$t_C = \frac{\int_{t_{start}}^{t_{end}} tA(t)dt}{u} \tag{15}$$

The standard deviation σ of the "growing days" from the center day of the growing season is:

$$\sigma = \left(\frac{\int_{t_{start}}^{t_{end}} \left(t - t_C\right)^2 A(t)dt}{u} \right)^{0.5} \tag{16}$$

The length of the growing season can then be measured by the scaled standard deviation:

$$L_E = 2\sqrt{3}\sigma \tag{17}$$

We name the scaled standard deviation the "Effective Growing Season Length" and denote it by L_E. The scaling factor $2\sqrt{3}\sigma$ is introduced so that L_E is exactly the width if the temporal pattern of $A(t)$ is a rectangle (Gu et al. 2003a). Gu et al. (2003a) defined the center day as the "center of gravity" of the curve $A(t)$. In the present paper, the center day is defined as a statistical mean and is thus more straightforward.

4.6   Effective Canopy Photosynthetic Capacity

From the carbon assimilation potential index and the effective growing season length, we can then define the seasonal "Effective Canopy Photosynthetic Capacity" (A_E) as:

$$A_E = \frac{u}{L_E} \tag{18}$$

4.7   Shape Parameters of the Seasonal Patterns of Plant Community Photosynthesis

The shape of the seasonal cycle of plant community photosynthesis often differs greatly among different sites. Borrowing two shape parameters from statistics, we define the Skewness (γ_S) and Kurtosis (γ_K) of the seasonal pattern of plant community photosynthesis as follows:

$$\gamma_S = \frac{1}{u}\int_{t_{start}}^{t_{end}} \left(\frac{t-t_C}{\sigma}\right)^3 A(t)dt \tag{19}$$

$$\gamma_K = \frac{1}{u}\int_{t_{start}}^{t_{end}} \left(\frac{t-t_C}{\sigma}\right)^4 A(t)dt - 3 \tag{20}$$

Figure 7 shows the temporal variation of the canopy photosynthetic capacity scaled by the carbon assimilation potential (i.e. A/u) with the values of skewness and kurtosis marked for the seven vegetation sites (the scaling makes the comparison among different curves easier). The skewness parameter is more negative if the photosynthetic activities are skewed to the end of the growing season (e.g. the cool-season vegetations, Fig. 7b). The kurtosis parameter is larger if the peak of the seasonal photosynthesis is sharper (e.g. the aspen forest vs. other warm-season vegetations, Fig. 7a). Different skewness and kurtosis may reflect adaptations of plant communities to specific climate conditions.

Fig. 7 The canopy photosynthetic capacity scaled for comparison in the shape of temporal variation among different sites.

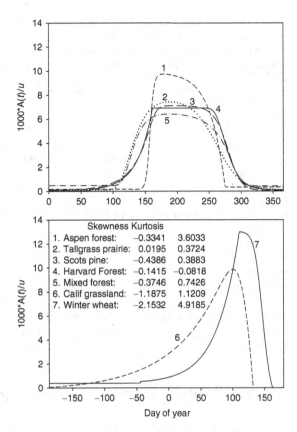

5   Application for Synthesis Across Sites

How useful is the framework described above? To answer this question, we need to examine how general the concepts and method of analysis developed are and more importantly, whether their application can lead to new scientific findings, questions, and testable hypotheses. Although the number of sites included in this study is limited (seven in total), the broad range of vegetation types covered indicates that the framework we developed can be widely applied. In the following, two synthesis examples are used to demonstrate that analyzing the dynamics of plant community photosynthesis based on the developed framework can produce fruitful scientific results. Table 1 summarizes the indices of photosynthetic cycles calculated for the seven vegetation sites in this study.

Table 1 Values of indices characterizing the dynamics of plant community photosynthesis at the seasonal time scale for the seven vegetation sites involved in this study. Study site abbreviations: *SP* scots pine, *AF* aspen forest, *HF* Harvard Forest, *TM* Tennessee mixed forest, *TP* tallgrass prairie, *CG* California grassland, *WW* winter wheat. Units of indices are given in the Appendix

Index	SP	AF	HF	TM	TP	CG	WW
Peak recovery rate	0.40	1.99	0.95	0.94	0.79	0.17	0.97
Peak recovery day	157	158	161	127	126	77	110
Peak senescence rate	−0.42	−1.02	−0.65	−0.72	−0.48	−1.20	−1.52
Peak senescence day	280	273	273	286	264	132	147
Carbon assimilation potential	2,473	2,589	3,460	5,267	4,671	1,666	2,773
Peak canopy photosynthetic capacity	17.58	25.23	23.91	33.76	34.80	16.41	36.03
Peak capacity day	193	178	193	188	188	101	112
Effective canopy photosynthetic capacity	14.40	17.72	20.73	25.53	27.60	8.12	12.65
Upturn day	125	151	136	111	108	361	73
Stabilization day	169	164	162	147	152	94	111
Downturn day	258	251	260	256	227	118	134
Recession day	300	276	297	303	299	132	158
Center day	206	208	209	200	199	52	88
Effective growing season length	172	146	167	206	169	205	219
Skewness	−0.44	−0.33	−0.14	−0.37	0.02	−1.19	−2.15
Kurtosis	0.39	3.60	−0.08	0.74	0.37	1.12	4.92

5.1   The Recovery–Senescence Relationship

The first synthesis example we present here is the relationship between the peak recovery rate and the peak senescence rate across sites (Fig. 8). For warm-season vegetation sites, the peak recovery rates are generally larger than the peak senescence rates. For cool-season vegetation sites, the opposite is true. But more interestingly, the relationship between the peak recovery and senescence rates seems linear across the warm-season vegetation sites. This close relationship between community photosynthetic recovery and senescence is unlikely due to the fitting function of choice since the function fits the data tightly, particularly for the linear features in the temporal variation of the canopy photosynthetic capacity (Figs. 5 and 6) and since a linear relationship was also reported by Gu et al. (2003a) who used a different fitting function. Therefore, the recovery–senescence linear relationship likely reflects a true conservative characteristic of plant community photosynthesis across sites. It may imply that the efficiency of a warm-season plant community to mobilize resources (nutrients and carbohydrates) to develop photosynthetic machinery in response to rapidly improving atmospheric conditions at the start of the growing season is closely related to its efficiency to withdraw and preserve crucial resources from leaves before abscission in response to deteriorating environmental conditions near the end of the growing season.

We have too few cool-season sites (only two) and thus cannot draw any firm conclusion regarding the recovery–senescence relationship for the cool-season vegetation

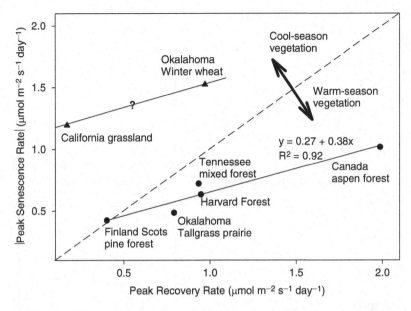

Fig. 8 The relationship between peak recovery and senescence rates across sites. The absolute values of peak senescence rates are used.

type. But this does not preclude us from pointing out the following observation: the line that passes through the two cool-season sites in the figure of peak recovery vs. senescence rate parallels the regression line for the warm-season vegetation type (Fig. 8). The probability that these two lines parallel each other by chance must be very low, considering that the sites involved were geographically separated and measurements used to derive these lines were independently acquired. If it is not due to chance, then there are grounds for making the following two hypotheses:

1. The relationship between the recovery and senescence of plant community photosynthesis is linear for the cool-season vegetation type as it is for the warm-season vegetation type.
2. The recovery-senescence relationship is invariant between the warm-season and cool season vegetation types, up to an offset in the intercept.

5.2 Factors Affecting the Carbon Assimilation Potential

The second example of using the developed methodology for synthesis across sites concerns the factors affecting the carbon assimilation potential. As we pointed out earlier, the carbon assimilation potential is an important measure of how much carbon dioxide can be assimilated in a year by a plant community under the constraint of climate. The carbon assimilation potential can be maximized by increasing the peak canopy photosynthetic capacity and/or the growing season length.

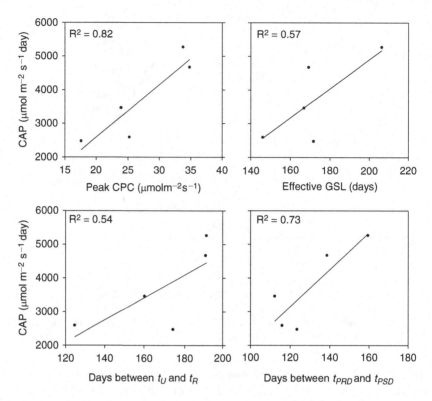

Fig. 9 The change of the carbon assimilation potential (CAP) with the peak canopy photosynthetic capacity (CPC) and with different measures of the growing season length (GSL). t_U, Upturn Day; t_R, Recession Day; t_{PRD}, Peak Recovery Day; t_{PSD} Peak Senescence Day. Only the warm-season vegetation sites are included.

Figure 9 compares, for the warm-season vegetation type, the relationship between the carbon assimilation potential and the peak canopy photosynthetic capacity with various measures of the growing season length. For this synthesis, the cool-season vegetation type is not included because the number of cool-season sites is too few and because the control for the carbon assimilation potential is obviously different between the warm and cool-season vegetation types. The comparison shown in Fig. 9 indicates the peak canopy photosynthetic capacity is a better predictor for the canopy carbon assimilation potential than are the measures of the growing season length. Although common ecological and environmental factors affect both peak canopy photosynthetic capacity and growing season length, different factors have variable degrees of influence on them. Peak canopy photosynthetic capacity should be strongly affected by leaf photosynthetic capacity and leaf area index of the canopy (Noormets et al., current volume). Leaf photosynthetic capacity and leaf area index are controlled primarily by nutrient and water availability at a site. In contrast, growing season length is determined mainly by climate conditions (i.e. temperature, photoperiod, etc). Thus the finding that peak canopy photosynthetic capacity is a better predictor of carbon assimilation potential than are measures of growing season length may indicate that ecophysiological conditions such as nutrient and water availability could be more important than the variation in climate conditions in controlling carbon dioxide assimilation at the seasonal time scale.

6 Discussion and Conclusions

In this chapter we continued the effort initiated in Gu et al. (2003a) to develop the methodology for analyzing the dynamics of plant community photosynthesis at the seasonal time scale based on eddy covariance flux measurements. We proposed a new function to represent the photosynthetic cycle of plant communities and suggested that the dynamics of plant community photosynthesis generally consist of five distinctive phases in sequence. These phases are pre-phase, recovery phase, stable phase, senescence phase, and termination phase; each phase results from the interactions between the inherent biological and ecological processes and the progression of climatic conditions and reflects unique functioning of plant communities at different stages of the growing season. We also improved the set of indices to characterize and quantify the transitions between phenophases in the dynamics of plant community photosynthesis.

We applied the improved framework of analysis to seven vegetation sites which ranged from evergreen and deciduous forests to crop to grasslands and include both cool-season and warm-season vegetation types. We found that for the warm-season vegetation type, the recovery of plant community photosynthesis at the beginning of the growing season was faster than the senescence at the end of the growing season while for the cool-season vegetation type, the opposite was true. Additionally, for the warm-season vegetation type, the recovery was closely correlated with the senescence such that a faster photosynthetic recovery was associated with speedier photosynthetic senescence and *vice versa*. We hypothesized that a similar close correlation could also exist for the cool-season vegetation type, and furthermore, the recovery–senescence relationship may be invariant between the warm-season and cool-season vegetation types up to an offset in the intercept. This hypothesis, which the present analysis aroused but didn't have enough data to confirm, awaits more studies. We also found that while the growing season length affected how much carbon dioxide could be potentially assimilated by a plant community over the course of a growing season, ecophysiological factors that affect leaf area/photosynthetic capacity development (e.g. nutrient and water availability) could be even more important at this scale. This implies that the climate warming-induced increase in the growing season length may have a limited enhancement effect on the terrestrial carbon uptake. These results and insights demonstrate that the proposed method of analysis and system of terminology can serve as a foundation for studying the dynamics of plant community photosynthesis and such studies can be fruitful.

Where should we go from here? The dynamics of plant community photosynthesis need to be studied at more eddy covariance flux sites, especially, Mediterranean or cool-season vegetation sites. A greatly expanded analysis would allow us to develop a comprehensive picture on how the photosynthetic phenological indices of plant community as well as the relationships among them change with vegetation types and climatic conditions. Cross-site emergent patterns such as the conserved relationship between peak recovery rate and peak senescence rate (Fig. 8) and the dominant control of carbon assimilation potential by peak canopy photosynthetic capacity (Fig. 9) could be confirmed or established. Mechanistic explanations of these emergent patterns may require development of new ecological theories and in-depth physiological and biochemical studies of underlying processes. These efforts could serve as the starting point for developing the science of community photosynthesis. Within this new scientific discipline, many outstanding questions could be pursued. For example, do different plant communities have their unique photosynthetic signatures? How do changes in climate and soil nutrient conditions drive photosynthetic cycle events? How are photosynthetic cycle events related to vegetation structural cycle events? Understanding of molecular and leaf photosynthesis will be necessary but not sufficient for developing answers to these questions just as the advantage of diffuse radiation at the canopy level cannot be explained based on molecular and leaf photosynthesis alone (Gu et al. 2002, 2003b). We will need to study how the characteristics of plant community photosynthesis are related to traits and adaptations of individual species in the community, how plant community as a whole shapes its photosynthetic adaptation and evolution under environmental changes, and how plant community photosynthetic cycles interact with soil nutrient and carbon pool dynamics. In particular, fruitful results could be obtained by investigating the recovery–senescence relationship and its physiological basis.

Acknowledgements We thank Drs. Rich Norby and Asko Noormets for commenting on the paper. This study was supported by the U.S. Department of Energy, Office of Science, Biological and Environmental Research Program, Environmental Science Division. Oak Ridge National Laboratory (ORNL) is managed by UT-Battelle, LLC, for the U.S. Department of Energy under the contract DE-AC05-00OR22725.

Appendix: List of Terms

Canopy photosynthetic capacity (A, μmol m^{-2} s^{-1}): the maximal gross photosynthetic rate at the canopy level that can be expected for a plant community at a given time of a year when the seasonal variation in climatic conditions is taken into account.

Carbon assimilation potential (u, μmol m^{-2} s^{-1} day): the integration of canopy photosynthetic capacity over a year (the area under the curve of canopy photosynthetic capacity in a plot of canopy photosynthetic capacity vs. day of year).

Center day (t_C, the number of days from 1st Jan.): the mean "growing day t of year" when t is treated as a random variable whose "probability density function" is $A(t)/u$ where A is the canopy photosynthetic capacity and u the carbon assimilation potential.

Downturn day (t_D, the number of days from 1st Jan.): the day on which the peak canopy photosynthetic capacity is predicted to occur based on the senescence line. Around the downturn day, canopy photosynthetic capacity often starts to decrease sharply.

Effective canopy photosynthetic capacity (A_E, μmol m^{-2} s^{-1}): the ratio of the carbon assimilation potential to the effective growing season length.

Effective growing season length (L_E, days): the scaled standard deviation of the "growing day t of year" when t is treated as a random variable whose probability density function is $A(t)/u$ where A is the canopy photosynthetic capacity and u the carbon assimilation potential.

Kurtosis (γ_k): a measure of the peakedness of the curve $A(t)$, the scaled and shifted fourth central moment of the "probability density function" $A(t)/u$.

Peak capacity day (t_p, the number of days from 1st Jan): the day on which the peak canopy photosynthetic capacity and thus the peak of the growing season occur.

Peak canopy photosynthetic capacity (A_p, µmol m^{-2} s^{-1}): the peak of the canopy photosynthetic canopy during the growing season.

Peak recovery day (t_{PRD}, the number of days from 1st Jan.): the day of the year on which the peak recovery rate occurs.

Peak recovery rate (k_{PRR}, µmol m^{-2} s^{-1} day^{-1}): the largest growth rate of canopy photosynthetic capacity during the growing season.

Peak senescence day (t_{PSD}, the number of days from 1st Jan.): the day of the year on which the peak senescence rate occurs.

Peak senescence rate (k_{PSR}, µmol m^{-2} s^{-1} day^{-1}): the most negative growth rate of canopy photosynthetic capacity during the growing season.

Pre-phase (Phase I): the initial stage of the seasonal cycle of plant community photosynthesis during which canopy photosynthetic capacity tends to increase slowly and often steadily.

Recession day (t_R, the number of days from 1st Jan): the day on which the senescence line intercepts with the *x*-axis.

Recovery line (RL): a line that closely approximates the linear feature within the recovery phase (Phase II) of the seasonal dynamics of plant community photosynthesis and is defined by the canopy photosynthetic capacity and its growth rate on the peak recovery day.

Recovery phase (Phase II): the second stage of the seasonal cycle of plant community photosynthesis during which the canopy photosynthetic capacity tends to increase rapidly and linearly.

Senescence line (SL): a line that closely approximates the linear feature during the senescence phase (Phase IV) of the seasonal dynamics of plant community photosynthesis and is defined by the canopy photosynthetic capacity and its growth (decline) rate (negative) on the peak senescence day.

Senescence phase (Phase IV): the fourth stage of the seasonal cycle of plant community photosynthesis during which canopy photosynthetic capacity tends to decline rapidly and linearly.

Skewness (γ_s): a measure of the asymmetry of the curve $A(t)$, the scaled third central moment of the 'probability density function' $A(t)/u$.

Stabilization day (t_s, the number of days from 1st Jan): the day on which the peak canopy photosynthetic capacity is predicted to occur based on the recovery line.

Stable phase (Phase III): the third stage of the seasonal cycle of plant community photosynthesis during which canopy photosynthetic capacity remains relatively stable.

Termination phase (Phase V): the final stage of the seasonal cycle of plant community photosynthesis during which canopy photosynthetic capacity is reduced to zero or approaches to zero slowly.

Upturn day (t_U, the number of days from 1st Jan.): the day on which the recovery line intercepts with the *x*-axis. Around the upturn day, the canopy photosynthetic capacity often starts to increase sharply.

References

Baldocchi DD, Falge E, Gu LH, Olson R, Hollinger D, Running S, Anthoni P, Bernhofer C, Davis KJ, Evans R, Fuentes JD, Goldstein AH, Katul G, Law BE, Lee X, Malhi Y, Meyers T, Munger W, Oechel W, Paw U KT, Pilegaard K, Schmid HP, Valentini R, Verma S, Vesala T, Wilson K, Wofsy S. 2001. FLUXNET: A new tool to study the temporal and spatial variability of ecosystem-scale carbon dioxide, water vapor, and energy flux densities. Bulletin of the American Meteorological Society 82:2415–34.

Baldocchi, D., Hicke, B.B. and Meyers, T.P. (1988) Measuring biosphere-atmosphere exchanges of biologically related gases with micrometeorological methods. Ecology 69, 1331–1340.

Baldocchi, D.D. (2003) Assessing the eddy covariance technique for evaluating carbon dioxide exchange rates of ecosystems: past, present and future. Global Change Biol. 9, 479–492.

Baldocchi, D.D., Xu, L.K. and Kiang, N. (2004) How plant functional-type, weather, seasonal drought, and soil physical properties alter water and energy fluxes of an oak–grass savanna and an annual grassland. Agric. For. Meteorol. 123, 13–39.

Black, T.A., Chen, W.J., Barr, A.G., Arain, M.A., Chen, Z., Nesic, Z., Hogg, E.H., Neumann, H.H. and Yang, P.C. (2000) Increased carbon sequestration by a boreal deciduous forest in years with a warm spring. Geophys. Res. Lett. 27, 1271–1274.

Bowling, D.R., Pieter, P.T. and Monson, R.K. (2001) Partitioning net ecosystem carbon exchange with isotopic fluxes of CO2. Global Change Biol. 7, 127–145.

Burba, G.G. and Verma, S.B. (2005) Seasonal and interannual variability in evapotranspiration of native tallgrass prairie and cultivated wheat ecosystems. Agric. For. Meteorol. 135, 190–201.

Campbell, J.E., Carmichael, G.R., Chai, T., Mena-Carrasco, M., Tang, Y., Blake, D.R., Blake, N.J., Vay, S.A., Collatz, G.J., Baker, I., Berry, J.A., Montzka, S.A., Sweeney, C., Schnoor, J.L. and Stanier, C.O. (2008) Photosynthetic control of atmospheric carbonyl sulfide during the growing season. Science 322, 1085–1088.

Falge, E., Baldocchi, D.D., Tenhunen, J., Aubinet, M., Bakwin, P.S., Berbigier, P., Bernhofer, C., Burba, G., Clement, R., Davis, K.J., Elbers, J.A., Goldstein, A.H., Grelle, A., Granier, A., Guddmundsson, J., Hollinger, D., Kowalski, A.S., Katul, G., Law, B.E., Malhi, Y., Meyers, T., Monson, R.K., Munger, J.W., Oechel, W., Paw U, K.T., Pilegaard, K., Rannik, Ü., Rebmann, C., Suyker, A., Valentini, R., Wilson, K. and Wofsy, S. (2002) Seasonality of ecosystem respiration and gross primary production as derived from FLUXNET measurements. Agric. For. Meteorol. 113, 53–74.

Goulden, M.L., Munger, J.W., Fan, S.M., Daube, B.C. and Wofsy, S.C. (1996) Measurements of carbon sequestration by long-term eddy covariance: methods and critical evaluation of accuracy. Global Change Biol. 2, 169–182.

Gu, L. and Baldocchi, D.D. (2002) Foreword to the Fluxnet special issue. Agric. For. Meteorol. 113, 1–2.

Gu, L., Post, W.M., Baldocchi, D., Black, T.A., Verma, S.B., Vesala, T. and Wofsy, S.C. (2003a). Phenology of vegetation photosynthesis. In: Schwartz, M.D. (Ed.) Phenology: An Integrated Environmental Science. Kluwer, Dordecht, pp. 467–485.

Gu, L.H., Baldocchi, D.D., Verma, S.B., Black, T.A., Vesala, T., Falge, E.M. and Dowty, P.R. (2002) Advantages of diffuse radiation for terrestrial ecosystem productivity. J. Geophys. Res. (D Atmos.) 107, art. no. 4050.

Gu, L.H., Baldocchi, D.D., Wofsy, S.C., Munger, J.W., Michalsky, J.J., Urbanski, S.P. and Boden, T.A. (2003b) Response of a deciduous forest to the Mount Pinatubo eruption: Enhanced photosynthesis. Science 299, 2035–2038.

Hikosaka, K. (2003) A model of dynamics of leaves and nitrogen in a plant canopy: An integration of canopy photosynthesis, leaf life span, and nitrogen use efficiency. Am. Nat. 162, 149–164.

Liu, Q., Edwards, N.T., Post, W.M., Gu, L., Ledford, J. and Lenhart, S. (2006) Temperature-independent diel variation in soil respiration observed from a temperate deciduous forest. Global Change Biol. 12, 2136–2145.

Niinemets, Ü., Kull, O. and Tenhunen, J.D. (2004) Within-canopy variation in the rate of development of photosynthetic capacity is proportional to integrated quantum flux density in temperate deciduous trees. Plant Cell Environ. 27, 293–313.

NIST/SEMATECH (2006) e-Handbook of Statistical Methods. http://www.itl.nist.gov/div898/handbook/.

Rannik, Ü., Aubinet, M., Kurbanmuradov, O., Sabelfeld, K.K., Markkanen, T. and Vesala, T. (2000) Footprint analysis for measurements over a heterogeneous forest. Bound.-Lay. Meteorol. 97, 137–166.

Routhier, M.C. and Lapointe, L. (2002) Impact of tree leaf phenology on growth rates and reproduction in the spring flowering species Trillium erectum (Liliaceae). Am. J. Bot. 89, 500–505.

Schwartz, M.D. (Ed.) (2003) Phenology: An Integrative Environmental Science. Kluwer, Dordrecht, pp. 592.

Suyker, A.E. and Verma, S.B. (2001) Year-round observations of the net ecosystem exchange of carbon dioxide in a native tallgrass prairie. Global Change Biol. 7, 279–289.

Wilson, K.B., Baldocchi, D.D. and Hanson, P.J. (2000) Spatial and seasonal variability of photosynthetic parameters and their relationship to leaf nitrogen in a deciduous forest. Tree Physiol. 20, 565–578.

The Phenology of Gross Ecosystem Productivity and Ecosystem Respiration in Temperate Hardwood and Conifer Chronosequences

Asko Noormets, Jiquan Chen, Lianhong Gu, and Ankur Desai

Abstract The relative duration of active and dormant seasons has a strong influence on ecosystem net carbon balance and its carbon uptake potential. While recognized as an important source of temporal and spatial variability, the seasonality of ecosystem carbon balance has not been studied explicitly, and still lacks standard terminology. In the current chapter, we apply a curve fitting procedure to define seasonal transitions in ecosystem gross productivity (GEP) and respiration (ER), and we show that the temporal changes in these two fluxes are not synchronous, and that the transition dates and rates of change vary both across sites and between years. Carbon uptake period (CUP), a common phenological metric, defined from ecosystem net carbon exchange (NEE), is related to these periods of activity, but the differential sensitivities of GEP and ER to environmental factors complicate the interpretation of variation in CUP alone. On a landscape scale, differences in stand age represent a major source of heterogeneity reflected in different flux capacities as well as microclimate. In the current study, we evaluate age-related differences in the phenological transitions of GEP and ER using hardwood and conifer chronosequences. While a significant portion of variability in GEP seasonality was explained with stand age, the influence of interannual climatic variability exceeded these, and was the predominant factor affecting ER seasonality. The length of the

A. Noormets (✉)
Department of Forestry and Environmental Resources,
North Carolina State University, Raleigh, NC, USA
e-mail: anoorme@ncsu.edu

J. Chen
Department of Environmental Sciences, University of Toledo, Toledo, OH, USA
e-mail: jiquan.chen@utoledo.edu

L. Gu
Environmental Sciences Division, Oak Ridge National Laboratory, Oak Ridge, TN, USA
e-mail: lianhong-gu@ornl.gov

A. Desai
Department of Atmospheric and Oceanic Sciences,
University of Wisconsin-Madison, WI, USA
e-mail: desai@aos.wisc.edu

A. Noormets (ed.), *Phenology of Ecosystem Processes*,
DOI 10.1007/978-1-4419-0026-5_3, © Springer Science+Business Media, LLC 2009

active season (ASL) varied more due to differences in the timing of the end rather than the start of the active period. ASL of GEP was consistently greater in conifers than hardwoods, but the opposite was true for ER.

Abbreviations

Phenological stages and dates

ASL^{AH} length of active season (days)
ASL^{BG} length of active season (days)
ASL^{x0} length of active season (days)
EOS   end of active season (DOY)
x0   day of half-maximum flux (DOY)
LFD   length of flux development period (days)
LFR   length of flux recession period (days)
LPF   length of peak flux period (days)
RD   rate of development (g C m^{-2} d^{-1})
RR   rate of recession (g C m^{-2} d^{-1})
SOS   start of active season (DOY)

Subscripts appended to any of the above

GEP   index or date referring to gross ecosystem productivity
ER   index or date referring to ecosystem respiration

Site abbreviations

IH03   intermediate hardwood, 2003
IP03   intermediate red pine, 2003
MH02   mature hardwood, 2002
MH03   mature hardwood, 2003
MP02   mature red pine, 2002
MP03   mature red pine, 2003
YH02   young hardwood, 2002
YP02   young red pine, 2002

1 Introduction

Climate warming has now been recognized even in the popular media (Gore 2006) and the complexities of the mechanisms contributing to it are increasingly understood (Hansen et al. 1997; Houghton et al. 1998; IPCC 2007). Cues from sources as diverse as paleoclimatic proxies, surface and airborne observations, satellite remote sensing and meteorological models all indicate that land surface temperatures have been increasing globally at a rate of 0.2–0.25°C per decade over the last

30–100 years (Bogaert et al. 2002; IPCC 2007), caused by both natural and anthropogenic forcing (Hegerl et al. 1996; Santer et al. 1996; Stott et al. 2000). The association between the earlier start of the active season (SOS) for biological processes and warmer winter temperatures is particularly strong in the eastern US and Northern Europe. There, the trend of increasing mean temperatures is increasingly reflected in the seasonal development of plants (Linderholm 2006), as well as the expansion of the ranges of species (Chuine and Beaubien 2001; Tape et al. 2006).

Given the greater increase in the late winter and early spring temperatures than late spring and early summer temperatures (Groisman et al. 1994), the advance of spring phenology has been greater than the delay of fall (Menzel and Fabian 1999; Penuelas et al. 2002; Schwartz and Chen 2002). Also, the advance in the flowering time has been greater in early- than late-flowering species (Fitter and Fitter 2002; Lu et al. 2006), with contrasting effects between prairie plants flowering before and after the peak temperature (Sherry et al. 2007). Globally, the spring events in plant development, including times of flowering, bud break and leaf expansion, marking the SOS have been advancing at a rate of about 1.5–3 (but up to 5.4) days per decade (Ahas et al. 2002; Beaubien and Freeland 2000; Chmielewski and Rötzer 2001; Menzel and Fabian 1999; Schwartz et al. 2006; Schwartz and Reiter 2000; Smith et al. 2004), associated with increasing mean temperature (Menzel 2003), earlier retreat of snow cover (Aurela et al. 2004; Groisman et al. 1994; Lu et al. 2006) and decreasing occurrence of late freeze events (Penuelas and Filella 2001). These trends have been more pronounced in Eurasia than North America (Smith et al. 2004), and stronger in trees than shrubs and grasses (Parmesan 2007).

The ground observations of earlier and longer vegetation activity have been confirmed through global analyses of the magnitude and timing of the annual cycle of CO_2 concentration in the atmosphere (Keeling et al. 1996; Myneni et al. 1997). Remote sensing models have identified significant trends towards increased net primary productivity (NPP) in recent decades (Cao et al. 2004), which in some estimates can be a significant proportion of the annual total. For example, Jackson et al. (2001) proposed that the lengthening of the growing season by 5–10 days may increase the annual NPP of forest systems by as much as 30%. Such a large response can only be possible if gross ecosystem productivity (GEP) responds more to higher temperatures than ecosystem respiration (ER). Indeed, it has been confirmed through ground measurements that the lengthening of the growing season by a certain number of days in spring stimulates ecosystem C uptake more than a lengthening by the same number of days in the fall (Kramer et al. 2000; Piao et al. 2008). This has been attributed to greater radiative inputs and longer days, as well as better moisture availability as the result of snow melt and relatively lower evaporative demand in spring than in fall (Barr et al. 2004; Black et al. 2000).

While it is true that ecosystem C balance tends to be more sensitive to early spring than late fall, not all forests respond similarly. Early spring was found to stimulate GEP more than ER in some boreal stands (Black et al. 2000). In others, delayed fall stimulated ER more than GEP, leading to smaller net C uptake (Hollinger et al. 2004). It is likely that the degree of sensitivity of ER to spring and fall warming depends on the duration of these transition periods. With long fall, leaf litter can decompose to a greater extent before dormancy, and respond to variations in temperature, whereas with short fall the litter primarily decays during the following growing season, exhibiting greater sensitivity to variations in spring temperature. In addition to direct effect on carbon fluxes, warm and late falls may have carry-over effects associated with plant development. Heide (2003) reported that delayed fall as the result of higher than average temperatures led to frost damage and subsequent delayed bud break the following spring, resulting in lower net carbon uptake. And then there are stands where prolonged GS in spring and fall has similar effect on ecosystem C balance (Goulden et al. 1996). In all, the trends are generally more consistent for spring phenology, whereas changes in the fall are more variable (Menzel 2002), varying by biome, species, and climatic variability (e.g. extreme events) during the preceding growing season (Walther et al. 2002). Later freeze (Smith et al. 2004) and leaf coloring (Menzel and Fabian 1999) have been reported primarily in the boreal forests in North America, but even there the trends in the fall are weaker than in the spring. Overall, recent trends in ecosystem CO_2 exchange and atmospheric fall-to-winter CO_2 build-up in the northern hemisphere suggest that as a general rule, warmer falls stimulate ER more than GEP, and lead to suppressed net productivity (Piao et al. 2008).

In addition to the spring-fall asymmetry, phenological stages close to one another may also have different sensitivity to increasing temperature. For example, in Central Europe the first leaf and last freeze dates have been advancing at about the same rate, whereas in Northern Eastern Europe the first leaf date has advanced faster than the last freeze date, and in East Asia, the last freeze date has advanced faster than the first leaf date (Schwartz et al. 2006). Such non-uniform responses are currently poorly understood, but may have major implications for plant vitality (Chuine and Beaubien 2001), synchrony with pollinators (Parmesan 2007) and ecosystem carbon balance (Oechel et al. 2000). Furthermore, the asymmetric changes in the ASL may also differentially affect GEP and ER. The different timing and magnitude of GEP and ER was invoked by Schaefer et al. (2005) to explain the increased amplitude of atmospheric CO_2 concentration (Myneni et al. 1997). It has recently been recognized that the feedback by vegetation on the physical and chemical properties of the atmosphere is not negligible, and the changes brought about in plant phenology may further trigger changes in the climate it responds to (Bonan et al. 2003; Lenton 2000; Meir et al. 2006). While the significance of feedback effects and interactive influences between temperature and other factors on phenology has been recognized (Norby et al. 2003), progress in this understanding is only beginning to be made. Despite the gaps in knowledge and the fact that the seasonality of ecosystem C exchange remains among the most significant uncertainties in land surface biogeochemical models (e.g. Kathuroju et al. 2007; Olesen et al. 2007), it is clear that it plays a major role in the interannual variability of ecosystem net carbon balance (e.g. Aurela et al. 2004; Barr et al. 2007; Carrara et al. 2003), and that global patterns in annual NEE strongly correlate with differences in growing season length (Fig. 1; Baldocchi et al. 2001).

As of now, there is no standard terminology for assessing the seasonality of C exchange (or any other ecosystem process). The term "growing season" has been adopted from traditional phenology where it may mark different phases in the duration of deciduous canopy – from bud break to bud set, bud break to leaf fall, or from fully

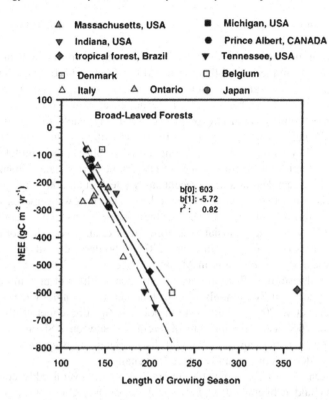

Fig. 1 Annual net ecosystem exchange of carbon (NEE) as a function of carbon uptake period (with permission, from Baldocchi et al. 2001). The b[0] and b[1] coefficients are the intercept and slope of the linear regression fitted to all sites, excluding the tropical forest in Brazil. The r^2 is the coefficient of determination, and the dashed lines are the 95% confidence intervals of the regression. Negative values indicate carbon uptake by vegetation.

expanded leaf to the beginning of leaf senescence. While all these dates are related, they may not always change similarly. Unlike traditional phenology of discrete events, the continuous scale of ecosystem biogeochemical fluxes (esp. ER) makes it more difficult to define the transition points. At this point, the science of phenology of ecosystem processes is still in the phase of trying to define the relevant transition points. However, the problem of identifying phases in a continuous data stream is not unique to ecology. Land surface phenology attempts to consolidate remotely sensed greenness indices (normalized difference vegetation index, NDVI, or enhanced vegetation index, EVI) with traditional phenological indices (Reed et al. 1994; White et al. 2005; White and Nemani 2003; Yang et al. 2007). Often, the transition from dormant to active periods has been defined by half-maximum greenness (Fisher et al. 2007; Schwartz et al. 2002). While there is little theoretical justification for this choice, especially as the relationship between satellite-based and ground observations of phenological changes is often poor (Badeck et al. 2004), this technique lends itself well to diverse streams of time series data. Finding the relevant transition points could be facilitated by existing gas exchange models that have proven relatively accurate for delineating seasonal dynamics (e.g. Mäkelä et al. 2004, 2008).

For studying the phenology of ecosystem processes, the most promising recent development is the numerical method proposed by Gu et al. (2003) and developed further in the current volume (chapter "Characterizing the Seasonal Dynamics of Plant Community Photosynthesis Across a Range of Vegetation Types"). It isolates individual transitions between two periods of relatively stable fluxes, and fits a cumulative Weibull distribution function to this transition (Eqn. 1 in Methods). Similar approach was proposed for analyzing the seasonality of the remotely sensed normalized difference vegetation index (NDVI) time series data (Zhang et al. 2003), but was overlooked in a more recent study (Beck et al. 2006) that proposed a double logistic function instead. While the latter approach was superior to other fitting algorithms currently used in land surface phenology (second-order Fourier transform and asymmetric Gaussian function), Beck et al. (2006) did not develop the information as rigorously as Gu et al. (2003), who derived a full suite of phenological indicators (more details in Methods). Beck et al. (2006) only used the fitted curve to derive the inflection points, equivalent to the half-maximum points from sigmoid curve fits, commonly used in land surface phenology (Fisher et al. 2007; Schwartz et al. 2002). The transition points defined by Gu et al. (2003) are also significant because they allow to differentiate between ASL and C uptake potential of ecosystems, which, as we will discuss later, is a step towards deeper mechanistic understanding of land surface C exchange.

As an alternative to the somewhat ambiguous ASL, several eddy covariance (EC) studies (Baldocchi et al. 2005; Falge et al. 2002) and EC-remote sensing syntheses (Churkina et al. 2005; White and Nemani 2003) have adopted carbon uptake period (CUP) as the basis of delineating seasonality and comparing different ecosystems. This is particularly convenient, because EC measures the net balance of vegetation-atmosphere carbon exchange, and provides an accurate tool for amongstands comparison. However, the asynchronicity of GEP and ER, due to their sensitivity to different environmental drivers, makes CUP difficult to predict. For example, Kramer et al. (2002) found that while several models predicted the seasonal dynamics and annual total NEE correctly, the predictions of GEP and ER were wrong. Falge et al. (2002) reported that in temperate deciduous and boreal conifer forests, the seasonality of ER was generally delayed in relation to GEP. Depending on the relative juxtaposition of the seasonalities of these two fluxes, the onset of CUP may be advanced or delayed. Furthermore, it may be difficult to define CUP in conifers because ecosystem C balance may be positive (gaining C) throughout the year. Even if arbitrarily chosen thresholds (e.g. 20% of minimum NEE, as in Suni et al. 2003a) correlate well with cumulative temperature or model estimates (as it did for Suni et al. 2003a), it is difficult to apply these at different sites, or justify them on a mechanistic basis. Separating individual processes also makes the data available for studies beyond those of net carbon balance. For example, long-term C storage is closely tied to heterotrophic respiration (or ER, loosely speaking) and soil C (Field et al. 2007; Richter et al. 1999; Schlesinger and Lichter 2001), and understanding its seasonality independent of other confounding fluxes would be of greater benefit than knowing the seasonality of NEE. For mechanistic understanding and prognostic power, we propose that the seasonality of ecosystem C exchange should be assessed on the basis of individual processes. In current study, working with eddy covariance data, we partition NEE to GEP and ER, acknowledging that ER itself represents a compound process and should ideally be partitioned further.

Variability in phenological changes occurs both globally, along broad environmental gradients (Baldocchi et al. 2001), as well as locally (Noormets et al. 2007). As a rule, ecosystem net productivity increases with the growing season length (Fig. 1) that generally decreases with latitude. However, forest stands may differ in their microclimate even when exposed to similar weather and soil conditions (Noormets et al. 2008), and these differences are expected to affect the seasonal development of photosynthetic and respiratory capacities. Across different forest types, there is a consistent trend for lower mean soil temperature in older stands with more closed canopy compared to younger open-canopied stands (Fig. 2; Bond-Lamberty et al. 2005a, b). Besides temperature, the differences also manifest in net radiation and vapor pressure deficit (VPD) (Amiro et al. 2006; Ewers et al. 2005). The age-related differences in annual carbon balance may be in the order of 500 g C m^{-2} (Noormets et al. 2007), and accounting for these is crucial for accurate representation of regional integrated carbon balance (Desai et al. 2008). Stand age may influence C exchange through the interacting influences of canopy cover, the amount of live and dead biomass, understory composition, radiation balance and microclimate. While there may be considerable variation among young stands due to diverse post-harvest management practices that control the amount of harvest residue (Devine and Harrington 2007), young stands typically tend to lose snow and warm up earlier than mature stands (Amiro 2001; Bergeron et al. 2008; Noormets et al. 2007). When the concept of chronosequence is treated more loosely, allowing for species succession throughout the stand development, like in a fire chronosequence (Goulden et al. 2006), the differences in phenology and carbon

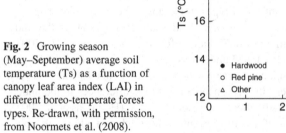

Fig. 2 Growing season (May–September) average soil temperature (Ts) as a function of canopy leaf area index (LAI) in different boreo-temperate forest types. Re-drawn, with permission, from Noormets et al. (2008).

exchange become even more pronounced, with CUP varying from as short as 65 days in young stands with primarily herbaceous vegetation to 130 days in old stands with coniferous vegetation. Understanding the mechanism of age-related differences in C fluxes could potentially provide an additional constraint to regional C cycle models by combining the information about the phenological dynamics of one stand with the age composition of stands in the area of interest. In earlier studies we have shown the significance of considering stand age and forest type-related variability for regional upscaling of C budgets (Desai et al. 2008). Thus, consistent differences with stand age could prove helpful in applications from continental data assimilation systems (e.g. AmeriFlux Data Assimilation System) to monitoring issues of practical interest (e.g. the Early Warning System; Hargrove and Hoffman 2005).

The goals of the current study were to (1) apply consistent criteria for defining the seasonality of GEP and ER, and (2) evaluate the phenology of these fluxes in relation to each other, and across two chronosequences – three deciduous and three coniferous stands – to evaluate the effect of different canopy duration and different microclimate on the seasonal development of C fluxes. We will also evaluate the relationship of the seasonality of GEP and ER with CUP, and discuss additional phenological metrics that affect ecosystem carbon balance.

2 Methods

2.1 Study Area and Site Characteristics

The study area is located in northern Wisconsin, USA (46°30′–46°45′N, 91°2′ –91°22′W) and belongs to the northern coniferous-deciduous biome. The climate is humid-continental with 30-year temperature normals from −16°C in January to 25°C in July, and annual precipitation over the same period ranged from 660 to 910 mm. The dominant vegetation types are second-growth hardwood stands with aspen, birch and maple as the predominant species, and red pine and jack pine plantations. We measured ecosystem carbon exchange in a pair of different-aged hardwood and a pair of conifer stands each in 2002 and 2003. The mature stands were measured in both years, whereas the young stands were sampled for only 1 year. The paired sampling scheme was designed to provide information about age-related differences in C exchange in the two predominant forest types in the area. The predominant species and main site characteristics are given in Table 1, and further details about the study area, measurement and gapfilling protocols can be found in Noormets et al. (2007). The mature stands had a well developed understory, and the young stands had a significant herbaceous-grassy component, whereas the intermediate-aged stands had very dense overstory canopy and very limited understory. While differences in the understory composition could potentially affect ecosystem C fluxes (Heijmans et al. 2004), we did not detect obvious signs of it in the current study and do not address this pos-

Table 1 Site properties, including dominant species, age, leaf area index (LAI), canopy cover and basal area. Further site information available in Noormets et al. (2008a). Site abbreviations: *MH* mature hardwood, *IH* intermediate-aged hardwood, *YH* young hardwood, *MP* mature red pine, *IP* intermediate-aged red pine, *YP* young red and jack pine

Site	Dominant species	Age (years)	LAI (m^2 m^{-2})	Canopy cover (%)	Basal area (m^2 ha^{-1})
MH	*Populus grandidentata, Betula papyrifera, Quercus rubra, Acer rubrum, Acer saccharum*	65, 66	3.9	97	33.5
IH	*Populus grandidentata, Populus tremuloides*	17	3.0	92	11.8
YH	*Acer rubrum, Populus grandidentata, Populus tremuloides*	3	1.2–1.4	2	1.5
MP	*Pinus resinosa, Populus grandidentata*	63, 64	2.5–2.8	73	26.9
IP	*Pinus resinosa*	21	2.8	60	18.2
YP	*Pinus banksiana, Pinus resinosa*	8	0.5	17	4.7

sibility further. However, as the understory phenology is often shifted in relation to overstory (Richardson and O'Keefe, current volume) it should be considered as a potential source of variation on a site-by-site basis.

As ER is extrapolated from nighttime data, and GEP is calculated as the sum of NEE and ER, it is possible (at least, in principle) that gapfilling protocols may affect the phenological parameters. While our gapfilling protocols include tests for bias and have been validated against other commonly used gapfilling approaches (Moffat et al. 2007), the procedures for producing defensible ER and GEP estimates are beyond the scope of this chapter. Please see Falge et al. (2001) and Moffat et al. (2007) for further details.

2.2 Phenological Transitions

To distinguish periods of relatively high activity from those of relatively low activity or dormancy, we use cumulative Weibull distribution function to delineate the "active season" for GEP and ER (Gu et al. 2003). It is a generalization of the commonly used "growing season". Gu et al. (2003) defined seasonal transitions in GEP by fitting cumulative Weibull distribution function to daily maximum photosynthesis. In the current study, we use daily flux totals for both GEP and ER. Monitoring changes in flux capacity (light-saturated) may often be preferential to integrals, but it also has limitations. For example, eddy covariance data, recorded every 30–60 min has limited number of observations per day, and sometimes estimating capacities from such datasets may be difficult (especially for ER). However, both scales reflect the same trends and could be analyzed, and both data types may be sensitive to extended periods of adverse weather suppressing the fluxes of interest. The Weibull cumulative distribution function is given by:

$$y = y_0 + \beta_1 \left[1 - e^{-\left(\frac{x - x_0 + \beta_2 \ln(2)^{\frac{1}{\beta_3}}}{\beta_2} \right)^{\beta_3}} \right] \tag{1}$$

where y is the daily integral of the flux of interest, y_0 is the base-value of y during the dormant season, x is day-of-year (DOY) for the first half of the year, and days until the end of the year for the second half of the year, x_0 is the DOY at half-maximum y (fitted), β_1 is the difference between peak and base y, β_2 is the difference between 75th and 25th percentiles of the time from base to peak y, and β_3 is a shape parameter. Parameters y_0, x_0, β_1, β_2 and β_3 are estimated in model fitting. The universality of Eqn. (1) allows defining relatively simple transitions between a single active period and a single dormant period per year, but would also work in situations where multiple active periods or multiple levels of activity are the norm (please see Gu et al., in current volume).

Once the model parameters for spring development and fall recession are determined, the first and second derivatives of predicted fluxes can be used to singularly identify the turning points. The maximum and minimum of the first derivative identify the midpoints for spring and fall as the days with greatest change in the fluxes of interest (points C and F in Fig. 3). The period CF corresponds to the active season length according to the half-maximum (crossing x_0, Eqn. 1) thresholding method used in remote sensing (see Sect. 1). The spring and fall maxima of the second derivative mark flux upturn date (point B, this and other names of the transition dates follow the nomenclature suggested by Gu et al. 2003) and flux retardation date (point G), whereas the minima mark the flux saturation date (point D) and flux downturn date (point E). Periods between BD, DE and EG mark the length of flux development, peak flux, and flux recession periods (LFD, LPF and LFR, respectively). Active season length can be defined in different ways, as the period between points A and H (e.g. as suggested by Gu et al. 2003), B and G, or C and F. In the current study we found greater and more consistent differences in the period defined by half-maximum points (C and F) and the extended active season (period BG) than the one defined by AH. We will denote the respective active season lengths as ASL[x0], ASL[BG], and ASL[AH]. The slopes of regressions fitted to the predicted fluxes during phases BD and EG give us the rate of flux development (RD) and the rate of flux retardation (RR), respectively. To illustrate the range of variability of various phenological turning points, the duration of different stages, and the differences between GEP and ER at the same stand, Fig. 4 depicts the seasonal dynamics of the fluxes, the fitted Weibull distribution function and its first and second derivatives in the conifer chronosequence.

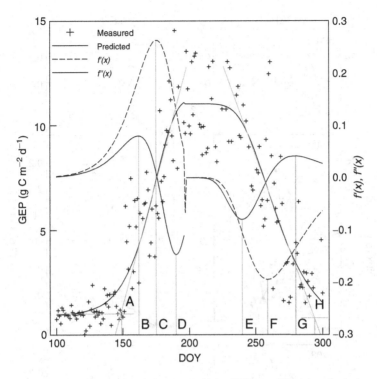

Fig. 3 Seasonal course of gross ecosystem productivity (GEP), and the phenological transition points (denoted with letters, A–H) as defined by the extremes of the first and second derivatives of the fitted cumulative Weibull functions Eqn. (1). The scale of the second derivative is enhanced tenfold for visual clarity. See text for details.

2.3   Requirements for Data Quality and Level of Integration

The seasonality of fluxes can be assessed from different types of daily data – daily maximum values, daily totals or normalized flux capacities (e.g. maximum photosynthetic capacity, P_{max}, or temperature-normalized base respiration, R_{10}). In the current study, we use daily integrals because of their greater robustness, whereas earlier studies have analyzed the changes in daily maximum values (Falge et al. 2002; Gu et al. 2003). Daily integrals require prior gapfilling and may potentially be affected by model assumptions, but such bias can be evaluated using established procedures (Moffat et al. 2007). Both data types may be sensitive to extended periods of adverse weather if it suppresses flux capacities. However, brief periods of a few days that do not reverse flux development, as well as periods of missing data do not significantly affect the phenological parameters. The parameters can be

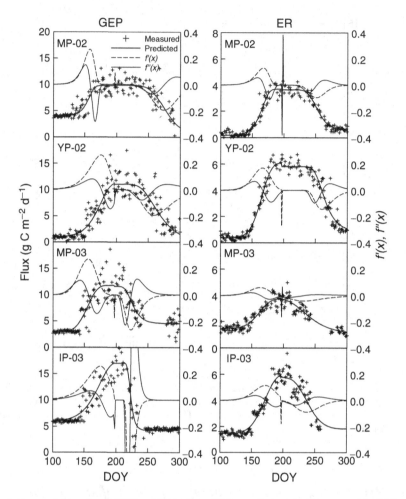

Fig. 4 Seasonal dynamics of GEP and ER in three different red pine stands, measured in 2002 and 2003. Overlaid with daily fluxes are the fitted Weibull function, and its 1st and 2nd derivatives.

estimated even with a sizeable proportion of missing data, as long as the transition points are represented.

Theoretically, the flux capacities should minimize the influence of environmental variability, providing the smoothest seasonal change curve. However, our tests of using normalized flux capacities (P_{max}, R_{10}) to evaluate the seasonality of ecosystem carbon exchange showed that these parameters were not robust. For example, estimating ASL from changes in R_{10} resulted in significantly shorter ASL (by nearly 2 months; data not shown) than when estimating it from daily total ER. Other weaknesses of the parameter-based approach include that R_{10} is not estimable independently, but is closely related to the temperature sensitivity parameter of the temperature response model (Noormets et al. 2008), and the data must often be grouped by more than a single day to achieve stable parameter estimates. Therefore, we can not recommend that the parameters of the light and temperature response models commonly used in gapfilling eddy covariance data be used for evaluating phenological transitions. It should also be emphasized that diurnal changes in fluxes may confound with detecting phenological changes, and their influence on the analysis should be minimized.

2.4 Robustness of Parameter Estimates

The robustness of the phenological parameter estimates was evaluated using Monte Carlo analysis. We treated the first fitted model as the ideal data, and added noise by sampling with replacement from the observed population of residuals. Fitting new Weibull functions to 1,000 of these synthetic datasets gave us 95% confidence intervals for the derived phenological parameters on the scale of 0.1–0.4 days. While tightly constrained, the parameters were sensitive to the temporal structure of the synthetic error. The variance in C fluxes increases with increasing flux magnitude (Hollinger and Richardson 2005) and therefore, overestimating the variance during low fluxes, i.e. during the dormant period, could lead to unrealistic seasonal dynamics in the synthetic data. To avoid this, the error must be sampled from the appropriate subpopulation. This, however, leads to a circular approach because we want to use the synthetic data to draw inferences about the seasonality, but we must define the seasons in order to sample the residuals.

2.5 Statistical Analyses

The parameters of the Weibull function were estimated with a nonlinear curve fitting procedure (PROC NLIN, SAS 9). The significance of forest type and year on the phenological parameters was evaluated with Tukey's post-hoc tests on least squares means in ANOVA framework (PROC MIXED), and the effects of stand age, leaf area index (LAI) and basal area (BA) were evaluated with regression analysis (PROC REG). All significant effects we defined at $p < 0.05$ level, unless noted otherwise. The relative effects of age, year and forest type were determined with a stepwise regression (PROC REG), with both entry and staying p-values at 0.15. It should be noted that measuring the younger sites in separate years in the face of large year-to-year variation restricts the level of confidence with which the conclusions can be drawn about age-related changes. However, some strong trends that emerge in the current study can be further tested on longer data series when they become available.

3 Results

3.1 Main Sources of Variation

The fitted model explained the spring and fall changes better for ER (average R^2 across sites 0.92 in spring and 0.93 in fall) than GEP (average $R^2 = 0.87$ (spring) and 0.89 (fall)). No difference was observed among age classes. The primary source of variability of phenological parameters among the eight site-years was interannual difference in weather conditions (Table 2), leading to 29 ± 10 days longer ASL^{x0}_{GEP} ($t_{1,6} = 3.0$, $P = 0.024$) and 24 ± 5 days longer ASL^{x0}_{ER} ($t_{1,6} = 4.65$, $P = 0.004$) in 2002 than 2003. This was attributed to warmer autumn and later leaf fall in the former, whereas the differences in SOS were small (2.0 ± 3.8 and 0.6 ± 5.9 days for SOS_{GEP} and SOS_{ER}, respectively) and not statistically significant (Fig. 5). Interestingly, significant differences between the 2 years were also observed in LPF and ASL^{BG}, whereas those in ASL^{AH} varied by site. The age-related trends were significant for LFD_{GEP} and LPF_{GEP}, indicating that mature stands developed faster and maintained peak GEP longer than the young stands. ASL_{ER}, however, started later and ended more gradually in mature than young stands (Table 2). Our initial hypothesis of longer ASL in young than mature stands did not prove true (Fig. 5), even though RD_{ER}, RR_{ER} and RR_{GEP} were greater in the young than mature stands. These change rates and ASL^{x0}_{ER} and EOS_{ER} were narrowly constrained in the mature but not young stands. The expected dichotomy between the deciduous and conifer stands was observed only for GEP (mean ASL^{x0}_{GEP} of 80 ± 3 days in hardwoods and 87 ± 3 days in conifers), whereas ASL^{x0}_{ER} was 15 ± 6 days longer in hardwood than conifer stands. No other consistent differences were detected between the forest types.

Partitioning the relative contributions of age and year using a stepwise regression showed that the most significant source of variation for GEP seasonality was stand age, whereas interannual weather variability dominated the seasonality of ER (Table 3). Most age-related differences were also significant when expressed on

Table 2 Probability of type I error of significant difference in phenological parameters among age classes and years (both direct and interactive effects). Model denominator degrees of freedom = 6

	GEP			ER		
	Age	Year	Age × Year	Age	Year	Age*Year
SOS	0.120	0.534	0.142	0.023	0.885	0.032
EOS	0.329	0.059	0.839	0.071	0.023	0.015
ASL^{x0}	0.055	0.012	0.281	0.875	0.001	0.009
ASL^{AH}	0.163	0.055	0.526	0.406	0.029	0.002
ASL^{BG}	0.087	0.024	0.414	0.669	0.009	0.006
LPF	0.043	0.012	0.525	0.298	0.0004	0.205
LFD	0.042	0.501	0.171	0.184	0.305	0.024
LFR	0.450	0.689	0.905	0.407	0.001	0.050
RD	0.704	0.218	0.198	0.094	0.466	0.206
RR	0.053	0.085	0.178	0.027	0.518	0.456

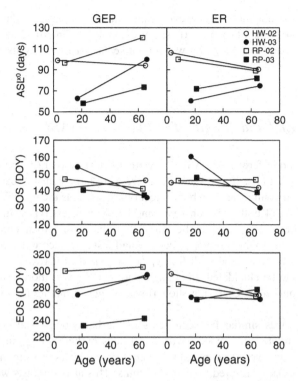

Fig. 5 Active season length (ASL^{x0}), start of active season (SOS), and end of active season (EOS) for gross ecosystem productivity (GEP) and ecosystem respiration (ER) in different-aged hardwood (HW) and conifer (RP) stands during 2 years (2002 and 2003). The 95% confidence intervals for all phenological indices were 0.1–0.4 days, as estimated with Monte Carlo analysis. That is, all visual differences on the figures are statistically significant in terms of precision of calculation. The significance of interannual and age-related differences, however, was derived from the mean parameters for every site-year.

Table 3 Adjusted coefficients of determination (and corresponding P-values) of stepwise regression models for the dependence of phenological parameters on age and year

	GEP		ER	
	Age	Year	Age	Year
SOS			0.35 (0.13)	
EOS	0.42 (0.08)			
ASL^{x0}				0.73 (<0.01)
ASL^{AH}	0.39 (0.10)			
ASL^{BG}		0.48 (0.06)		0.38 (0.10)
LPF	0.21 (0.10)	0.52 (0.04)		0.91 (<0.01)
LFD	0.42 (0.08)			
LFR				0.86 (<0.01)
RD				
RR	0.98 (<0.01)		0.59 (0.03)	

LAI or BA basis (not shown). However, sometimes age integrated the site characteristics better than the other properties. For example, the intermediate and mature
stands had comparable LAI, and it was a poor scalar for resolving their functional
differences.

3.2 Start, End and Length of Active Season (ASL)

The among-stand differences in ASL^{x0}_{GEP} were greatest by year (in conifers over 50
days) and by forest type (Figs. 4 and 5). There were large year-to-year differences
in the ASL^{x0}_{ER} in young stands in both forest types, whereas the mature stands were
relatively consistent both between years and between forest types. In both years,
SOS_{ER} (but not SOS_{GEP}) was earlier in mature hardwood than conifer stands.
Whether this is due to the more open canopy and faster surface warming or greater
metabolic activity associated with foliage development in hardwood than conifer
stands remains to be elucidated. In young stands, SOS_{ER} was more consistent than
SOS_{GEP}. The contiguous CUP (not shown) was longer in mature than young hardwood stands.

The EOS_{GEP} was similar between years in the hardwood stands, and showed
consistent age-related differences, with 5–10 days earlier EOS_{GEP} in young than
mature stands. However, the contrast between different years was big in the conifer
stands, where EOS_{GEP} differed by over 2 months. This also contrasts with the consistency of EOS_{ER} in all mature stands. EOS_{ER} was within 5 days between the 2
years in both mature hardwood and conifer stands. Overall, SOS was earlier and
EOS later (with the exception of EOS_{ER} in 2002) in mature than young stands.

3.3 The Duration and Rate of Change of Flux Transitions

The length of the flux development period decreased with age for GEP (LFD_{GEP};
Fig. 6), whereas the age-related trends in LFD_{ER} varied by year (Table 2). Both GEP
and ER had significantly longer LPF in 2002 than 2003, and LPF_{GEP} also increased
with stand age (Table 2). The length of flux recession period (LFR) exhibited
significant interannual differences for ER, but not for GEP (Table 2), whereas the
rate of recession (RR) differed among years in GEP. Both RR_{GEP} and RR_{ER} were
greater in young than mature stands.

The rates of flux development in spring were relatively constant for both GEP
(RD_{GEP}) and ER (RD_{ER}; Fig. 7; Table 2), with the exception of IH which had nearly
threefold higher RD than the other stands. The rates of recession (RR) were consist-
ently greater in the young than mature stands. The |RD| were greater than |RR| in
the mature but not in the young stands. Both RD and RR were about twofold greater
for GEP than ER, likely because of the greater change of magnitude from winter- to
summertime fluxes. However, if this was the sole reason, the pattern should have

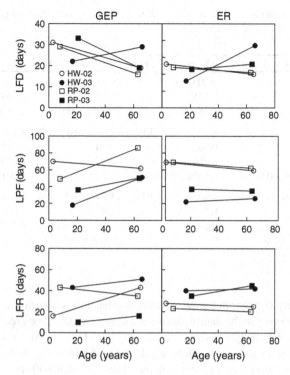

Fig. 6 Lengths of flux development (LFD), peak flux (LPF) and flux recession (LFR) periods of GEP and ER in different-aged hardwood (HW) and conifer (RP) stands during 2 years (2002 and 2003).

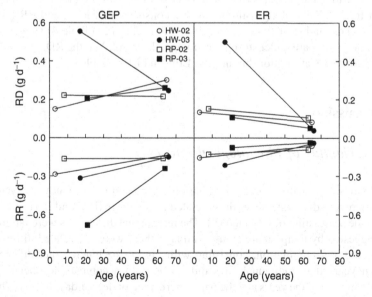

Fig. 7 Rates of development (RD, spring) and recession (RR, fall) of GEP and ER in different-aged- hardwood (HW) and conifer (RP) stands during 2 years (2002 and 2003).

Table 4 Regression parameters of ASL^{x0} vs. SOS, ASL^{x0} vs. EOS, RD vs. LFD, RR vs. LFR, LFR vs. LFD and RR vs. RD relationships

	GEP		ER	
	$ASL = \alpha + \beta \times SOS$	$ASL = \alpha + \beta \times EOS$	$ASL = \alpha + \beta \times SOS$	$ASL = \alpha + \beta \times EOS$
P-value	0.525	0.009	0.511	0.016
R^2	0.071	0.705	0.075	0.649
β	-0.95 ± 1.41	0.68 ± 0.18	-0.48 ± 0.69	1.14 ± 0.34
	$RD = \alpha + \beta \times LFD$	$RR = \alpha + \beta \times LFR$	$RD = \alpha + \beta \times LFD$	$RR = \alpha + \beta \times LFR$
P-value	0.376	0.062	0.094	0.685
R^2	0.132	0.466	0.397	0.029
β	-0.0070 ± 0.0073	0.0079 ± 0.0034	-0.015 ± 0.008	0.0011 ± 0.0026
	$LFR = \alpha + \beta \times LFD$	$RR = \alpha + \beta \times RD$	$LFR = \alpha + \beta \times LFD$	$RR = \alpha + \beta \times RD$
P-value	0.542	0.935	0.392	0.005
R^2	0.065	0.001	0.124	0.764
β	-0.62 ± 0.96	0.05 ± 0.60	0.54 ± 0.59	-0.37 ± 0.08

been consistently stronger in hardwood than conifer stands, which was not the case (Fig. 7). It is curious that the intermediate-aged stands exhibited the greatest deviation from otherwise relatively conservative RD and RR (Fig. 7). We hypothesize that the rapid change in the intermediate-aged stands could be related to their high homogeneity in species composition compared to other stands (Table 1), leading to highly synchronized on- and offset of vegetation activity.

The expected inverse relationship between the duration of the transition periods (LFD and LFR) and the rates of change (RD and RR) was seen for GEP (Table 4), but not for ER. This, along with the notable consistency of RD_{ER} and RR_{ER} in the mature stands, and their tight relationship with each other (Table 4) suggests that these change rates are under strong physiological control. Yet, the RD_{GEP} and RR_{GEP} did not exhibit clear relationship nor did LFD and LFR (Table 4).

4 Discussion

4.1 Variability of Phenological Phases

It is obvious that of the sources of variation considered in this study, the year-to-year differences dominated over those related to stand age (LAI and BA) and forest type on the seasonality of GEP and ER. The interannual differences were far greater than suggested by temperature sums. Although there were age-related differences in degree-day accumulation, consistent with Fig. 2, the much larger differences between years than individual sites did not lead to a comparable phenological response in each of the years and the parameterization of degree-day models (Chuine

2000) across years proved difficult (data not shown). The large interannual variability corroborates the observations of large year-to-year variability (up to 30 days) of SOS (Suni et al. 2003b) and ASL (Zha et al. 2008) in boreal forests. However, in the current study, among-stand differences in ASL^{x0} were more related to differences in EOS than SOS (Table 4). It has been found, however, that earlier SOS tends to stimulate GEP more than ER (e.g. Black et al. 2000; Jackson et al. 2001), whereas later EOS has the opposite effect (Piao et al. 2008), thus increasing and decreasing net productivity, respectively. Although the effect of EOS on ecosystem C balance may be smaller than that of SOS on per day basis, its effect can still be significant when the interannual variability of EOS exceeds that in SOS (Fig. 5). Furthermore, given that ASL_{ER} was more consistent in the mature than young stands (Fig. 5), whereas the magnitude of ER decreased with age in these systems (Noormets et al. 2007), the effect on net ecosystem C balance may be determined primarily by the changes in ASL_{GEP}. It is important to note that the expected differences in the SOS and ASL between the deciduous and conifer stands were not observed.

Although typically classified as northern temperate forests, the spring and fall transition dynamics of both GEP and ER suggest greater similarity with boreal than temperate forests. For temperate forests, LFD_{GEP} is usually found to be greater than LFR_{GEP} (Falge et al. 2002; Morecroft et al. 2003). For example, Morecroft et al. (2003) observed that the increase in leaf photosynthetic capacity (P_{max}) in an oak forest took longer (50–70 days after bud break) than the decline in the fall (14 days). In a study of forests from different climate zones, Falge et al. (2002) observed a similar pattern among temperate forests, whereas in the boreal zone LFD_{GEP} < LFR_{GEP}. Likewise, in the current study the mean LFD < LFR for both GEP and ER (LFR = 32.2 ± 3.1 and LFD = 24.4 ± 1.5 days, $F_{1,38}$ = 2.29, p = 0.029), although for a given site LFD and LFR were not well correlated (Table 4). Furthermore, the often proposed (e.g. DeForest et al. 2006; Falge et al. 2002) later cessation of ER in the fall compared to GEP was not observed in this study (Fig. 5). Likewise, the |RD| and |RR| did not systematically differ although there were individual stands exhibiting seasonal asymmetry in flux development (e.g. IP; Fig. 4). While RD_{ER} and RR_{ER} were highly correlated (Table 4), RD_{GEP} and RR_{GEP} were not. The latter contrasts with the findings of Gu et al. (2003), and might be related to the current study reflecting local variability, whereas Gu et al. addressed a broader latitudinal gradient. Comparing the relatively large differences in the flux change rates (RD and RR) in the young stands to the relative stability in the duration of the transition periods (LFD and LFR) suggests that any observed correlation might be mostly a reflection of the difference between the dormant season base flux and peak flux, as the LFD and LFR were strongly controlled by weather conditions (Table 2). It is unclear what controls the notable consistency of RD and RR as well as EOS_{ER} and ASL^{x0}_{ER} in the mature stands. Further research in this area is warranted, as ER plays a key role in affecting ecosystem net carbon balance, and its response to climate variability can be expected to be greater in the early stages of stand development (Fig. 5; Noormets et al. 2007).

4.2   Differences Between GEP and ER

The temporal dynamics of GEP and ER development were related, but clearly responded to different cues. The phenological indices of GEP exhibited consistent age-related variability, whereas differences in ER seasonality were mostly explained with those in weather (Table 3). The hypothesized systematic difference between ASL_{GEP} and ASL_{ER} due to later EOS_{ER} was not observed, because the temporary increase in ER following leaf fall did not affect points E and G (Fig. 3) and RR_{ER} derived from them. However, RD and RR were consistently about twofold greater for GEP than ER. This could partly be an artifact due to the greater absolute value of the GEP, but the longer LFD_{GEP} than LFD_{ER} suggested that this was not the only cause (Fig. 6). The variability of observed fluxes during the transition period was smaller in ER than GEP, but could be related to the considerable gapfilling of night-time data (Noormets et al. 2008).

4.3   Various Definitions of ASL

The variety and vagueness of the definitions of "growing season" could easily lead to the comparison of studies that use different approaches for delineating seasonality. In the current study, we have presented four different measures of ASL for each flux (ASL^{x0}, ASL^{AH}, ASL^{BG}, LPF). Some additional possibilities are described by Barr et al. in chapter "Climatic and Phenological Controls of the Carbon and Energy Balances of Three Contrasting Boreal Forest Ecosystems in Western Canada" in the current volume. Of the different measures of ASL, ASL^{x0} explained the most variation among stands, offering encouragement that the popular measures of half-maximum NDVI or EVI in remote sensing studies (Fisher et al. 2007; Schwartz et al. 2002) might also suit well for analyzing surface carbon fluxes.

Figure 8 shows that the different measures of ASL were generally well correlated, but the small-scale age-related differences did not always transfer. The carbon uptake period (CUP), the most commonly used measure of ASL in eddy covariance studies (Baldocchi et al. 2005; Churkina et al. 2005; Falge et al. 2002), was 79 (YH02), 133 (IH03), 131 (MH02) and 140 (MH03) days, and similar in scale to ASL^{AH}. However, YH02 had much shorter CUP than ASL^{AH}_{GEP}, because of large ER and alternating sink and source periods during the ASL (see Noormets et al. 2007). The inclusion or exclusion of this site significantly affected the correlation coefficients between CUP and other measures of ASL (Table 5). The change in R^2 was smallest for LPF_{ER} indicating that the among-site differences in sink strength were strongly affected by the duration of peak respiration. We have discussed this in greater detail in the past (Noormets et al. 2007, 2008). As the discontinuity of CUP makes comparisons with ASL (which is continuous by definition) difficult to interpret, it is important that uniform standards be adopted for delineating the seasonality of different processes that would allow comparisons among studies. Currently, it is often not clearly defined if CUP includes source days (NEE>0) during otherwise active canopy.

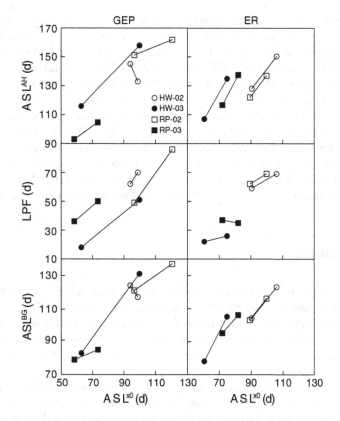

Fig. 8 Relationships between different measures of growing season length.

Table 5 Correlation coefficients between the contiguous carbon uptake period (CUP) and four measures of active season length for GEP and ER. Correlations were calculated for all hardwood stands and without YH02

	GEP		ER	
	All deciduous	w/o YH	All deciduous	w/o YH
ASL^{x0}	−0.32	0.44	−0.80	−0.24
ASL^{AH}	0.26	0.58	−0.70	0.52
ASL^{BG}	−0.38	0.31	−0.76	−0.25
LPF	−0.56	0.07	−0.76	−0.60

4.4 Predicting Responses to Climate Change

The timing of phenological transitions and phases could affect ecosystem C budget when ER and GEP are affected differentially by the changing weather dynamics. The greatest proportion of variability was explained by interannual climate differences, far exceeding the influence of stand age and forest type. Given that the influence of year was greater on the seasonality of ER than of GEP, it appears that the immediate implications for the regional carbon balance may be constrained by the stability of ASL_{ER}. While the literature suggests strong negative relationship between SOS and annual NEP (i.e. the later the SOS, the lower the NEP), in the current study most of the variability in ASL originated from EOS. Yet, detecting even the more conservative 7–10 day difference observed in the mature stands between the 2 years may require 10–12 year data record, assuming the 3.3–3.7 days per decade advance rate observed for trees (Estrella et al. 2007; Parmesan 2007). While several flux sites have attained such record, the analysis is further facilitated by the coordination and standardization under the global FLUXNET monitoring network. Given that interannual variation exceeds local among-sites differences, a network of sites with shorter individual record length is still well suited for the analysis of trends.

5   Conclusions

The largest proportion of variation in GEP and ER seasonality was explained by interannual variations in weather. The among-site differences in ASL were due to different EOS rather than SOS, with the latter being more consistent among sites, and less responsive to interannual climatic differences. However, this does not contradict the earlier conclusions of greater effect of unit change in SOS than EOS on annual carbon balance.

The seasonalities of GEP and ER differed in both timing and sensitivity to external factors. The age-related differences were stronger in the seasonality of GEP, whereas that of ER was dominated by interannual weather variability. The development and recession rates of fluxes were proportional to the difference between winter base and summer peak fluxes (i.e. $RD_{GEP} \approx 2 \times RD_{ER}$ and $RR_{GEP} \approx 2 \times RR_{ER}$).

The differences between deciduous and evergreen forest types were minimal, with marginally longer ASL_{GEP} in evergreen than deciduous stands, whereas ASL_{ER} was longer in the latter. The phenology of gas exchange suggests that these forests, typically classified as northern temperate stands, actually resemble boreal forests in terms of their flux development and recession dynamics.

Acknowledgements   This work was supported by USDA Forest Service Southern Global Change Program (AN, JC) and National Science Foundation (JC). We thank Ben Bond-Lamberty, Eero Nikinmaa and Andrew Richardson for constructive comments.

References

Ahas, R., Aasa, A., Menzel, A., Fedotova, V.G. and Scheifinger, H. (2002) Changes in European spring phenology. Int. J. Climatol., 22, 1727–1738.
Amiro, B.D. (2001) Paired-tower measurements of carbon and energy fluxes following disturbance in the boreal forest. Global Change Biol., 7, 253–268.

Amiro, B.D., Orchansky, A.L., Barr, A.G., Black, T.A., Chambers, S.D., Chapin III, F.S., Goulden, M.L., Litvak, M., Liu, H.P. and McCaughey, J.H. (2006) The effect of post-fire stand age on the boreal forest energy balance. Agric. For. Meteorol., 140, 41–50.

Aurela, M., Laurila, T. and Tuovinen, J.P. (2004) The timing of snow melt controls the annual CO_2 balance in a subarctic fen. Geophys. Res. Lett., 31, art. no. L16119.

Badeck, F.W., Bondeau, A., Bottcher, K., Doktor, D., Lucht, W., Schaber, J. and Sitch, S. (2004) Responses of spring phenology to climate change. New Phytol., 162, 295–309.

Baldocchi, D.D., Black, T.A., Curtis, P.S., Falge, E., Fuentes, J.D., Granier, A., Gu, L., Knohl, A., Pilegaard, K., Schmid, H.P., Valentini, R., Wilson, K., Wofsy, S., Xu, L. and Yamamoto, S. (2005) Predicting the onset of net carbon uptake by deciduous forests with soil temperature and climate data: a synthesis of FLUXNET data. Int. J. Biometeorol., 49, 377–387.

Baldocchi, D.D., Falge, E., Gu, L.H., Olson, R., Hollinger, D., Running, S., Anthoni, P., Bernhofer, C., Davis, K.J., Evans, R., Fuentes, J.D., Goldstein, A.H., Katul, G., Law, B.E., Lee, X., Malhi, Y., Meyers, T., Munger, W., Oechel, W., Paw U, K.T., Pilegaard, K., Schmid, H.P., Valentini, R., Verma, S., Vesala, T., Wilson, K. and Wofsy, S. (2001) FLUXNET: A new tool to study the temporal and spatial variability of ecosystem-scale carbon dioxide, water vapor, and energy flux densities. Bull. Am. Meteorol. Soc., 82, 2415–2434.

Barr, A.G., Black, T.A., Hogg, E.H., Griffis, T.J., Morgenstern, K., Kljun, N., Theede, A. and Nesic, Z. (2007) Climatic controls on the carbon and water balances of a boreal aspen forest, 1994-2003. Global Change Biol., 13, 561–576.

Barr, A.G., Black, T.A., Hogg, E.H., Kljun, N., Morgenstern, K. and Nesic, Z. (2004) Inter-annual variability in the leaf area index of a boreal aspen-hazelnut forest in relation to net ecosystem production. Agric. For. Meteorol., 126, 237–255.

Beaubien, E.G. and Freeland, H.J. (2000) Spring phenology trends in Alberta, Canada: links to ocean temperature. Int. J. Biometeorol., 44, 53–59.

Beck, P.S.A., Atzberger, C., Hogda, K.A., Johansen, B. and Skidmore, A.K. (2006) Improved monitoring of vegetation dynamics at very high latitudes: A new method using MODIS NDVI. Remote Sens. Environ., 100, 321–334.

Bergeron, O., Margolis, H.A., Coursolle, C. and Giasson, M.-A. (2008) How does forest harvest influence carbon dioxide fluxes of black spruce ecosystems in eastern North America? Agric. For. Meteorol., doi:10.1016/j.agrformet.2007.10.012, 148, 537–548.

Black, T.A., Chen, W.J., Barr, A.G., Arain, M.A., Chen, Z., Nesic, Z., Hogg, E.H., Neumann, H.H. and Yang, P.C. (2000) Increased carbon sequestration by a boreal deciduous forest in years with a warm spring. Geophys. Res. Lett., 27, 1271–1274.

Bogaert, J., Zhou, L., Tucker, C.J., Myneni, R.B. and Ceulemans, R. (2002) Evidence for a persistent and extensive greening trend in Eurasia inferred from satellite vegetation index data. J. Geophys. Res. (D Atmos.), 107, 10.1029/2001JD001075.

Bonan, G.B., Levis, S., Sitch, S., Vertenstein, M. and Oleson, K.W. (2003) A dynamic global vegetation model for use with climate models: concepts and description of simulated vegetation dynamics. Global Change Biol., 9, 1543–1566.

Bond-Lamberty, B., Gower, S.T., Ahl, D.E. and Thornton, P.E. (2005a) Reimplementation of the Biome-BGC model to simulate successional change. Tree Physiol., 25, 413–424.

Bond-Lamberty, B., Wang, C.K. and Gower, S.T. (2005b) Spatiotemporal measurement and modeling of stand-level boreal forest soil temperatures. Agric. For. Meteorol., 131, 27–40.

Cao, M.K., Prince, S.D., Small, J. and Goetz, S.J. (2004) Remotely sensed interannual variations and trends in terrestrial net primary productivity 1981–2000. Ecosystems, 7, 233–242.

Carrara, A., Kowalski, A., Neirynck, J., Janssens, I.A., Yuste, J.C. and Ceulemans, R. (2003) Net ecosystem CO_2 exchange of mixed forest in Belgium over 5 years. Agric. For. Meteorol., 119, 209–227.

Chmielewski, F.M. and Rötzer, T. (2001) Response of tree phenology to climate change across Europe. Agric. For. Meteorol., 108, 101–112.

Chuine, I. (2000) A unified model for budburst of trees. J. Theor. Biol., 207, 337–347.

Chuine, I. and Beaubien, E.G. (2001) Phenology is a major determinant of tree species range. Ecol. Lett., 4, 500–510.

Churkina, G., Schimel, D., Braswell, B.H. and Xiao, X. (2005) Spatial analysis of growing season length control over net ecosystem exchange. Global Change Biol., 11, 1777–1787.

DeForest, J.L., Noormets, A., Tenney, G., Sun, G., McNulty, S.G. and Chen, J. (2006) Phenophases in an oak-dominated forest alter the soil respiration-temperature relationship. Int. J. Biometeorol., 51, 135–144.

Desai, A.R., Noormets, A., Bolstad, P.V., Chen, J., Cook, B.D., Davis, K.J., Euskirchen, E.S., Gough, C., Martin, J.M., Ricciuto, D.M., Schmid, H.P., Tang, J. and Wang, W. (2008) Influence of vegetation type, stand age and climate on carbon dioxide fluxes across the Upper Midwest, USA: Implications for regional scaling of carbon flux. Agric. For. Meteorol., 148, 288–308.

Devine, W.D. and Harrington, C.A. (2007) Influence of harvest residues and vegetation on microsite soil and air temperatures in a young conifer plantation. Agric. For. Meteorol., 145, 125–138.

Estrella, N., Sparks, T.H. and Menzel, A. (2007) Trends and temperature response in the phenology of crops in Germany. Global Change Biol., 13, 1737–1747.

Ewers, B.E., Gower, S.T., Bond-Lamberty, B. and Wang, C.K. (2005) Effects of stand age and tree species on canopy transpiration and average stomatal conductance of boreal forests. Plant Cell Environ., 28, 660–678.

Falge, E., Baldocchi, D., Olson, R., Anthoni, P., Aubinet, M., Bern-hofer, C., Burba, G., Ceulemans, R., Clement, R., Dolman, H., Granier, A., Gross, P., Grünwald, T., Hollinger, D., Jensen, N.-O., Katul, G., Keronen, P., Kowalski, A., Lai, C.T., Law, B.E., Meyers, T., Moncrieff, J., Moors, E., Munger, J.W., Pilegaard, K., Rannik, Ü., Rebmann, C., Suyker, A., Tenhunen, J., Tu, K., Verma, S., Vesala, T., Wilson, K. and Wofsy, S. (2001) Gap filling strategies for defensible annual sums of net ecosystem exchange. Agric. For. Meteorol., 107, 43–69.

Falge, E., Baldocchi, D.D., Tenhunen, J., Aubinet, M., Bakwin, P.S., Berbigier, P., Bernhofer, C., Burba, G., Clement, R., Davis, K.J., Elbers, J.A., Goldstein, A.H., Grelle, A., Granier, A., Guddmundsson, J., Hollinger, D., Kowalski, A.S., Katul, G., Law, B.E., Malhi, Y., Meyers, T., Monson, R.K., Munger, J.W., Oechel, W., Paw U, K.T., Pilegaard, K., Rannik, Ü., Rebmann, C., Suyker, A., Valentini, R., Wilson, K. and Wofsy, S. (2002) Seasonality of ecosystem respiration and gross primary production as derived from FLUXNET measurements. Agric. For. Meteorol., 113, 53–74.

Field, C.B., Lobell, D.B., Peters, H.A. and Chiariello, N.R. (2007) Feedbacks of terrestrial ecosystems to climate change. Annu. Rev. Environ. Res., 32, 1–29.

Fisher, J.I., Richardson, A.D. and Mustard, J.F. (2007) Phenology model from surface meteorology does not capture satellite-based greenup estimations. Global Change Biol., 13, 707–721.

Fitter, A.H. and Fitter, R.S.R. (2002) Rapid Changes in Flowering Time in British Plants. Science, 296, 1689–1691.

Gore, A. (2006) An Inconvenient Truth: The Planetary Emergency of Global Warming and What We Can Do About It. Rodale Books: New York.

Goulden, M.L., Munger, J.W., Fan, S.M., Daube, B.C. and Wofsy, S.C. (1996) Exchange of carbon dioxide by a deciduous forest: Response to interannual climate variability. Science, 271, 1576–1578.

Goulden, M.L., Winston, G.C., McMillan, A.M.S., Litvak, M.E., Read, E.L., Rocha, A.V. and Elliot, J.R. (2006) An eddy covariance mesonet to measure the effect of forest age on land-atmosphere exchange. Global Change Biol., 12, 2146–2162.

Groisman, P.Y., Karl, T.R. and Knight, R.W. (1994) Observed impact of snow cover on the heat balance and the rise of continental spring temperatures. Science, 263, 198–200.

Gu, L., Post, W.M., Baldocchi, D., Black, T.A., Verma, S.B., Vesala, T. and Wofsy, S.C. (2003). Phenology of vegetation photosynthesis. In: Schwartz, M.D. (Ed.) Phenology: An Integrated Environmental Science. Kluwer; Dordecht, pp. 467–485.

Hansen, J., Sato, M., Lacis, A. and Ruedy, R. (1997) The missing climate forcing. Philosophical Transactions of the Royal Society of London Series B-Biological Sciences, 352, 231–240.

Hargrove, W.W. and Hoffman, F.M. (2005) Potential of multivariate quantitative methods for delineation and visualization of ecoregions. Environ. Manage., 34, S39–S60.

Hegerl, G.C., vonStorch, H., Hasselmann, K., Santer, B.D., Cubasch, U. and Jones, P.D. (1996) Detecting greenhouse-gas-induced climate change with an optimal fingerprint method. J. Clim., 9, 2281–2306.

Heide, O.M. (2003) High autumn temperature delays spring bud burst in boreal trees, counterbalancing the effect of climatic warming. Tree Physiol., 23, 931–936.

Heijmans, M.M.P.D., Arp, W.J. and Chapin, F.S. (2004) Carbon dioxide and water vapour exchange from understory species in boreal forest. Agric. For. Meteorol., 123, 135–147.

Hollinger, D.Y., Aber, J.D., Dail, B., Davidson, E.A., Goltz, S.M., Hughes, H., Leclerc, M.Y., Lee, J.T., Richardson, A.D., Rodrigues, C., Scott, N.A., Achuatavarier, D. and Walsh, J. (2004) Spatial and temporal variability in forest-atmosphere CO_2 exchange. Global Change Biol., 10, 1689–1706.

Hollinger, D.Y. and Richardson, A.D. (2005) Uncertainty in eddy covariance measurements and its application to physiological models. Tree Physiol., 25, 873–885.

Houghton, R.A., Davidson, E.A. and Woodwell, G.M. (1998) Missing sinks, feedbacks, and understanding the role of terrestrial ecosystems in the global carbon balance. Global Biogeochem. Cycles, 12, 25–34.

IPCC (2007). Climate Change 2007: Synthesis Report. Contribution of Working Groups I, II and III to the Fourth Assessment Report of the Intergovernmental Panel on Climate Change.Pachauri, R.K. and Reisinger, A. (Eds.), IPCC, Geneva, Switzerland, pp.104.

Jackson, R.B., Lechowicz, M.J., Li, X. and Mooney, H.A. (2001). Phenology, growth, and allocation in global terrestrial productivity. In: Saugier, B., Roy, J. and Mooney, H.A. (Eds.), Terrestrial Global Productivity: Past, Present, and Future. Academic: San Diego, CA, pp. 61–82.

Kathuroju, N., White, M.A., Symanzik, J., Schwartz, M.D., Powell, J.A. and Nemani, R.R. (2007) On the use of the advanced very high resolution radiometer for development of prognostic land surface phenology models. Ecol. Model., 201, 144–156.

Keeling, C.D., Chin, J.F.S. and Whorf, T.P. (1996) Increased activity of northern vegetation inferred from atmospheric CO_2 measurements. Nature, 382, 146–149.

Kramer, K., Leinonen, I., Bartelink, H.H., Berbigier, P., Borghetti, M., Bernhofer, C., Cienciala, E., Dolman, A.J., Froer, O., Gracia, C.A., Granier, A., Grünwald, T., Hari, P., Jans, W., Kellomäki, S., Loustau, D., Magnani, F., Markkanen, T., Matteucci, G., Mohren, G.M.J., Moors, E., Nissinen, A., Peltola, H., Sabate, S., Sanchez, A., Sontag, M., Valentini, R. and Vesala, T. (2002) Evaluation of six process-based forest growth models using eddy-covariance measurements of CO_2 and H_2O fluxes at six forest sites in Europe. Global Change Biol., 8, 213–230.

Kramer, K., Leinonen, I. and Loustau, D. (2000) The importance of phenology for the evaluation of impact of climate change on growth of boreal, temperate and Mediterranean forests ecosystems: an overview. Int. J. Biometeorol., 44, 67–75.

Lenton, T.M. (2000) Land and ocean carbon cycle feedback effects on global warming in a simple Earth system model. Tellus B, 52, 1159–1188.

Linderholm, H.W. (2006) Growing season changes in the last century. Agric. For. Meteorol., 137, 1–14.

Lu, P., Yu, Q., Liu, J. and Lee, X. (2006) Advance of tree-flowering dates in response to urban climate change. Agric. For. Meteorol., 138, 120–131.

Mäkelä, A., Hari, P., Berninger, F., Hänninen, H. and Nikinmaa, E. (2004) Acclimation of photosynthetic capacity in Scots pine to the annual cycle of temperature. Tree Physiol., 24, 369–376.

Mäkelä, A., Pulkkinen, M., Kolari, P., Lagergren, F., Berbigier, P., Lindroth, A., Loustau, D., Nikinmaa, E., Vesala, T. and Hari, P. (2008) Developing an empirical model of stand GPP with the LUE approach: analysis of eddy covariance data at five contrasting conifer sites in Europe. Global Change Biol., 14, 92–108.

Meir, P., Cox, P. and Grace, J. (2006) The influence of terrestrial ecosystems on climate. Trends Ecol. Evol., 21, 254–260.

Menzel, A. (2002) Phenology: Its importance to the global change community – An editorial comment. Clim. Change, 54, 379–385.

Menzel, A. (2003) Plant phenological anomalies in Germany and their relation to air temperature and NAO. Clim. Change, 57, 243–263.

Menzel, A. and Fabian, P. (1999) Growing season extended in Europe. Nature, 397, 659–659.

Moffat, A.M., Papale, D., Reichstein, M., Hollinger, D.Y., Richardson, A.D., Barr, A.G., Beckstein, C., Braswell, B.H., Churkina, G., Desai, A.R., Falge, E., Gove, J.H., Heimann, M., Hui, D., Jarvis, A.J., Kattge, J., Noormets, A. and Stauch, V.J. (2007) Comprehensive comparison of gap filling techniques for eddy covariance net carbon fluxes. Agric. For. Meteorol., 147, 209–232.

Morecroft, M.D., Stokes, V.J. and Morison, J.I.L. (2003) Seasonal changes in the photosynthetic capacity of canopy oak (*Quercus robur*) leaves: the impact of slow development on annual carbon uptake. Int. J. Biometeorol., 47, 221–226.

Myneni, R.B., Keeling, C.D., Tucker, C.J., Asrar, G. and Nemani, R.R. (1997) Increased plant growth in the northern high latitudes from 1981 to 1991. Nature, 386, 698–702.

Noormets, A., Chen, J. and Crow, T.R. (2007) Age-dependent changes in ecosystem carbon fluxes in managed forests in northern Wisconsin, USA. Ecosystems, 10, 187–203.

Noormets, A., Desai, A.R., Cook, B.D., Euskirchen, E.S., Ricciuto, D.M., Davis, K.J., Bolstad, P.V., Schmid, H.P., Vogel, C.V., Carey, E.V., Su, H.B. and Chen, J. (2008) Moisture sensitivity of ecosystem respiration: Comparison of 14 forest ecosystems in the Upper Great Lakes Region, USA. Agric. For. Meteorol., 148, 216–230.

Norby, R.J., Hartz-Rubin, J.S. and Verbrugge, M.J. (2003) Phenological responses in maple to experimental atmospheric warming and CO_2 enrichment. Global Change Biol., 9, 1792–1801.

Oechel, W.C., Vourlitis, G.L., Hastings, S.J., Zulueta, R.C., Hinzman, L. and Kane, D. (2000) Acclimation of ecosystem CO_2 exchange in the Alaskan Arctic in response to decadal climate warming. Nature, 406, 978–981.

Olesen, J.E., Carter, T.R., Diaz-Ambrona, C.H., Fronzek, S., Heid-mann, T., Hickler, T., Holt, T., Quemada, M., Ruiz-Ramos, M., Rubaek, G.H., Sau, F., Smith, B. and Sykes, M.T. (2007) Uncertainties in projected impacts of climate change on European agriculture and terrestrial ecosystems based on scenarios from regional climate models. Clim. Change, 81, 123–143.

Parmesan, C. (2007) Influences of species, latitudes and methodologies on estimates of phenological response to global warming. Global Change Biol., 13, 1860–1872.

Penuelas, J. and Filella, I. (2001) Responses to a Warming World. Science, 294, 793–794.

Penuelas, J., Filella, I. and Comas, P. (2002) Changed plant and animal life cycles from 1952 to 2000 in the Mediterranean region. Global Change Biol., 8, 531–544.

Piao, S.L., Ciais, P., Friedlingstein, P., Peylin, P., Reichstein, M., Luyssaert, S., Margolis, H., Fang, J.Y., Barr, A., Chen, A.P., Grelle, A., Hollinger, D.Y., Laurila, T., Lindroth, A., Richardson, A.D. and Vesala, T. (2008) Net carbon dioxide losses of northern ecosystems in response to autumn warming. Nature, 451, 49–53.

Reed, B.C., Brown, J.F., VanderZee, D., Loveland, T.L., Merchant, J.W. and Ohlen, D.O. (1994) Measuring phenological variability from satellite imagery. J. Veg. Sci., 5, 703–714.

Richter, D.D., Markewitz, D., Trumbore, S.E. and Wells, C.G. (1999) Rapid accumulation and turnover of soil carbon in a re-establishing forest. Nature, 400, 56–58.

Santer, B.D., Taylor, K.E., Wigley, T.M.L., Johns, T.C., Jones, P.D., Karoly, D.J., Mitchell, J.F.B., Oort, A.H., Penner, J.E., Ramaswamy, V., Schwarzkopf, M.D., Stouffer, R.J. and Tett, S. (1996) A search for human influences on the thermal structure of the atmosphere. Nature, 382, 39–46.

Schaefer, K., Denning, A.S. and Leonard, O. (2005) The winter Arctic oscillation, the timing of spring, and carbon fluxes in the Northern Hemisphere. Global Biogeochem. Cycle, 19.

Schlesinger, W.H. and Lichter, J. (2001) Limited carbon storage in soil and litter of experimental forest plots under increased atmospheric CO_2. Nature, 411, 466–469.

Schwartz, M.D., Ahas, R. and Aasa, A. (2006) Onset of spring starting earlier across the Northern Hemisphere. Global Change Biol., 12, 343–351.

Schwartz, M.D. and Chen, Z.Q. (2002) Examining the onset of spring in China. Clim. Res., 21, 157–164.

Schwartz, M.D., Reed, B.C. and White, M.A. (2002) Assessing satellite-derived start-of-season measures in the conterminous USA. Int. J. Climatol., 22, 1793–1805.

Schwartz, M.D. and Reiter, B.E. (2000) Changes in North American spring. Int. J. Climatol., 20, 929–932.

Sherry, R.A., Zhou, X., Gu, S., Arnone, J.A., III, Schimel, D.S., Verburg, P.S., Wallace, L.L. and Luo, Y. (2007) Divergence of reproductive phenology under climate warming. Proceedings of the National Academy of Sciences, 104, 198–202.

Smith, N.V., Saatchi, S.S. and Randerson, J.T. (2004) Trends in high northern latitude soil freeze and thaw cycles from 1988 to 2002. J. Geophys. Res., 109.

Stott, P.A., Tett, S.F.B., Jones, G.S., Allen, M.R., Mitchell, J.F.B. and Jenkins, G.J. (2000) External control of 20th century temperature by natural and anthropogenic forcings. Science, 290, 2133–2137.

Suni, T., Berninger, F., Markkanen, T., Keronen, P., Rannik, Ü. and Vesala, T. (2003a) Interannual variability and timing of growing-season CO_2 exchange in a boreal forest. J. Geophys. Res., 108, No. D9, 4265, doi:10.1029/2002JD002381, 2.1–2.8.

Suni, T., Berninger, F., Vesala, T., Markkanen, T., Hari, P., Mäkelä, A., Ilvesniemi, H., Hanninen, H., Nikinmaa, E., Huttula, T., Laurila, T., Aurela, M., Grelle, A., Lindroth, A., Arneth, A., Shi-bistova, O. and Lloyd, J. (2003b) Air temperature triggers the recovery of evergreen boreal forest photosynthesis in spring. Global Change Biol., 9, 1410–1426.

Tape, K., Sturm, M. and Racine, C. (2006) The evidence for shrub expansion in Northern Alaska and the Pan-Arctic. Global Change Biol., 12, 686–702.

Walther, G.-R., Post, E., Convey, P., Menzels, A., Parmesan, C., Beebee, T.J.C., Fromentin, J.-M., Hoegh-Guldberg, O. and Bairlein, F. (2002) Ecological responses to recent climate change. Nature, 416, 389–395.

White, M.A., Hoffman, F., Hargrove, W.W. and Nemani, R.R. (2005) A global framework for monitoring phenological responses to climate change. Geophys. Res. Lett., 32, Art. No. L04705.

White, M.A. and Nemani, A.R. (2003) Canopy duration has little influence on annual carbon storage in the deciduous broad leaf forest. Global Change Biol., 9, 967–972.

Yang, F., Ichii, K., White, M.A., Hashimoto, H., Michaelis, A.R., Votava, P., Zhu, A.X., Huete, A., Running, S.W. and Nemani, R.R. (2007) Developing a continental-scale measure of gross primary production by combining MODIS and AmeriFlux data through Support Vector Machine approach. Remote Sens. Environ., 110, 109–122.

Zha, T., Barr, A.G., Black, T.A., McCaughey, H., Bhatti, J.S., Hawthorne, I., Krishnan, P., Kidston, J., Saigusa, N., Shashkov, A. and Nesic, Z. (2008) Carbon sequestration in boreal Jack pine stands following harvesting. Global Change Biol., 15, 1475–1487.

Zhang, X.Y., Friedl, M.A., Schaaf, C.B., Strahler, A.H., Hodges, J.C.F., Gao, F., Reed, B.C. and Huete, A. (2003) Monitoring vegetation phenology using MODIS. Remote Sens. Environ., 84, 471–475.

Phenological Differences Between Understory and Overstory: A Case Study Using the Long-Term Harvard Forest Records

Andrew D. Richardson and John O'Keefe

Abstract The timing of phenological events varies both among species, and also among individuals of the same species. Here we use a 12-year record of spring and autumn phenology for 33 woody species at the Harvard Forest to investigate these differences. Specifically, we focus on patterns of leaf budburst, expansion, coloration and fall, in the context of differences between canopy and understory species, and between canopy and understory individuals of the same species. Many understory species appear to adopt a strategy of "phenological escape" in spring but not autumn, taking advantage of the high-light period in spring before canopy development. For all but a few of these species, the spring escape period is very brief. Relationships between canopy and understory conspecifics (i.e. individuals of the same species) varied among species, with leaf budburst and leaf fall occurring earlier in understory individuals of some species, but later in other species. We fit standard models of varying complexity to the budburst time series for each species to investigate whether biological responses to environmental cues differed among species. While there was no clear consensus model, Akaike's Information Criterion (AIC) indicated that a simple two-parameter "Spring warming" model was best supported by the data for more than a third of all species, and well supported for two-thirds of all species. More highly-parameterized models involving various chilling requirements (e.g., Alternating, Parallel or Sequential chilling) were less well supported by the data. Species-specific model parameterization suggested that responses to both chilling and forcing temperatures vary among species. While there were no obvious differences in this regard between canopy and understory species, or between early- and late-budburst species, these results imply that species can be expected to differ in their responses to future climate change.

A.D. Richardson (✉)
Department of Organismic and Evolutionary Biology, Harvard University,
26 Oxford Street, Cambridge MA 02138
e-mail: andrewr@solo.sr.unh.edu

J. O'Keefe
Harvard Forest, 324 North Main Street, Petersham, MA 01366
e-mail: jokeefe@fas.harvard.edu

A. Noormets (ed.), *Phenology of Ecosystem Processes*,
DOI 10.1007/978-1-4419-0026-5_4, © Springer Science+Business Media, LLC 2009

1   Introduction

Plants exhibit both phenotypic plasticity and genotypic adaptation to different growth environments (Schlichting 1986; Sultan 1995). In terms of phenology, these responses can be manifest as differences in spring budburst and autumn senescence within species across environmental gradients at varying scales, from regional (e.g., across the native range of a species; see Lechowicz 1984; Raulier and Bernier 2000) to local (e.g., along an elevational gradient: see Richardson et al. 2006) and even microsite (e.g., with regard small-scale topographic variation and cold air drainage, see Fisher et al. 2006). These patterns are commonly interpreted in the context of differences in temperature regime among different growth environments.

However, just as differences in light environment between understory and canopy result in well-known differences in leaf structure (Boardman 1977; Lichtenthaler et al. 1981), so too can plants respond to the ambient light environment by adopting different phenological strategies. For example, many herbaceous understory species in temperate deciduous forests have been shown to adopt "phenological escape" (Crawley 1997) strategies of shade avoidance in order to increase the seasonal integral of potential photosynthesis and to capitalize on other seasonally-limited resources for growth (e.g., Muller 1978). By emerging earlier in the spring than canopy species (or by delaying autumn senescence), shade-avoiding understory plants can take advantage of a short-lived high-light growth environment (e.g., Sparling 1967; Mahall and Bormann 1978; Crawley 1997). As demonstrated by greenhouse studies, early-emerging individuals not only have more time to grow, but also grow more quickly and accumulate more resources than individuals that emerge later (Rathcke and Lacey 1985). Textbook examples of herbaceous species adopting the escape strategy include *Hyacinthoides non-scripta* and *Anemone nemorosa* (Crawley 1997). Such a strategy is not without tradeoffs, however: early-emerging individuals often have a lower chance of survival, but much higher reproductive success, than individuals that emerge later (Rathcke and Lacey 1985). Earlier emergence increases susceptibility to spring frost damage which may not only damage photosynthetic machinery but also lead to foliar necrosis (e.g., Gu et al. 2008). Also, plants that emerge under high-light conditions may not be morphologically well-adapted to low light conditions. For example, work by Sparling (1967) demonstrated that herbs emerging in spring, prior to canopy closure, had photosynthetic characteristics of shade intolerant species, whereas species emerging in mid-summer exhibited shade tolerant characteristics (see also Hull 2002). However, Rothstein and Zak (2001) reported that acclimation of individual leaves could occur as the growing season progressed and the below-canopy light environment changed: leaves of *Viola pubescens* emerged prior to canopy closure with sun leaf characteristics, but as the canopy closed, both chemical (e.g., N partitioning to Rubisco vs. chlorophyll) and structural (leaf mass to area ratio) acclimation resulted in the same leaves becoming more like shade leaves, and thus better suited to a low-light environment.

While phenological escape is well-documented for herbaceous species, much less is known about the degree to which this strategy is adopted by woody plants, either shrub/small tree species restricted to the understory, or juvenile/suppressed individuals of canopy species. Previous work in this regard has generally been focused on a relatively limited number of species, and results have not been entirely consistent among studies. While the overall pattern is one of understory species and juveniles of canopy species leafing out earlier in spring, but not necessarily dropping their leaves later in autumn, than mature canopy trees (Gill et al. 1998; Seiwa 1999a, b; Augspurger and Bartlett 2003; Barr et al. 2004; Augspurger et al. 2005), important questions remain. For example, how widespread is the phenological escape strategy among woody understory plants? Or, what traits do species adopting this strategy have in common? And, are there differences between understory species and juveniles of canopy species in terms of how this strategy is expressed?

In the sections that follow, we begin by providing a brief review of how the understory growth environment varies from season to season. We then discuss evidence for phenological variation within and among species, first broadly and then specifically with regard to differences between understory and canopy. This is followed by an evaluation of potential fitness consequences of different phenological strategies. We then use data from the long-term phenology records maintained at the Harvard Forest as a case study to investigate the following questions in the context of patterns of spring (budburst and leaf expansion) and autumn (leaf coloration and leaf fall) phenology for an entire forest community, comprising 33 different woody species:

1. Are there differences in phenology between understory tree and shrub species and canopy tree species (i.e., those that have the potential to grow into the canopy)?
2. Within the class of canopy species, are there differences in phenology between suppressed individuals growing in the understory, and dominant or co-dominant individuals in the overstory?
3. Given that year-to-year variation in weather (e.g., warm vs. cool spring) gener- ally results in corresponding shifts in the timing of phenological events (Hunter and Lechowicz 1992), how do species differ in their sensitivities to climatic vari- ability? For example:

(a) Is the sequence in which species reach particular phenophases consistent from year to year?
(b) Are there differences among species (particularly with regard to canopy vs. understory species) in terms of which temperature-based model to predict budburst is best supported by the data?

1.1   Seasonal Variation in the Understory Growth Environment

The seasonal patterns of light availability for various canopy strata in a temperate deciduous forest depend largely on the development and senescence of the overstory canopy (throughout this remainder of this chapter we use "canopy" and "overstory" as

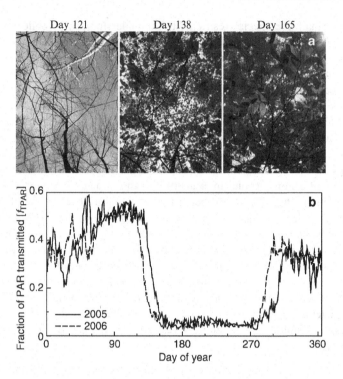

Fig. 1 Rapid changes in the state of the canopy (**a**) affect the below-canopy light environment (**b**), as measured by $f_{TPAR} = Q_{transmitted}/Q_{incident}$, the fraction of incident photosynthetically active radiation that is transmitted through the canopy. Data from the Bartlett Experimental Forest, White Mountain National Forest, New Hampshire. Photo credit: Chris Costello.

synonyms), with changing zenith angle and overall solar irradiance acting as secondary factors (Baldocchi et al. 1984). As an example, data from the Bartlett Experimental Forest, a northern hardwood (maple-beech-birch) forest in central New Hampshire, indicate a rapid spring decrease in the fraction of photosynthetically active radiation transmitted through the canopy (f_{TPAR}) as the developmental trajectory progresses from a leafless (day 121 in Fig. 1a) to a fully-foliated (day 165 in Fig. 1a) canopy. Over a period of roughly 1 month, f_{TPAR} declines from ~50% to ~5% (Fig. 1b, see also Baldochi et al. 1986; Gill et al. 1998). During this time the understory light environment is extremely heterogeneous, as canopy species leaf out at different times and leaf elongation proceeds at different rates (Kato and Komiyama 2002). With canopy closure, sunflecks become ever more important for understory photosynthesis (Chazdon and Pearcy 1991). In the autumn, the reverse pattern occurs, over a similarly brief period, with the onset of leaf senescence and, ultimately, abscission. It is important to note that the physical presence of the canopy does not necessarily mean the canopy is physiologically active (Sakai et al. 1997): particularly in the autumn, canopy photosynthesis is reduced with the onset of senescence, but shading of the understory continues until canopy leaves are actually dropped.

Beyond fluxes of photosynthetically active radiation, the presence or absence of a full canopy affects other aspects of energy and mass transfer, with implications for various factors affecting growth and development of understory plants, including: leaf temperatures and saturation deficit; precipitation throughfall as well as soil surface heat flux and evaporation; and diurnal ranges of air and soil temperatures.

1.2 Phenological Differences Within and Among Species

For a given tree species, climatic gradients (particularly latitude but also elevation) are a dominant control on phenology, as budburst by higher-latitude and higher-elevation individuals generally follows that of lower-latitude and lower-elevation individuals (e.g., Lechowicz 1984; Richardson et al. 2006; Fisher et al. 2007). This can be attributed to varying rates of spring warming, as temperature is considered the primary trigger for spring budburst, at least in temperate, boreal, and arctic ecosystems. However, provenance trials indicate that there is genetic variation among populations in terms of responsiveness to environmental cues. For example, in common-garden experiments, northern populations have generally been shown to leaf out in advance of southern populations, although the reverse is true in some species (Lechowicz 1984).

In temperate forests, coexisting tree species commonly leaf out at different times, from the typically early emergence of *Betula* and *Populus*, to the late emergence of *Quercus* and *Fraxinus*. Among these species, budburst dates vary by 3 weeks or more in a temperate deciduous forest in Quebec, although a similar ordering of related species is observed both elsewhere in North America and also in Europe (Lechowicz 1984). At the same time, the rate of canopy development also differs among species, with leaf elongation proceeding rapidly in some species (*Prunus* and *Acer*) but more slowly in other species (*Juglans* and *Carya*) (Lechowicz 1984). Perhaps surprisingly, the timing of budburst is not correlated with ecological niche (e.g., shade tolerance) or with phylogenetic history (both advanced and primitive families can be either early or late leafing out, and within families, and even within genera, there is just as much variation as among families); however, noting that many late-leafing tree species tended to be ring-porous (i.e., woody species in which the xylem vessels laid down at the beginning of the growing season are much larger in diameter than those produced later in the growing season), Lechowicz (1984) hypothesized that ring-porous species leaf out later than diffuse-porous species because large diameter vessels are more prone to cavitation caused by freeze-thaw xylem embolism. Thus in ring-porous species budburst can only occur after new xylem has been formed each spring, as otherwise there would be little or no capacity to transport water to developing leaves. In diffuse-porous species, on the other hand, the hydraulic conductivity is less affected by winter freezing, and leaf development can occur in advance of, or concurrently with, the formation of new xylem.

1.3   Phenological Differences Between Understory and Overstory

Woody understory shrubs, small trees, and juvenile seedlings and saplings of canopy tree species might all be expected to exhibit phenological escape strategies similar to those of herbaceous understory species, but previous research has provided mixed results. Here we review some of this earlier work.

Comparing the phenology of woody understory plants (three understory species, and juveniles of two canopy tree species) with the development and senescence of the forest canopy, Augspurger et al. (2005) found that for four of the five species studied, leaf expansion preceded canopy development. The first of these four, *Aesculus glabra*, leafed out more than 3 weeks before the canopy and received approximately 97% of its annual irradiance during this spring high-light period. Budburst and foliar development of the fifth species, *Asimina tribola*, was slow and lagged behind canopy development. Senescence and leaf drop of *Aesculus* occurred roughly 2 months before the canopy, but for the other four species the timing of autumn senescence was similar to that of the canopy. Gill et al. (1998) reported earlier leaf expansion and later autumn senescence in *Viburnum alnifolium*, an understory shrub, compared to the northern hardwood canopy species. But, in a deciduous boreal forest, foliar development of the early-flowering but comparatively late-leafing hazelnut (*Corylus cornuta*) understory lagged that of the *Populus tremuloides* canopy by 3 weeks or more. Here, aspen were found to require only about half as many degree-days as hazelnut to reach the same level of development (Barr et al. 2004). This study also reported greater interannual variability in both spring and autumn phenology for canopy, compared to understory species. The woody species *Aesculus sylvatica*, a small deciduous tree native to the southeastern United States has an unusual spring-green phenology similar to that of many understory herbs. Its leaves emerge in mid-March, only to senesce in late May when the canopy above is beginning to close (dePamphilis and Neufeld 1989). For this understory species, the growing season is very short and is restricted to the high-light period of spring.

Juvenile and mature individuals of canopy tree species can also differ phenologically. Seiwa (1999a, b) reported that phenology changed with ontogeny in both *Acer mono* and *Ulmus davidiana*. In both species, there was a general pattern of later leaf emergence with increasing tree age and tree height (from small seedlings to saplings to small trees and large trees). Similar results were reported by Augspurger and Bartlett (2003), who compared the phenology of juvenile and adult individuals of 13 different deciduous species. Budburst of juveniles was significantly earlier (average ≈8 days) than that of conspecific (i.e., of the same species) adults (see also Gill et al. 1998) in ten species; for none of the species was the opposite pattern observed.

There is a wide range of variability in phenological differences between juveniles and adults in autumn. For example, whereas there was no significant age-related difference in the timing of autumn leaf drop for *Acer* (Seiwa 1999a), seedlings of *Ulmus* retained their leaves longer than mature trees (Seiwa 1999b). Similarly mixed results were presented by Augspurger and Bartlett (2003), who reported that for most of the 13 species studied, the timing of onset of senescence and completion of leaf fall did not vary between juveniles and adults. Where significant differences were observed, the pattern of variation was not consistent among species: juveniles retained their leaves longer in some instances (e.g. *Acer saccharum*), but shorter in others (e.g. *Aesculus glabra*). In contrast to these results, Gill et al. (1998) found that understory *Acer saccharum* and *Fagus grandifolia* retained green leaves later in autumn than overstory trees of the same species.

On the basis of the springtime patterns, Seiwa (1999a, b) suggested that the overall net benefits of earlier leaf-out must therefore decrease with increasing tree height, reflecting a shifting balance between the advantages (increased carbon gain and reduced susceptibility to herbivory, e.g., Crawley 1997) and the disadvantages (risk of frost damage, e.g., Lechowicz 1984) of early leaf out. This would seem to be a logical explanation: for individuals deep under the canopy, leafing out early offers the largest rewards. By comparison, for dominant canopy trees, the competition for light has already been "won", and leafing out early is of much less benefit. Thus, one would predict that leaf expansion "progress from the base of the canopy upward" but autumn senescence "[progress] from the top of the canopy downward" (Gill et al. 1998).

1.4 Consequences of Different Phenological Strategies

Studies of phenological differences among individuals, either among species or within species, have sought to explain patterns of phenological variation by considering how different phenological strategies could work to enhance an individual's fitness. This might be achieved through maximization of carbon gain, minimization of the likelihood of frost damage, exploitation of seasonally limited resources such as light, water or nutrients, or reduction of herbivory by completing foliar development (including lignification and production of secondary metabolites) before insect emergence.

As suggested above, the potential photosynthetic gains to understory plants from leafing out prior to canopy development are large. Data from the Bartlett Experimental Forest indicate that the accumulated below-canopy flux of photosynthetically active radiation (PAR) increases at a rate of ≈ 15 mol photons m^{-2} d^{-1} from snowmelt through the onset of canopy development (i.e., when fTPAR ≈ 0.50 in Fig. 1b), whereas from the completion of canopy development to the onset of senescence, accumulated transmitted PAR increases at a rate of only ≈ 1.5 mol photons m^{-2} d^{-1} (fTPAR ≈ 0.05), a tenfold difference. Between snowmelt ($\approx$day 105) and the end of the year, the accumulated transmitted PAR equals ≈ 900 mol m^{-2}; of this total, roughly half is received between snowmelt and the midpoint of canopy development ($\approx$day 130). These data help to explain how, as reported in the literature, opportunistic species such as the herbaceous perennial *Erythronium americanum*, which emerges almost immediately after snowmelt but is senescent soon after canopy closure (Mahall and Bormann 1978), can essentially complete their life cycle before the passing of the summer solstice (e.g., Muller 1978).

Similar gains might be expected of understory plants that retain their leaves in autumn longer than canopy trees. However, in terms of potential photosynthetic carbon assimilation, a high-light day at the beginning of the growing season provides greater stimulation of photosynthesis than a high-light day at the end of the growing season. There are at least three reasons for this. First, the maximum incident solar radiation flux is higher in the spring than the autumn, and the zenith angle is larger; both of these translate to considerably more light being transmitted through to the understory in the spring compared to autumn. Second, spring temperatures are generally warmer and more favorable to photosynthesis than autumn temperatures, and water tends to be less limiting in spring (especially following snowmelt) than in autumn (Chen et al. 1999). Third, seasonal declines in both leaf area and photosynthetic capacity mean that as senescence approaches, whole-plant rates of uptake tend to be lower than in spring (Gill et al. 1998). However, this is not universal: in some species the photosynthetic capacity of elongating leaves is quite low, which may prevent efficient exploitation of high irradiances in late spring. For example, Morecroft et al. (2003) reported that after budburst, *Quercus robur* took 2 months to reach peak photosynthetic capacity.

The potential increases in plant carbon gain resulting from phenological escape have been evaluated both through observational as well as modeling studies. Harrington et al. (1989) compared the phenology and photosynthesis of two native and two exotic shrubs growing in a Wisconsin forest and found that the exotic species had leaf life spans that were almost 2 months longer than those of the native species. Further analysis showed that the exotic shrubs accumulated roughly 30% of their annual carbon gain in the spring, and 10% of their annual carbon gain in the fall, during the period when the competing native shrub *Cornus racemosa* was leafless. Similarly, Jolly et al. (2004) used an ecosystem model to investigate how changes in the length of leaf display (extending both the start and the end of the growing season by 1 or 2 weeks) might affect the net productivity of canopy and understory species. Whereas productivity of canopy trees was hardly increased under either experimental scenario (+2% and +4% for scenarios 1 and 2, respectively), very large productivity increases were predicted for the understory (+32% and +53%, respectively). One reason for the modest increase in overstory productivity was that increased respiration during mid-summer largely offset any additional photosynthesis in spring and autumn. On the other hand, the understory productivity was enhanced directly by the increased light interception in spring and autumn, and indirectly because earlier leaf-out resulted in the production of more foliage relative to the base scenario, which further increased light interception throughout the entire growing season.

With the above literature review providing a background, we now use long-term phenological data from the Harvard Forest to investigate phenological differences between understory and overstory species in a temperate deciduous broadleaf forest.

2 Harvard Forest Case Study

2.1 Site Description

The Harvard Forest (42.54°N, 72.18°W, el. 220–410 m ASL) is located in central Massachusetts, about 100 km west of Boston. The climate is cool and moist temperate, with a mean July temperature of 20°C and mean January temperature of −7°C. Mean annual precipitation is 1,100 mm, and is distributed evenly across the seasons. The soils are predominantly sandy loams derived from glacial till, and are generally moderately to well drained, and acidic. Forests are dominated by transition hardwoods: red oak (*Quercus rubra*, 36% of basal area) and red maple (*Acer rubrum*, 22% of basal area) with other hardwoods (including black oak, *Quercus velutina*, white oak, *Quercus alba*, and yellow birch, *Betula alleghaniensis*) together accounting for 14% of the total basal area. Conifers include eastern hemlock (*Tsuga canadensis*, 13% of basal area), red pine (*Pinus resinosa*, 8% of basal area) and white pine (*Pinus strobus*, 6% of basal area). Canopy leaf area index (LAI) is ≈5 m^2 m^{-2}.

2.2 Phenology Observations

Since 1990, springtime phenology observations have been made (by J.O'K.) at 3–7 day intervals from April through June. Bud break, leaf development, flowering, and fruit development have been monitored on three or more individuals (a total of 115 permanently marked trees or shrubs) of 33 woody species. Budburst is defined as when 50% of the buds on an individual have recognizable leaves emerging from them. Near-complete leaf development is defined as the date when at least 50% of the leaves on an individual have reached 75% of their final (mature) size. Autumn phenology observations have been made since 1991 (excepting 1992). Weekly observations of percent leaf coloration and percent leaf fall begin in September and continue through complete abscission. Here we focus on two key dates: leaf coloration (date when 50% of leaves on an individual are colored) and leaf fall (date when 50% of leaves on an individual have fallen). All dates were determined by linearly interpolating between adjacent observation periods.

The species selection (Table 1) includes both overstory trees and understory trees and shrubs. All individuals are located within 1.5 km of the Harvard Forest headquarters at elevations between 335 and 365 m, in habitats ranging from closed forest, through forest-swamp margins, to dry, open fields. Beginning in 2002, the number of species observed in spring was reduced. For nine species (red maple, sugar maple, striped maple, yellow birch, American beech, white ash, witch hazel, red oak, and white oak), complete spring observations are still conducted, and the same observation schedule maintained. For an additional eight species, only budburst is monitored. At the same time, the number of species observed in autumn was reduced to 14 (red maple, sugar maple, striped maple, shadbush, yellow birch,

Table 1 Woody species monitored in spring and autumn phenology surveys at Harvard Forest since 1990. "Canopy potential" column indicates whether species is potentially an overstory canopy species, or whether it is restricted to the understory. "Canopy Position" columns (Dominant, Codominant, Intermediate, and Suppressed) indicate the number of individuals monitored, for each species, within each canopy stratum (as determined in 1991). The "Ongoing" column indicate the species which continue to be monitored (as of 2007; data for other species available only through 2001): *S* complete spring observations, *s* spring observations, budburst only, *F* complete fall observations, *S/F* complete spring and fall observations

	Code	Latin binomial	Common name	Canopy potential	D	C	I	S	Ongoing
					Canopy Pos. (# obs.)				
1	ACPE	*Acer pensylvanicum*	striped maple	Under				4	*S/F*
2	ACRU	*Acer rubrum*	red maple	Over		3	2		*S/F*
3	ACSA	*Acer saccharum*	sugar maple	Over	1			2	*S/F*
4	AMSP	*Amelanchier canadensis*	shadbush	Under			1	3	*s/F*
5	ARSP	*Aronia sp.*	chokeberry	Under				3	
6	BEAL	*Betula alleghaniensis*	yellow birch	Over			3		*S/F*
7	BELE	*Betula lenta*	black birch	Over		2		1	*s/F*
8	BEPA	*Betula papyrifera*	paper birch	Over	1	2		1	*s/F*
9	BEPO	*Betula populifolia*	grey birch	Over		1	1	2	
10	CADE	*Castanea dentata*	chestnut	Under				3	
11	COAL	*Cornus alternifolia*	alt-leaf dogwood	Under				3	*s*
12	CRSP	*Crataegus sp.*	hawthorn	Under				3	*s*
13	FAGR	*Fagus grandifolia*	american bech	Over		1	2	1	*S/F*
14	FRAM	*Fraxinus americana*	white ash	Over	1		3	1	*S/F*
15	HAVI	*Hamamelis virginiana*	witch hazel	Under				3	*S*
16	ILVE	*Ilex verticillata*	winterberry	Under				4	
17	KAAN	*Kalmia angustifolia*	sheep laurel	Under				3	
18	KALA	*Kalmia latifolia*	mountain laurel	Under				3	
19	LYLI	*Lyonia ligustrina*	maleberry	Under				3	
20	NEMU	*Nemopanthus mucronata*	mountain holly	Under				3	
21	NYSY	*Nyssa sylvatica*	black gum	Over		1	1	1	*F*
22	PIST	*Pinus strobus*	white pine	Over	1	1	1	1	
23	POTR	*Populus tremuloides*	trembling aspen	Over		1	2		*s*
24	PRSE	*Prunus serotina*	black cherry	Over		2		2	*s/F*
25	QUAL	*Quercus alba*	white oak	Over		2		1	*S/F*
26	QURU	*Quercus rubra*	red oak	Over		3	1		*S/F*
27	QUVE	*Quercus velutina*	black oak	Over		2		2	*s/F*
28	RHSP	*Rhododendron sp.*	azalea	Under				3	
29	SAPU	*Sambucus pubens*	red elderberry	Under				4	
30	TSCA	*Tsuga canadensis*	hemlock	Over			1	2	
31	VACO	*Vaccinium corymbosum*	highbush blueberry	Under				4	
32	VIAL	*Viburnum alnifolium*	hobblebush	Under				3	
33	VICA	*Viburnum cassinoides*	witherod	Under				3	

black birch, paper birch, American beech, white ash, black gum, black cherry, white oak, red oak, and black oak). Our analysis here focuses on the years up to 2001, since the widest range of species are available for these years. However, when evaluating the complexity of budburst model structure that can be supported by the data, we conduct a separate analysis for the 17 species for which data since 2001 are also available.

Phenology observations are ongoing and the complete dataset is available online (http://harvardforest.fas.harvard.edu/data/p00/hf003/hf003.html).

Complete on-site climatological data are available for the period of study from the Shaler (1964–2002) and Fisher (2001–) meteorological stations, located near the forest's administrative buildings. For this analysis, we use daily mean air temperature (°C), calculated from recorded daily maxima and minima.

2.3 Phenology Models

A variety of different models have been used to model spring phenology of temperate species (e.g., Hänninen 1995; Schwartz 1997; Chuine et al. 1998; Chuine 2000; Schaber and Badeck 2003; Richardson et al. 2006; note that models to predict autumn phenology are comparatively less-well developed, see Schaber and Badeck 2003). To date, there is no consensus on which modeling approach is best. There are a number of reasons for this. First, models that give the best fit for one data set may perform the worst when validated against an external data set (Chuine et al. 1998, 1999). Second, the model that works best for one species may perform poorly for other species (Hunter and Lechowicz 1992; Chuine et al. 1998, 1999). Third, models that are considered physiologically realistic (see discussion by Hänninen 1995) have sometimes been found to perform no better than simple, empirical models (Hunter and Lechowicz 1992). Fourth, a range of different model structures can often provide equally good fits to the available data (Hänninen 1995; Schaber and Badeck 2003), and studies with synthetic data show that biologically "incorrect" models can be parameterized so as to provide good fits, and even to make satisfactory predictions (Hunter and Lechowicz 1992).

Here we use the long-term Harvard Forest budburst and daily weather data to evaluate a range of models that have been previously presented in the literature. The models provide a context for interpreting observed differences in phenology among species and from year-to-year. A key question of interest is whether there are differences among species (particularly with regard to overstory vs. understory species) in terms of which model is best supported by the data at hand; from this it may be possible to learn about how species vary in relation to temperature sensitivities and thresholds.

The models we use are largely based on (but not necessarily identical to) those presented by Chuine et al. (1999), and our description and nomenclature follows this earlier work (see Table 2 for a list of symbols). Model parameters are fit separately for each species, allowing for species-specific biological responses to environmental cues; the parameters to be fit depend on the model structure (see Tables 2 and 3).

Table 2 List of symbols used in budburst models, after Chuine et al. (1999)

Symbol	Definition
Fit parameters	
t_1	Time step at which accumulation of chilling units begins (not used in Spring warming models; fit parameter in all other models)
C^*	Chilling state at which transition from rest to quiescence occurs (fit parameter in Sequential and Parallel1 models; not used in other models)
t_2	Time step at which accumulation of forcing units begins (fit parameter in Spring warming models; equal to t_1 in Alternating and Parallel2 models; date when $S_c = C^*$ in Sequential and Parallel1 models)
F^*	Forcing state at which transition from quiescence to budburst occurs (fit parameter in Spring warming and Sequential models; function of S_c in Alternating, Parallel1 and Parallel2 models)
T_{chill}	Critical temperature for chilling function $R_c(t)$ (not used in Spring warming models; fit parameter in all other models)
T_{force}	Critical temperature for forcing function $R_f(t)$ (fit parameter in all models)
a, b	Model constants ($a > 0$, $b < 0$) relating F^* to S_c, i.e. $F^* = a \exp(b\, S_c(t))$ at $t = y$ (not used in Spring warming, Alternating or Sequential models; fit parameter in all other models)
Model states and drivers	
t	Current time step
$x(t)$	Daily mean temperature (°C) at time t
$S_c(t)$	Cumulative chilling achieved at time t
$S_f(t)$	Cumulative forcing achieved at time t
$R_c(t)$	Increment in state of chilling at time t
$R_f(t)$	Increment in state of forcing at time t
y	Predicted budburst date (where $S_f = F^*$)

Table 3 Phenology models fit to Harvard Forest phenology data. Parameters are defined in Table 2. Spring warming, Sequential, Alternating and Parallel model structures are described in text. CF1 and CF2 refer to different functional forms for forcing and chilling rates, as described in text

Model name	Fit parameters
Spring warming CF1	3 (t_2, T_{force}, F^*)
Spring warming CF2	2 (t_2, F^*)
Alternating CF1	4 (t_1, T_{force}, a, b); $t_2 = t_1$; $T_{chill} = T_{force}$
Sequential CF1	5 (t_1, T_{chill}, T_{force}, C^*, F^*)
Sequential CF2	4 (t_1, T_{chill}, C^*, F^*)
Parallel1 CF1	6 (t_1, T_{chill}, T_{force}, C^*, a, b)
Parallel1 CF2	5 (t_1, T_{chill}, C^*, a, b)
Parallel2 CF1	5 (t_1, T_{chill}, T_{force}, a, b); $t_2 = t_1$
Parallel2 CF2	4 (t_1, T_{chill}, a, b); $t_2 = t_1$

In general, budburst is predicted to occur only when some combination of chilling and accumulated warming ("forcing") have been achieved (i.e., $S_c(t) \geq C^*$ and $S_f(t) \geq F^*$). The state of chilling, S_c, is the time integral (from t_1) of the rate of chilling, R_c, which is specified as a function of daily mean air temperature, $x(t)$, i.e. Eqn. (1):

$$S_c(t) = \sum_{t_1} R_c(x(t)) \tag{1}$$

The cumulative state of forcing, S_f, similarly represents the time integral (from t_2, which, depending on model structure, either equals t_1 or requires that a chilling threshold be met, i.e. where $S_c \geq C^*$) of the rate of forcing, R_f, which is also a function of air temperature, i.e. Eqn. (2):

$$S_f(t) = \sum_{t_2} R_f(x(t)) \tag{2}$$

Models differ depending on whether or not a period of chilling is strictly required, either prior to or concurrently with forcing. In "spring warming" models (Cannell and Smith 1983; Hunter and Lechowicz 1992), there are no chilling requirements and only forcing temperatures affect the timing of budburst. On the other hand, three types of chilling can be specified:

Sequential: In the sequential model, forcing has no effect until all chilling requirements have been met. During a "period of rest", the state of chilling accumulates from day t_1 until the state of chilling reaches a threshold value, C^*. At this point (day t_2) there is a transition to a "period of quiescence" and accumulation of forcing units begins and continues until a threshold value, F^*, is reached, triggering budburst.

Alternating: The state of chilling and the state of forcing both advance together over time from day t_1 (t_2 is set to t_1): above a threshold temperature, forcing degree-days are accumulated; below the threshold temperature, chilling days are accumulated (thus forcing occurs whenever chilling is not occurring, and vice versa). Requirements for chilling and forcing thresholds C^* and F^* are not specified explicitly: rather, as more chilling is accumulated, the forcing required for budburst is reduced (when $b < 0$, here as in Eqn. (3):

$$F^* = a\exp\left(bS_c(y)\right) \tag{3}$$

Parallel: Similar to the alternating model, in the parallel model forcing and chilling advance together over time, and as more chilling is accumulated, the forcing requirement for budburst is reduced. However, unlike the alternating model, forcing need not occur whenever chilling is not occurring. We distinguish two versions of parallel chilling. In the first (Parallel1), a threshold value of chilling, C^*, must first be reached before forcing units are accumulated (i.e., the model requires a transition from rest to quiescence). In the second (Parallel2), chilling and forcing both accumulate from t_1 (i.e., t_2 is set to t_1). (Note that as described here, both sequential and alternating chilling are essentially restricted versions of the parallel model. For more a more thorough treatment of chilling, see: Cannell and Smith 1983; Hunter and Lechowicz 1992; Kramer 1994; Chuine et al. 1998, 1999).

For each of these different model structures (Spring warming, Sequential, Alternating or Parallel), various functional forms of the equations for R_c and R_f are possible. Here we consider two variants. In one approach (here denoted CF1), the state of chilling is specified in terms of "chilling days," which are accumulated as $R_c = 1$ where $x(t) < T_{chill}$ and $R_c = 0$ otherwise, and the state of forcing is specified in terms of "forcing degree-days," which are accumulated as $R_f = x(t)\text{-}T_{force}$ where $x(t) > T_{force}$ and $R_f = 0$ otherwise. In another approach (here denoted CF2), rates of chilling and forcing are both specified as nonlinear functions of $x(t)$ (Sarvas 1974 in Chuine et al. 1999). More specifically, in CF2, (unitless) chilling is accumulated according to the triangular function in Eqn. (4):

$$R_c = 0 \qquad \text{where } x(t) \le -3.4 \text{ or } x(t) \ge 10.4 \qquad (4a)$$

$$R_c = \frac{x(t) + 3.4}{T_{chill} + 3.4} \qquad \text{where } -3.4 < x(t) \le T_{chill} \qquad (4b)$$

$$R_c = \frac{x(t) - 3.4}{T_{chill} - 10.4} \qquad \text{where } T_{chill} < x(t) \le 10.4 \qquad (4c)$$

In CF2, the rate of forcing is a sigmoid function of $x(t)$, and (unitless) forcing is accumulated as in Eqn. (4) for $x(t) > 0$:

$$R_f = \frac{28.4}{1 + \exp(-0.185(x(t) - 18.4))} \qquad (5)$$

The way in which these submodels are combined, and the parameters that are optimized for each model, are listed in Table 3. The resulting models vary both in complexity and in their underlying assumptions about the nature of the physiological processes involved, as manifest in terms of general model structure (e.g., the nature of chilling requirements and tradeoffs between S_c and F^*), functional form (CF1 vs. CF2), and the number of free parameters to be optimized.

2.4   Model Parameterization

Nonlinearities and discontinuities in phenology models (e.g., degree day accumulation begins on a particular day, above a particular temperature) mean there is a very real possibility of model parameter sets which yield only locally, and not globally, optimal agreement between model and data (e.g., Chuine et al. 1998). Our parameter optimization method was based on simulated annealing-type routines using Monte Carlo techniques (Metropolis et al. 1953) as described by Press et al. (1992) and used previously to estimate phenology model parameters by Chuine et al. (1998). FORTRAN code for the models and optimization algorithm is available from A.D.R.

2.5   Model Selection Criteria

An objective method is needed to select the most appropriate model from among a range of competing structures. While possible options include F-tests and within-sample or out-of-sample cross-validation methods, alternative approaches based on information theory are becoming popular in many fields, including ecology. For example, Akaike's (1973) criterion is rigorously based on the expected Kullback-Liebler information of each model (for a full overview of Akaike's method, as well as many examples from the ecological literature, see Burnham and Anderson 2002). Akaike's Information Criterion (AIC) quantifies how well the data at hand support various candidate models. Assuming Gaussian errors with constant variance, AIC is typically calculated as in Eqn. (6):

$$AIC = n\log(\sigma^2) + 2p \tag{6}$$

Here, n is the number of observations, p is the number of fit parameters plus one, and σ^2 is the residual sum of squares divided by n. The model with the lowest AIC is considered the best model, given the data at hand, but it is important to note that AIC is not in any sense a formal or statistical hypothesis test (Anderson et al. 2000). For a given data set and a set of candidate models, AIC effectively balances improving explanatory power (lower σ^2) against increasing complexity (larger p). In this way, AIC selects against models with an excessive number of parameters.

Alternatives to AIC have been developed (e.g., Schwartz 1978, Hurvich and Tsai 1990), and while we will not discuss these here, we will apply a correction factor which has been developed for cases where n is small relative to p. The small-sample corrected criterion, AIC_C, is calculated (Burnham and Anderson 2002; Motulsky and Christopoulos 2003) as in Eqn. (7):

$$AIC_C = AIC + \frac{2p(p+1)}{n-p-1} \tag{7}$$

The absolute difference in AIC_C scores between two models can be used to evaluate the weight of evidence in support of one model (the model with the lower AIC_C) over another model (Burnham and Anderson 2002): if the difference, ΔAIC_C, is small or zero, then both models are essentially equally likely to be the best model. If $\Delta AIC_C \approx 2.0$, then the model with the lower AIC_C is almost three times more likely to be best. If, however, $\Delta AIC_C \approx 6.0$, then the model with the lower AIC_C is about 20 times more likely to be best. Here we calculate ΔAIC_C relative to the "best" model (lowest AIC_C) for each species, and express all AIC-based results in terms of this metric.

3 Results

3.1 Patterns of Variation Among Species

For both canopy and understory species, budburst dates varied by roughly 6 weeks among species (Fig. 2; four-letter species codes are reported here in the text to facilitate figure interpretation; full species information is listed in Table 1). Among canopy species, black cherry (PRSE) had the earliest mean budburst date (day 110), while white pine (PIST) had the latest (day 157). Among understory species, red

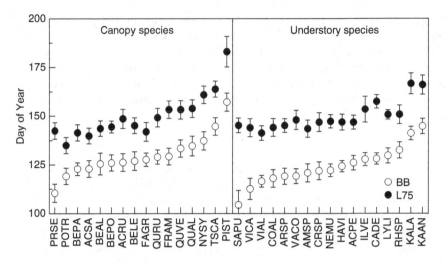

Fig. 2 Spring phenology differences across species, and in relation to potential canopy position (canopy vs. understory species). Budburst (BB) is date when 50% of leaf buds have burst (*hollow circles*); L75 is date when 50% of leaves have reached 75% of final size (*filled circles*). X-axis is sorted by date of budburst (four-letter species codes as in Table 1). Error bars indicate ±1 standard deviation of the mean-adjusted phenophase date anomalies.

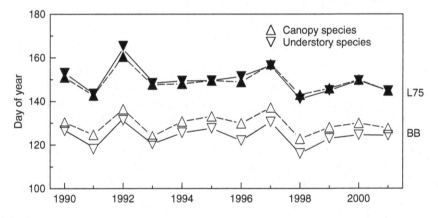

Fig. 3 Spring phenology of understory and canopy species at the Harvard Forest, 1990–2001. Budburst (BB) is date when 50% of leaf buds have burst (*hollow symbols*); L75 is date when 50% of leaves have reached 75% of final size (*filled symbols*). Each data point is the average across all species in each class (16 canopy species, 17 understory species).

elderberry (SAPU) had the earliest mean budburst date (day 105) and sheep laurel (KAAN) the latest (day 144). The date when leaves reached 75% of final size varied by a similar amount for canopy species (48 days, from day 134 for trembling aspen, POTR, to day 182 for white pine). However, the range of variation was only half as great for understory species (25 days, from day 141 for hobblebush, VIAL, to day 166 for mountain laurel, KALA).

The time required for leaf expansion (i.e., the number of days between budburst and when leaves reached 75% of final size) varied by roughly twofold. For canopy species, expansion took between 14 days (American beech, FAGR) and 32 days (black cherry), compared to between 18 days (azalea, RHSP) and 40 days (red elderberry) for understory species. For understory ($r = -0.67$, $P < 0.01$), but not canopy ($r = 0.01$, $P = 0.97$), species, the time required for expansion was negatively correlated with budburst date; in other words, leaf development progressed more slowly in early-leafing species than late-leafing species, which may be a result of degree-days accumulating more slowly in early spring compared to late spring.

There was a general tendency for budburst of understory species to be earlier than that of canopy species. In fact, budburst of 13 of 17 understory species occurred prior to the mean canopy budburst date. The only exceptions were the relatively late-leafing species maleberry, azalea, mountain laurel, and sheep laurel, two of which are ever-green species (mountain laurel and sheep laurel). The average budburst date for understory species was day 124, compared to day 130 for canopy species (day 120 and 126, respectively, when only deciduous species are considered). This pattern was consistent over time, with the difference always 3 days or greater (Fig. 3), and a paired (i.e., mean for understory species vs. mean for canopy species, paired by year) t-test indicated that this difference was statistically significant ($P < 0.001$; results were unchanged when only deciduous species considered).

By comparison, the date when leaves reached 75% of final size differed little between canopy and understory species. The average date at which leaves reached this phenophase was day 150 for understory species, which was within a day of that for the canopy species. A t-test indicated no statistically significant difference between canopy and understory species ($P = 0.30$; the same pattern was observed when analysis was limited to the deciduous species).

The date of leaf coloration varied by less than 3 weeks among canopy species, but by almost a full week more among understory species (Fig. 4). Among canopy species, red maple (ACRU) had the earliest mean leaf coloration date (day 274), while trembling aspen (POTR) the latest (day 292). Among understory species, azalea (RHSP) had the earliest mean leaf coloration date (day 272) while winter-berry (ILVE) had the latest (day 295). Thus both the earliest and latest species in terms of leaf coloration date tended to be understory species, although this was not the case in every year (e.g., POTR was in a number of years the latest species).

Whereas in the spring there was a strong tendency, particularly among understory species, to change rank order between budburst and when leaves reached 75% of final size (Fig. 2), in autumn the species order of leaf coloration and leaf fall was more consistent. For example, the earliest and latest species for leaf fall were the same as those for leaf coloration for both groups of species. Furthermore, the rank correlation between dates of leaf coloration and leaf fall was higher (Spearman rank correlation, $\rho = 0.95$) than between budburst date and the date when leaves reached 75% of final size ($\rho = 0.81$).

There was a slight tendency for dates of leaf coloration and leaf fall of understory species to be later than that of canopy species. The average leaf coloration date for canopy species was day 280, compared to day 281 for understory species; the

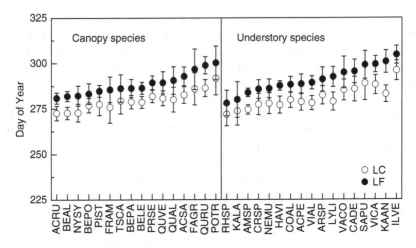

Fig. 4 Autumn phenology differences across species, and in relation to potential canopy position (canopy vs. understory species). Leaf coloration (LC) is date when 50% of leaves are colored (*hollow circles*); leaf fall (LF) is date when 50% of leaves have fallen (*filled circles*). X-axis is sorted by date of budburst (four-letter species codes as in Table 1). Error bars indicate ±1 standard deviation of the mean-adjusted phenophase date anomalies.

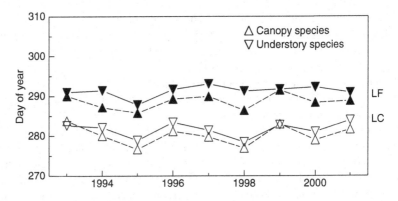

Fig. 5 Autumn phenology of understory and canopy species at the Harvard Forest, 1993–2001. Leaf coloration (LC) is date when 50% of leaves are colored (*hollow symbols*); leaf fall (LF) is date when 50% of leaves have fallen (*filled symbols*). Each data point is the average across all species in each class (16 canopy species, 17 understory species).

average leaf fall dates were days 289 and 291, respectively. Although small, these differences were consistent from year to year (Fig. 5; date of leaf coloration difference statistically significant at P = 0.01 by paired t-test, date of leaf fall difference statistically significant at P < 0.001; comparable results, except both differences significant at P < 0.001, were observed when analysis was limited to the deciduous species). However, for only 7 of 17 understory was leaf coloration date after the mean date for canopy species, and for only nine understory species was leaf fall date after the mean date for canopy species. Thus, while these patterns are statistically significant, they may not have great biological or ecological significance, and certainly do not indicate large differences in autumn phenology between broad groups of canopy and understory species. Rather, the differences are more pronounced among individual species within each group.

3.2 Differences Between Canopy and Conspecific Understory Trees

For six species (sugar maple, ACSA; black birch, BELE; paper birch, BEPA; gray birch, BEPO; American beech, FAGR; and black cherry, PRSE; see Table 1), one or more individuals of each was classified as either dominant or codominant (here, lumped together as "canopy trees"), and one or more individuals was classified as suppressed (here, "understory trees"). For each of these species we calculated, by year, the difference in date at which the canopy and understory trees reached each phenological stage, and then conducted a simple two-tailed *t*-test on these differences.

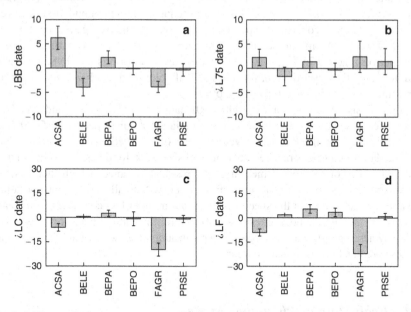

Fig. 6 Difference in dates at which canopy and understory individuals of six species reach different phenophases at the Harvard Forest. Species are sugar maple (ACSA), black birch (BELE), paper birch (BEPA), gray birch (BEPO), American beech (FAGR) and black cherry (PRSE). A positive value indicates that canopy individuals reach that phenophase at a later date than understory individuals. (**a**) Budburst (BB); (**b**) leaves reached 75% of final size (L75); (**c**) leaf coloration (LC); (**d**) leaf fall (LF). Error bars indicate 95% confidence intervals on the mean difference, with each year treated as a replicate.

These results indicate that whereas understory sugar maple and paper birch reached budburst significantly (P < 0.05) earlier than canopy trees of the same species, the reverse was true for black birch and American beech (Fig. 6a). These differences largely disappeared as foliar development proceeded: only for sugar maple was there still a significant (P < 0.05) difference between canopy and understory conspecifics in the date when leaves reached 75% of final size (Fig. 6b).

In autumn, dates of leaf coloration and leaf fall for below-canopy sugar maple and American beech were both significantly (P < 0.05) later than canopy trees of the same species, whereas understory black birch, paper birch, and gray birch all dropped their leaves significantly (P < 0.05) earlier than canopy trees of the same species (Fig. 6c, d). For black cherry, there was no consistent difference in spring or autumn phenology between canopy and understory individuals.

3.3 Interannual Variation and Consistency of Species Ordering

The above patterns of phenological variation among species were largely consistent from year-to-year, particularly for budburst. For example, there was a very strong correlation in budburst date between all pairs of years (Pearson correlation, average r = 0.94; range r = 0.87 to r = 0.98), and the sequence in which budburst occurred out was also highly consistent (Spearman rank correlation, average ρ = 0.92, range ρ = 0.77–0.98). A similar degree of consistency showed up comparing the date when leaves reached 75% of final size for early and late springs (r = 0.98, ρ = 0.95), and, to a slightly lesser degree, for dates of leaf coloration and leaf fall comparing early and late autumns (both r = 0.93, ρ = 0.92).

An analysis of variance on the 1990–2001 data set, with "species" and "year" as main effects, suggested that there is more variation among species (~65–80% of total variance) than there is across years (~20% of total variance for budburst date and the date when leaves reached 75% of final size, ~5% for dates of leaf coloration and leaf fall). For the spring measures, the residual variance (which includes the unaccounted for species × year interaction effect) was small (<10%) corroborating the above conclusion that the species patterns are more or less consistent from year-to-year. On the other hand, the residual variance for the autumn measures was large, ~25%, which suggests that the species patterns in autumn are somewhat less consistent (or less predictable) from year-to-year.

3.4 Evaluation of Phenology Models

AIC_C indicated the greatest support for the two-parameter Spring warming CF2 model, which was identified as the best model for 13 out of 33 species (ΔAIC_C = 0 in Table 4), despite having the highest root mean square error of all candidate models. A range of other models were identified as the best model for a more limited

Table 4 ΔAIC_C (difference between a particular model's AIC_C and the lowest AIC_C across all candidate models for a species) values for a range of different models (see text and Tables 2 and 3 for additional information) fit to Harvard Forest budburst data (1990–2001). Species codes are as given in Table 1. The best model, based on Akaike's Information Criterion corrected for small samples (AIC_C) has $\Delta AIC_C = 0$ and is indicated by *dark shading* and *bold type*; models with $\Delta AIC_C \leq 2$ indicated by *dark shading*, models with $\Delta AIC_C \leq 6$ indicated by *light shading*

	Spring warning CF1	Spring warning CF2	Alternating CF1	Sequential CF1	Sequential CF2	Parallel1 CF1	Parallel1 CF2	Parallel2 CF1	Parallel2 CF28
ACPE	0.4	1.9	3.5	**0.0**	7.8	13.5	11.0	8.9	8.5
ACRU	5.6	2.1	**0.0**	10.7	8.5	7.7	14.7	3.1	5.2
ACSA	2.7	4.9	4.5	**0.0**	14.7	4.6	10.6	11.0	7.8
AMSP	12.9	14.4	13.9	**0.0**	22.9	9.6	19.9	19.5	16.3
ARSP	5.3	0.2	**0.0**	10.9	8.8	13.2	10.4	6.1	4.7
BEAL	2.5	1.0	**0.0**	12.1	9.8	16.4	5.9	6.5	5.4
BELE	1.9	**0.0**	0.2	13.5	6.7	13.6	10.3	7.7	7.0
BEPA	1.3	1.7	**0.0**	12.0	8.6	17.5	3.4	6.9	2.8
BEPO	4.5	2.0	**0.0**	13.1	9.1	20.0	15.5	7.1	6.9
CADE	3.3	3.7	**0.0**	10.5	10.0	7.4	3.6	5.4	9.6
COAL	3.5	**0.0**	8.0	17.7	9.9	21.9	13.9	16.9	8.9
CRSP	1.2	0.5	**0.0**	0.5	9.3	10.0	4.4	5.4	2.4
FAGR	5.5	11.3	6.7	**0.0**	14.7	8.7	8.3	16.2	11.7
FRAM	2.0	**0.0**	3.0	12.3	10.0	12.0	8.6	11.0	6.4
HAVI	2.1	**0.0**	4.5	6.7	7.0	13.9	10.6	10.6	7.7
ILVE	1.6	**0.0**	3.1	10.9	8.4	14.9	10.9	7.3	7.3
KAAN	1.4	0.7	1.0	11.9	5.6	14.5	**0.0**	7.1	22.0
KALA	1.9	**0.0**	5.7	0.4	9.1	9.1	7.9	12.2	8.0
LYLI	3.9	1.1	**0.0**	15.2	7.9	16.6	10.2	11.9	7.9
NEMU	9.2	3.6	3.5	11.8	11.7	15.4	8.3	**0.0**	7.3
NYSY	1.3	1.5	5.5	7.4	6.4	15.1	**0.0**	11.7	7.0
PIST	1.1	**0.0**	0.7	6.9	0.3	7.2	1.0	12.9	3.1
POTR	2.7	**0.0**	6.8	10.5	2.7	14.9	9.2	12.0	8.0
PRSE	4.6	**0.0**	7.2	15.8	9.1	15.1	11.2	13.0	5.8
QUAL	1.9	4.6	2.6	**0.0**	8.9	9.4	9.8	8.3	9.1
QURU	1.6	1.0	0.3	**0.0**	9.2	7.7	1.7	7.4	3.5
QUVE	5.6	8.5	8.5	**0.0**	13.1	14.5	12.7	14.6	11.9
RHSP	5.5	4.4	**0.0**	13.2	10.5	20.9	17.9	10.8	13.0
SAPU	3.6	**0.0**	3.4	5.5.	2.5	9.3	9.6	14.6	8.1
TSCA	3.9	**0.0**	3.5	18.9	2.5	26.6	1.0	19.3	26.6
VACO	7.8	**0.0**	8.9	17.4	8.3	22.3	9.1	15.4	7.4
VIAL	0.7	6.5	0.6	**0.0**	14.8	7.2	9.4	5.5	3.7
VICA	6.1	**0.0**	10.8	16.0	7.4	18.5	13.9	20.5	7.6
All Species									
Best model	0	13	9	8	0	0	2	1	0
$\Delta AIC_C \leq 2$	13	22	14	10	1	0	5	1	0
$\Delta AIC_C \leq 6$	29	29	25	11	5	1	9	5	9

number of species: for example, the Alternating CF1 and Sequential CF1 models were indicated to be the best model for eight and nine species, respectively. However, the most highly parameterized model, Parallel1 CF1, with six parameters and the lowest RMSE, was not identified by AIC_c to be the best model for any of the 33 species. Simply put, with time series of only 12 years in length, a model with six parameters cannot be justified, because the penalty associated with additional parameters is larger than the associated improvement in model fit (even with 17 years of data, a model with five parameters would have to fit the data twice as well as a model with two parameters for the models to have the same AIC_c value).

Models with ΔAIC_c less than 2 are still considered to be reasonably well-supported by the data (Burnham and Anderson 2002). For the Spring warming CF2 model, ΔAIC_c was ≤2 for 22 species (and ≤6, which might be considered the limit at which a claim could be made that the data give any support for the model, for 29 species) (Table 4). No other model had ΔAIC_c ≤ 2 for more than 14 species; but the Spring warming CF1 and Alternating CF1 models both had ΔAIC_c ≤ 6 for more than three quarters of all species.

These patterns did not differ markedly between canopy and understory species; in both instances, the Spring warming CF2 model was identified as the best model for at least six species, and the patterns of variation in support for other models was comparable for both groups of species (summary results not shown). There was no obvious relationship between the timing of budburst (i.e., early vs. late leafing species) and the way in which models were ranked by AIC_c.

For the subset of species for which budburst observations have continued since 2001, we conducted a similar analysis on the 17-year data set (1990–2006) to investigate whether additional model complexity could be justified if longer time series were used for model fitting. Again, the Spring warming CF2 model had the lowest AIC_c for the largest number of species (7 of 17), and the Spring warming CF1 and Alternating CF1 models were at least marginally supported by the data for most species (Table 5). However, the five-parameter Parallel1 CF2 model and, to an even greater degree, the four-parameter Parallel2 CF2 model (ΔAIC_c ≤ 2 for 10 of 17 species), both emerged as viable candidate models in this analysis. This anaysis indicated greater support for the CF2, compared to CF1, formulations for rates of chilling and forcing: for the Spring warming, Sequential, Parallel1 and Parallel2 models, AIC_c values were consistently (75% of all species, on average) lower for CF2 than CF1 versions.

4 Discussion

4.1 Do Canopy and Understory Species Differ Phenologically?

The Harvard Forest phenology data show (Figs. 2 and 4) that in both spring and fall, there is a continuum of phenological activity, from "early" to "late" species. Interestingly, the range of variation is not markedly different for understory and canopy species. Moreover, on the surface, we do not see strong evidence of any

Table 5 AIC_c values for a range of different models (see text and Tables 2 and 3 for additional information) fit to Harvard Forest budburst data (1990–2006). Species codes are as given in Table 1. The best model, based on Akaike's Information Criterion corrected for small samples (AIC_c) has $\Delta AIC_c = 0$ and is indicated by *dark shading* and *bold type*; models with $\Delta AIC_c \leq 2$ indicated by *dark shading*, models with $\Delta AIC_c \leq 6$ indicated by *light shading*

	Spring warming CF1	Spring warming CF2	Alternating CF1	Sequential CF1	Sequential CF2	Parallel1 CF1	Parallel1 CF2	Parallel2 CF1	Parallel2 CF2
ACPE	**0.0**	0.6	0.5	6.8	4.4	1.8	6.1	4.6	0.8
ACRU	6.4	4.7	**0.0**	12.5	11.2	12.1	8.1	11.1	5.7
ACSA	1.6	5.0	7.6	6.5	11.9	3.2	4.4	5.4	**0.0**
AMSP	2.7	4.5	1.5	3.7	12.0	2.7	**0.0**	5.2	0.5
BEAL	4.5	1.9	**0.0**	10.6	5.1	12.4	7.7	10.5	4.4
BELE	2.6	**0.0**	3.4	11.7	0.6	11.9	8.0	4.8	5.1
BEPA	6.5	4.1	0.7	13.6	10.5	14.9	18.2	3.4	**0.0**
COAL	4.1	**0.0**	4.7	12.2	7.0	13.7	3.3	7.3	1.6
CRSP	0.4	4.2	4.0	6.0	11.3	7.6	0.4	6.7	**0.0**
FAGR	3.0	0.4	2.5	8.0	4.2	3.3	3.1	6.1	**0.0**
FRAM	4.5	**0.0**	5.4	12.6	5.2	14.9	19.1	13.5	7.0
HAVI	3.0	**0.0**	5.9	9.0	1.4	8.7	2.8	8.6	4.1
POTR	3.6	**0.0**	5.0	7.0	1.8	10.9	7.4	9.4	3.9
PRSE	4.0	**0.0**	4.0	9.5	7.6	4.0	1.8	12.7	1.2
QUAL	2.6	**0.0**	2.8	6.1	2.3	0.2	3.7	5.9	2.6
QURU	7.5	4.7	12.6	12.3	5.2	16.6	**0.0**	11.8	1.1
QUVE	**0.0**	4.8	2.4	2.7	9.6	8.9	1.3	10.9	1.8
All species									
Best model	2	7	2	0	0	0	2	0	4
$\Delta AIC_c \leq 2$	4	10	5	0	3	2	5	0	10
$\Delta AIC_c \leq 6$	14	17	15	3	9	6	10	6	16

clearly-defined groups of species (either understory or canopy) that exhibit dramatically different phenological patterns in either spring or fall.

However, budburst of 13 of 17 understory species occurred prior to the mean budburst date of the 16 canopy species (day 130). These differences between understory and canopy species had largely disappeared by the date when leaves reached 75% of their final size. In the autumn, both leaf coloration and leaf fall tended to consistently occur later in understory species than canopy species, but the difference in mean dates between groups of species was sufficiently small (1–2 days) as to be of questionable ecological significance. These patterns are broadly consistent with previously published studies (e.g., Gill et al. 1998; Augspurger et al. 2005; but see Barr et al. 2004).

In 10 of 17 understory species, budburst occurred prior to day 125 (species SAPU, red elderberry, through HAVI, witch hazel, in Fig. 2), whereas none of the five most dominant canopy species reach budburst until this date (yellow birch is the first, on day 125; but the most abundant species, red oak, does not reach budburst until day 129). Thus, these ten species are candidates for being classified as having adopted strategies of springtime phenological escape – but clearly the success with which this strategy is adopted (in terms of the potential increases in light interception and photosynthetic assimilation) varies among species, with the benefit potentially being considerably larger to the earliest species, such as red elderberry, compared to late budburst species, such as witch hazel. As noted above, however, the realized gains depend on the rate at which photosynthetic capacity is developed following leaf-out, and note that understory species tended to progress more slowly from budburst to 75% of final size, which could imply slow development of photosynthetic capacity as well.

Three understory species (striped maple, winterberry, and chestnut) leaf out between day 125 and day 130, i.e. concurrently with the majority of canopy species, and appear not to rely on the escape strategy. Special mention should be made of striped maple, which is known to be extremely shade tolerant, and chestnut, which was a dominant overstory tree across much of the eastern United States until the arrival of the chestnut blight in the early 1900s. Chestnut is now essentially restricted to root-sprouted saplings and small trees in the forest understory. It could be argued that chestnut would more appropriately be classified as a canopy species, but this would not change our overall interpretation of the patterns reported here.

Two of the four understory species with especially late budburst, mountain laurel (KALA) and sheep laurel (KAAN), are evergreen species that would presumably benefit little from early budburst or delayed senescence, since they retain most of their foliage throughout the entire year. In this manner, evergreen understory species are still able to exploit the spring and autumn high-light periods, and photosynthesis at these times of the year is considered important for their survival (e.g., Lassoie et al. 1983). Interestingly, while these species are both characterized by late budburst in spring, they are among the earliest (e.g., mountain laurel) and latest (sheep laurel) species to drop their leaves in autumn (Fig. 4).

In the autumn, the single most abundant species, red oak (QURU, accounting for 36% of basal area), is the third-last species to drop its leaves in this community (Fig. 4), which would make an autumn strategy of phenological escape relatively difficult for most understory species. On the other hand, red maple (ACRU) and yellow birch (BEAL) are both among the five most dominant canopy species but drop their leaves very early (Fig. 4). These examples highlight the fact that during the spring and autumn transition periods, the developmental state of the canopy is highly variable across space. Thus the amount of light reaching any point in the understory depends on the phenology of neighboring species (Kato and Komiyama 2002). In this regard, the degree to which an individual plant in the understory is able to opportunistically take advantage of spring and autumn high light periods largely depends on the species under which that plant grows. Detailed measurements to quantify the light environment experienced by individual understory plants, e.g., vertical profiles of leaf area index and clumping indices (and how these, and thus f_{TPAR}, vary seasonally), would offer insights into the degree to which a strategy of phenological escape is being successfully adopted.

Previous studies comparing phenology of understory and canopy species (e.g. Gill et al. 1998, Augspurger et al. 2005) have limited their analysis to a more restricted set of species than was considered here; our survey gives a broad overall view of phenological patterns within a forest community. We see that in springtime, there is a loosely-defined subset of understory species that appear to adopt a strategy of phenological escape; most of the non-evergreen species followed this strategy to some degree. However, the difference between understory and canopy species at Harvard Forest appears to be smaller than has been reported in previous studies. For example, only two understory species, red elderberry and witherod (SAPU and VICA in Fig. 2), had budburst that was more than 10 days in advance of budburst by any of the dominant canopy species. For the remaining 11 species that tended to leaf out in advance of the mean canopy budburst date, the period of phenological escape was generally very short, only on the order of several days, and thus the functional significance of the strategy (and the role of phenological escape in structuring the community) is not clear. This is a surprising result because it suggests that the costs (or associated risks) of earlier emergence at Harvard Forest may be larger than in other ecosystems where similar research has been previously conducted.

There are even less pronounced differences between canopy and understory species in the autumn. Both competitive and abiotic factors could explain this. For example, because the most dominant canopy species, red oak, is among the latest to drop its leaves, much of the understory remains shaded until very late autumn, by which time cooler temperatures, shorter days, and lower peak irradiances substantially reduce the potential photosynthetic gains of autumn phenological escape.

4.2   Do Canopy and Understory Conspecifics Differ Phenologically?

Previous studies have provided relatively consistent evidence of earlier budburst, but not necessarily later abscission, by understory individuals of canopy species (e.g., Augspurger and Bartlett 2003). Here, our evaluation of phenological differences between canopy trees and understory individuals of the same species indicates surprisingly mixed results in both spring and fall. For example, in spring, understory individuals of sugar maple and paper birch tended to reach budburst significantly earlier (by more than 5 days in the case of sugar maple) than canopy individuals, whereas the reverse was true for yellow birch and American beech (Fig. 6a). The pattern for sugar maple and paper birch is consistent with what would be expected based on previous studies, which generally show that within a species, seedlings, saplings and small trees leaf out in advance of mature canopy trees (Gill et al. 1998; Seiwa 1999a, b; Augspurger and Bartlett 2003), presumably because the potential carbon gains from earlier budburst decrease with increasing height.

The pattern for yellow birch and American beech is counter-intuitive, as this did not occur in any of the 13 species studied by Augspurger and Bartlett (2003), and Gill et al. (1998) reported exactly the opposite for American beech growing in a northern hardwood forest. The fact that differences between understory and canopy individuals had largely disappeared by the date when leaves reached 75% of final size (Fig. 6b) is surprising, as Augspurger and Bartlett (2003) reported that full leaf expansion also tended to occur earlier in juveniles than adults.

In autumn, leaf coloration and leaf fall of sugar maple and American beech understory individuals was much later (20 days in the case of American beech) than canopy individuals (comparable to patterns reported for these same species by Gill et al. 1998), whereas smaller but significant differences in the opposite direction were seen for paper birch (Fig. 6c, d). Similarly mixed results have been reported previously in the literature for autumn phenology (Seiwa 1999a, b; Augspurger and Bartlett 2003).

Within species, phenological differences between understory and canopy individuals have been attributed to vertical temperature gradients, with warmer ground-level temperatures in spring promoting earlier emergence (Gill et al. 1998; Augspurger 2004). If spring phenology is indeed under environmental, rather than developmental, control (Augspurger 2004), then these apparently species-specific patterns in the Harvard Forest data may be the result of microclimatic differences among understory individuals due to variation in topography (Fisher et al. 2006), canopy openness (Augspurger and Bartlett 2003), or even the phenology of neighboring species.

4.3   Do Species Respond Differently to Climatic Variation?

Our results indicated that the ordering of species in terms of phenological sequence was relatively consistent in spring (i.e., for budburst date and the date when leaves reached 75% of final size; see also Lechowicz 1984) from year-to-year, but somewhat less so in autumn (i.e., for dates of leaf coloration and leaf fall). This may be partially attributed to inherent uncertainties in the autumn measures, which are considerably more subjective than well-defined phenophases, such as budburst. The difference in consistency between spring and autumn ordering may also be a function of differences in species-level responses to environmental signals. For example, we might hypothesize that whereas all species are responding to similar temperature cues in the springtime, in autumn some species might be responding to various temperature thresholds, whereas other species are responding to changes in day length. In addition, drought, as well as wind and rain (major autumn storms tend to bring down a large number of leaves), are factors that affect autumn senescence and abscission, but possibly result in more stochastic and less predictable patterns of autumn phenology.

Akaike's Information Criterion indicated that the support for different phenological model structures was mixed and varied somewhat among species (Table 4). Although there was no clear consensus model that consistently worked best (see also Chuine et al. 1998) for all 33 study species (1990–2001 data), the data for most species gave reasonably strong support for three models, with the Spring warming CF2 model clearly the top choice. Both Hunter and Lechowicz (1992) and Chuine et al. (1998, 1999) also reported that a spring warming model performed as well, if not better than, more complex models involving chilling requirements. However, whereas results of Hunter and Lechowicz (1992) indicated support for a model with sequential chilling but not parallel chilling, results of Chuine et al. (1998) indicate support for a model with alternating chilling but not sequential or parallel chilling. Our results were mixed, differing depending on whether the shorter 1990–2001 (33 species) or longer 1990–2006 (17 species) dataset was being analyzed: the Sequential CF1 model was well-supported by 14 of 33 species in the shorter data set, but none of the 17 species in the longer data set, whereas the Parallel2 CF2 model was well supported by none of 33 species in the shorter data set, but 10 of 17 species in the longer data set. A possible explanation for this unusual pattern is that climatic conditions (and the phenological response to those conditions) for one or more years between 2002 and 2006 may have been sufficiently different from those between 1990 and 2001 to be incompatible with the Sequential CF1 model. However, as noted by Hänninen (1995), rigorous testing of mechanistic phenology models of the type studied here really requires experimental, rather than observational, data sets.

We saw no evidence of differences between overstory and understory species, or between evergreen and deciduous species, in terms of which model structure was preferred. And, while Schaber and Badeck (2003) reported that the best-fitting models differed between species with early and late budburst dates in Germany, we did not see evidence of a similar pattern in the Harvard Forest data. A possible explanation for this difference is that none of our models explicitly accounted for photoperiod (although, in spring warming models, the date at which forcing starts to accumulate could be interpreted as a photoperiod cue), whereas Schaber and Badeck (2003) found that day length improved model performance for species with late budburst (but see Cannell and Smith 1983).

Because similar model structures were indicated for most species, it could be difficult to argue that phenological responses are driven by species-specific cues (cf. Schaber and Badeck 2003). For example, there was no evidence that one group of species required sequential chilling, while another required parallel chilling, while a third group responded just to spring warming. On the other hand, looking at the species-specific model parameterization, certain species stand out as occupying distinct regions of multivariate parameter space. For example, for the Spring warming CF2 model, the understory evergreen species sheep laurel and mountain laurel, although late leafing, did not appear to require more forcing than other species, but rather appeared insensitive to forcing until early May, whereas many other species respond to forcing beginning in late March. For the Parallel2 CF2 model, there were large and significant differences among species in the parameters a and b, which control the degree to which the accumulated chilling reduces the forcing requirement (strongly in some species, almost not at all in other species). Similarly, the optimal chilling temperature, and the date at which chilling begins to accumulate, varied substantially among species. These differences in parameterization give insights into biologically-based species-specific responses to the environment (Hunter and Lechowicz 1992), and thus why the sequence in which budburst occurs might vary slightly from year-to-year. Furthermore, in the context of climate change, these differences have implications for how species will respond to future climates (Murray et al. 1989; Kramer 1994; see also Chuine et al. 2000 for an historical analysis), and thus imply that changes in the order that species leaf out can be expected. As evidence of this, we note that the species ordering was largely consistent between "early" and "late" spring years (Spearman rank correlation, P = 0.98), but whereas budburst of gray birch was more than 9 days earlier in "early" spring years, for hemlock it was only 3 days earlier. Although Kramer (1994) reported that species with late-spring budburst tend to exhibit less interannual variation (and thus less sensitivity to climatic variation) than those that leaf out in early spring (see also Murray et al. 1989), our results give only weak support for this hypothesis in general. The correlation between mean (by species) budburst date and the standard deviation (by species, across years) of budburst dates was in the expected direction (negative) but significantly different from zero only for understory ($r = -0.54$, $P = 0.02$) species. For canopy ($r = 0.10$, $P = 0.72$) species, or all species together ($r = -0.14$, $P = 0.44$), the correlation was not significantly different from zero.

5   Summary and Conclusions

We conducted an analysis of the long term Harvard Forest phenology record to investigate differences in spring and autumn phenology both among (understory species vs. canopy species) and within species (understory vs. canopy conspecifics). Budburst of most understory species occurred prior to budburst by any of the dominant canopy species, suggesting a strategy of phenological escape, although the period of escape was limited to just a few days for most understory species. Phenological differences between understory and canopy species had largely disappeared by the time leaves approached full expansion. Furthermore, in autumn, differences between understory and canopy species were less clear-cut; late abscission by the dominant canopy species, red oak, would tend to make phenological escape in this community a less viable strategy in autumn than in spring. Comparing understory and conspecific canopy individuals, we found that the patterns depended on the species in question: both earlier and later budburst, and accelerated or delayed senescence, were observed. An evaluation of a range of different models to predict budburst indicated support for both Spring warming models, with no chilling requirement, as well as models featuring Alternating and Parallel chilling. Models with Sequential chilling were not supported by the data. Although these patterns were broadly consistent among species, analysis of model parameterization indicated some differences in how species respond to environmental forcing. These differences help to explain the small year-to-year differences observed in the sequence of budburst among species.

This study, particularly the contrasting patterns of support for different models depending on whether the shorter (12 year) 33-species dataset or the longer (17 year) 17-species dataset was used, highlights the value of long-term, multi-species phenological data sets, with which it is possible to ask both ecological (how does phenology vary among and within species?) and biological (what environmental cues drive phenological events such as budburst?) questions. With ongoing efforts to link diverse "types" of phenology – e.g., to understand connections between phenology of above- (e.g., leaves) and below-ground (e.g., roots) growth, development and senescence, and how these relate to the phenology of ecosystem processes (particularly those related to the biogeochemical cycling of carbon and mineral nutrients, as discussed in many of the papers in this volume) as well as feedbacks to the climate system (surface albedo and partitioning of available energy to latent and sensible heat; Morisette et al. 2009) – mechanistic understanding of the causes of phenological differences both within and among species is of critical importance. Evidence documenting phenological changes in response to recent warming trends (Schwartz et al. 2006) suggests that the value of these long-term monitoring data will only continue to increase over time (Lovett et al. 2007).

Acknowledgements The National Science Foundation Long-Term Ecological Research (LTER) program supported the research at Harvard Forest. A.D.R. acknowledges support from the Northeastern States Research Cooperative. We thank Jeremy Fisher, Brenden McNeil, Alan Barr, Abraham Miller-Rushing, Jake Weltzin, and Asko Noormets for helpful comments on drafts of this chapter. This is a contribution of the Northeast Regional Phenology Network (NE-RPN).

References

Akaike, H. (1973) Information theory and an extension of the maximum likelihood principle. In: B.N. Petrov and F. Csaki (Eds.), *Proceedings of the 2nd International Symposium on Information Theory*, Akademiai Kiado, Budapest, pp. 267–281. (Reproduced in: S. Kotz and N.L. Johnson (2003), *Breakthroughs in Statistics, Vol. I, Foundations and Basic Theory*. Springer, New York, pp. 610–624.)

Anderson, D.R., Burnham, K.P. and Thompson, W.L. (2000) Null hypothesis testing: problems, prevalance and an alternative. J. Wildlife Manage. 64, 912–923.

Augspurger, C.K. (2004) Developmental versus environmental control of early leaf phenology in juvenile Ohio buckeye (*Aesculus glabra*). Can. J. Bot. 82, 31–36.

Augspurger, C.K. and Bartlett, E.A. (2003) Differnces in leaf phenology between juvenile and adult trees in a temperate deciduous forest. Tree Physiol. 23, 517–525.

Augspurger, C.K., Cheeseman, J.M. and Salk, C.F. (2005) Light gains and physiological capacity of understorey woody plants during phenological avoidance of canopy shade. Funct. Ecol. 19, 537–547.

Baldocchi, D., Hutchison, B., Matt, D. and McMillen, R. (1984) Seasonal variations in the radiation regime within an oak-hickory forest. Agric. For. Meteorol. 33, 177–191.

Baldochi, D., Hutchison, B., Matt, D. and McMillen, R. (1986) Seasonal variation in the statistics of photosynthetically active radiation penetration in an oak-hickory forest. Agric. For. Meteorol. 36, 343–361.

Barr, A.G., Black, T.A., Hogg, E.H., Kljun, N., Morgenstern, K. and Nesic, Z. (2004) Inter-annual variability in the leaf area index of a boreal aspen-hazelnut forest in relation to net ecosystem production. Agric. For. Meteorol. 126, 237–255.

Boardman, N.K. (1977) Comparative photosynthesis of sun and shade plants. Ann. Rev. Plant Physiol. 28, 355–377.

Burnham, K.P. and Anderson, D.R. (2002) *Model selection and multimodel inference: a practical information-theoretic approach (2nd ed.)*. Springer, New York.

Cannell, M. and Smith, R. (1983) Thermal time, chill days and prediction of budburst in *Picea sitchensis*. J. Appl. Ecol. 20, 951–963.

Chazdon, R.L. and Pearcy, R.W. (1991) The importance of sunflecks for forest understory plants. BioScience 41, 760–766.

Chen, W.J., Black, T.A., Yang, P.C., Barr, A.G., Neumann, H.H., Nesic, Z., Blanken, P.D., Novak, M.D., Eley, J., Ketler, R.J. and Cuenca, R. (1999) Effects of climatic variability on the annual carbon sequestration by a boreal aspen forest. Global Change Biol. 5, 41–53.

Chuine, I. (2000) A unified model for budburst of trees. J. of Theor. Biol. 207, 337–347.

Chuine, I., Cambon, G. and Comtois, P. (2000) Scaling phenology from the local to the regional level: advances from species-specific phenological models. Global Change Biol. 6, 943–952.

Chuine, I., Cour, P. and Rousseau, D.D. (1998) Fitting models predicting dates of flowering of temperate-zone trees using simulated annealing. Plant Cell Environ. 21, 455–466.

Chuine, I., Cour, P. and Rousseau, D.D. (1999) Selecting models to predict the timing of flowering of temperate trees: implications for tree phenology modelling. Plant Cell Environ. 22, 1–13.

Crawley, M.J. (1997) Life history and environment. In: M.J. Crawley (Ed.) Plant Ecology. Blackwell Science, Oxford, pp. 73–131.

dePamphilis, C.W. and Neufeld, H.S. (1989) Phenology and ecophysiology of *Aesculus sylvatica*, a vernal understory tree. Can. J. Bot. 67, 2161–2167.

Fisher, J.I., Mustard, J.F. and Vadeboncoeur, M.A. (2006) Green leaf phenology at Landsat resolution: Scaling from the field to the satellite. Remote Sens. Environ. 100, 265–279.

Fisher, J.I., Richardson, A.D. and Mustard, J.F. (2007) Phenology model from surface meteorology does not capture satellite-based greenup estimations. Global Change Biol. 13, 707–721.

Gill, D.S., Amthor, J.S. and Bormann, F.H. (1998) Leaf phenology, photosynthesis, and the persistence of saplings and shrubs in a mature northern hardwood forest. Tree Physiol. 18, 281–289.

Gu, L., Hanson, P.J., Post, W.M., Kaiser, D.P., Yang, B., Nemani, R., Pallardy, S.G., Meyers, T., 2008. The 2007 eastern US spring freeze: Increased cold damage in a warming world? BioScience, 58, 253–262.

Hänninen, H. (1995) Effects of climatic change on trees from cool and temperate regions: an ecophysiological approach to modeling of bud burst phenology. Can. J. Bot. 73, 183–199.

Harrington, R.A., Brown, B.J. and Reich, P.B. (1989) Ecophysiology of exotic and native shrubs in Southern Wisconsin. Oecologia 80, 356–367.

Hull, J.C. (2002) Photosynthetic induction dynamics to sunflecks of four deciduous forest understory herbs with different phenologies. Int. J. Plant Sci. 163, 913–924.

Hunter, A.F. and Lechowicz, M.J. (1992) Predicting the timing of budburst in temperate trees. J. Appl. Ecol. 29, 597–604.

Hurvich, C.M. and Tsai, C.L. (1990) Model selection for least absolute deviations regression in small samples. Stat. Prob. Lett. 9, 259–265.

Jolly, W.M., Nemani, R. and Running, S.W. (2004) Enhancement of understory productivity by asynchroouse phenology with overstory competitiors in a temperate deciduous forest. Tree Physiol. 24, 1069–1071.

Kato, S. and Komiyama, A. (2002) Spatial and seasonal heterogeneity in understory light conditions caused by differential leaf flushing of deciduous overstory trees. Ecol. Res. 17, 687–693.

Kramer, K. (1994) A modelling analysis of the effects of climatic warming on the probability of spring frost damage to tree species in the Netherlands and Germany. Plant Cell Environ. 17, 367–377.

Lassoie, J.P., Dougherty, P.M., Reich, P.B., Hinckley, T.M., Metcalf, C.M. and Dina, S.J. (1983) Ecophysiological investigations of understory eastern red cedar in central Missouri. Ecology 63, 1355–1366.

Lechowicz, M.J. (1984) Why do temperate deciduous trees leaf out at different times? Adaptation and ecology of forest communities. Am. Nat. 124, 821–842.

Lichtenthaler, H.K., Buschmann, C., Döll, M., Fietz, H.J., Bach, T., Kozel, U., Meier, D. and Rahmsdorf, U. (1981) Photosynthetic activity, chloroplast ultrastructure and leaf characteristics of high-light and low-light plants and of sun and shade leaves. Photosynth. Res. 2, 115–141.

Lovett, G.M., Burns, D.A., Driscoll, C.T., Jenkins, J.C., Mitchell, M.J., Rustad, L., Shanley, J.B., Likens, G.E. and Haeuber, R. (2007) Who needs environmental monitoring? Front. Ecol. Environ. 5, 253–260.

Mahall, B.E. and Bormann, F.H. (1978) A quantitativ description of the vegetative phenology of herbs in a northern hardwood forest. Bot. Gaz. 139, 467–481.

Metropolis, N., Rosenbluth, A.W., Rosenbluth, M.N., Teller, A.H. and Teller, E. (1953) Equations of state calculations by fast computing machines. J. Chem. Phys. 21, 1087–1092.

Morecroft, M.D., Stokes, V.J. and Morison, J.I.L. (2003) Seasonal changes in the photosynthetic capacity of canopy oak (*Quercus robur*) leaves: the impact of slow development on annual carbon uptake. Int. J. Biometeor. 47, 221–226.

Morisette, J.T., Richardson, A.D., Knapp, A.K., Fisher, J.I., Graham, E., Abatzoglou, J., Wilson, B.E., Breshears, D.D., Henebry, G.M., Hanes, J.M. and Liang, L. (2009) Tracking the rhythm of the seasons in the face of global change: phenological research in the 21st Century. Front. Ecol. Environ., doi: 10.1890/070217.

Motulsky, H.J. and Christopoulos, A. (2003) *Fitting models to biological data using linear and nonlinear regression. A practical guide to curve fitting.* GraphPad Software, Inc., San Diego, CA.

Muller, R.N. (1978) The phenology, growth and ecosystem dynamics of *Erythronium americanum* in the northern hardwood forest. Ecol. Monogr. 48, 1–20.

Murray, M.B., Cannell, M.G.R. and Smith, R.I. (1989) Date of budburst of fifteen tree species in Britain following climatic warming. J. Appl. Ecol. 26, 693–700.

Press, W.H., Teukolsky, S.A., Vetterling, W.T. and Flannery, B.P. (1992) *Numerical recipes in Fortran 77: The art of scientific computing.* Cambridge UP, New York.

Rathcke, B. and Lacey, E.P. (1985) Phenological patterns of terrestrial plants. Ann. Rev. Ecol. Syst. 16, 179–214.

Raulier, F. and Bernier, P.Y. (2000) Predicting the date of leaf emergence for sugar maple across its native range. Can. J. For. Res. 30, 1429–1435.

Richardson, A.D., Bailey, A.S., Denny, E.G., Martin, C.W. and O'Keefe, J. (2006) Phenology of a northern hardwood forest canopy. Global Change Biol. 12, 1174–1188.

Rothstein, D.E. and Zak, D.R. (2001) Photosynthetic adaptation and acclimation to exploit seasonal periods of direct irradiance in three temperate, deciduous-forest herbs. Funct. Ecol. 15, 722–731.

Sakai, R.K., Fitzjarrald, D.R. and Moore, K.E. (1997) Detecting leaf area and surface resistance during transition seasons. Agric. For. Meteorol. 84, 273–284.

Sarvas, R. (1974) Investigations on the annual cycle of development of forest trees. II. Autumn dormancy and winter dormancy. Communicationes Instituti Forestalis Fenniae 84, 1–101.

Schaber, J. and Badeck, F.W. (2003) Physiology-based phenology models for forest tree species in Germany. Int. J. Biometeor. 47, 193–201.

Schlichting, C.D. (1986) The evolution of phenotypic plasticity in plants. Ann. Rev. Ecol. Syst. 17, 667–693.

Schwartz, G.D. (1978) Estimating the dimension of a model. Ann. Stat. 6, 461–464.

Schwartz, M.D. (1997) Spring index models: An aproach to connecting satellite and surface phenology. In: H. Lieth and M.D. Schwartz (Eds.), *Phenology in Seasonal Climates.* Backhuys Publishers, Leiden, pp. 23–38.

Schwartz, M.D., Ahas, R. and Aasa, A. (2006) Onset of spring starting earlier across the Northern Hemisphere. Global Change Biol. 12, 343–351.

Seiwa, K. (1999a) Changes in leaf phenology are dependent on tree height in Acer mono, a deciduous broad-leaved tree. Ann. Bot. 83, 355–361.

Seiwa, K. (1999b) Ontogenetic changes in leaf phenology of *Ulmus davidiana var. japonica*, a deciduous broad-leaved tree. Tree Physiol. 19, 793–797.

Sparling, J.H. (1967) Assimilation rates of some woodland herbs in Ontario. Bot. Gaz. 128, 160–168.

Sultan, S.E. (1995) Phenotypic plasticity and plant adaptation. Acta Bot. Neerlandica 44, 363–383.

Phenology of Forest-Atmosphere Carbon Exchange for Deciduous and Coniferous Forests in Southern and Northern New England: Variation with Latitude and Landscape Position

Julian L. Hadley, John O'Keefe, J. William Munger,
David Y. Hollinger, and Andrew D. Richardson

Abstract We used ecosystem carbon exchange measurements at five sites in New England to examine how interannual variation in leaf development and leaf abscission, as well as latitude and landscape position, affected the phenology of carbon exchange in recent years. We studied three deciduous forest sites, two in southern and one in northern New England, at latitudes of about 42.54 and 44.28°N with carbon exchange records of 3–15 years, and also two coniferous forests, one also at about 42.54°N and the other at 45.25°N, with records of 4 and 11 years, including 3 years of concurrent data. In the southern New England deciduous forest with 15 years of data, the time at which carbon uptake increased in spring was significantly correlated with observed leaf development, but the cessation of carbon uptake was not significantly correlated with observed leaf abscission, which does not quickly follow leaf senescence in the dominant species, red oak (*Quercus rubra*). A measure of canopy greenness appears necessary for accurately estimating or predicting cessation of carbon uptake by this species. Differences between two southern New England deciduous forests in landscape position, slope aspect (northwest vs. east), and the degree of dominance by red oak vs. other deciduous and coniferous

J.L. Hadley(✉) and J. O'Keefe
Harvard Forest, Harvard University, 324 N. Main St., Petersham,
Cambridge, MA 01366, USA
e-mail: jhadley@fas.harvard.edu and jokeefe@fas.harvard.edu

J.W. Munger
Department of Earth and Planetary Sciences, Harvard University,
Cambridge, MA 01238, USA
e-mail: jwmunger@seas.harvard.edu

D.Y. Hollinger
USDA Forest Service, NE Research Station, 271 Mast Road, Durham, NH 03824, USA
e-mail: davidh@hypatia.unh.edu

A.D. Richardson
Complex Systems Research Center, University of New Hampshire,
Durham, NH 03824, USA
e-mail: andrew.richardson@unh.edu

A. Noormets (ed.), *Phenology of Ecosystem Processes*,
DOI 10.1007/978-1-4419-0026-5_5, © Springer Science + Business Media, LLC 2009

trees had effects on the annual time course of carbon exchange which were similar in magnitude to the effects of the 1.75° difference in latitude between these two deciduous forests and one further north. We hypothesize that a change in dominant tree species from red oak, which has ring-porous wood and must form new xylem each year prior to leaf growth, to diffuse porous trees (*Acer*, *Betula* and *Fagus* spp.), which lack this requirement, further north could enable the timing of leaf development to remain relatively early in the more northerly location, despite a cooler climate. A coniferous forest in southern New England (latitude 42.54°) showed two annual peaks in carbon uptake: a large one in spring before maximum carbon uptake by deciduous forests, and a smaller peak in autumn. In contrast, in a more northerly coniferous forest (45.25°N), the autumn peak was not observed. Significant late-winter (March) carbon uptake also occurred only in the more southerly conifer forest when early soil thawing occurred in 2006.

1 Introduction

Recent papers (Barr et al. 2004, 2007; Urbanski et al. 2007) have shown that inter-annual climate variation can strongly affect annual carbon exchange of temperate and boreal forests, through climatic influences on the development and senescence of the forest canopy, which can also be termed canopy phenology. In this chapter we will examine variation in the phenology of carbon (C) assimilation by forests in relation to forest canopy phenology at a long-term C exchange measurement site in New England, USA. We will also examine differences in phenology of C exchange at five New England sites in relation to two geographic and topographic variables, latitude and slope aspect, that influence either forest macroclimate or microclimate. Three of the sites are in predominantly deciduous forests and two in coniferous forests, allowing us to look at differences between these two fundamentally different forest types. Two of the deciduous forest sites are very close to each other at the same latitude, but with differences in slope aspect and wind exposure. In addition, there are nearly continuous records of net carbon exchange for more than a decade at two of the sites, creating a data set that is useful to at least begin an analysis of carbon exchange phenology.

The questions we will address in this chapter include:

What are the relationships between observed spring leaf development and autumnal leaf abscission and the beginning and end of forest ecosystem carbon uptake each year?

How do the timing and rate of change of the annual transitions from forest carbon loss to forest carbon uptake and *vice-versa* differ between forests of differing latitude, wind exposure, and species composition within the northeast U.S.?

What climatic and microclimatic variables are linked to differences between forests in the timing and rate of these annual carbon exchange transitions?

Do differences between the dominant tree species in different forests also affect the annual time course of carbon uptake and carbon loss?

2 Methods

In our analysis we use data from two sites with carbon flux data covering 10 or more years: a predominantly deciduous portion of the Harvard Forest in southern New England, containing about 25% conifers (HF-EMS; see Wofsy et al. 1993; Goulden et al. 1996; Barford et al. 2001; Urbanski et al. 2007) and a coniferous northern New England forest near Howland, Maine (Hollinger et al. 1999, 2004). In addition we will present data from a second deciduous area of the Harvard Forest, first measured in 2002 (HF-LPH; see Hadley et al. 2008) that has fewer conifers, less water and eastern hemlock stand (HF-Hemlock) with measurements in 2000–2001 and 2004 onward (Hadley and Schedlbauer 2002; Hadley et al. 2008). Finally, we will use measurements from a deciduous forest about 175 km further north than the Harvard forest, at the Bartlett Experimental Forest (BEF) in northern New Hampshire, with data from 2004 onward. Relevant data for all of these forests is given in Table 1, and some details of the eddy flux measurement

Table 1 Characteristics of the five New England forests. *HF-EMS* Harvard Forest Environmental Measurement Site, *HF-LPH* Harvard Forest Little Prospect Hill, *HF-Hemlock* Harvard Forest hemlock stand. Abbreviations of genera: *A. Acer, B. Betula, F. Fraxinus, Q. Quercus, T. Tsuga.* Mean annual temperatures are for 2004–2006

Site	HF-EMS	HF-LPH	HF-Hemlock	Bartlett	Howland
Latitude	42.538°N	42.540°N	42.539°N	44.28°N	45.25°N
Longitude	72.171°W	72.180°W	72.175°W	71.05°W	68.73°W
Elevation (m)	340	390	360	210	60
Tree age range (years)	65–100	45–100	100–200	70–120	60–190
Forest type	Deciduous	Deciduous	Coniferous	Deciduous	Coniferous
Dominant species (in decreasing abundance)	*Q. rubra* *A. rubrum* *T. canadensis* *Pinus rubra*	*Q. rubra* *A. rubrum*	*T. canadensis* *Pinus strobus*	*A. rubrum* *F. americana* *B. papyrifera* *A. saccharum*	*Picea rubens* *T. canadensis* *A. rubrum* *B. papyrifera*
Canopy height (m)	25	17	22	24	20
Aboveground carbon (t ha⁻¹)	95	40	105	105	120
Tree density (ha⁻¹)	660	900	600	625	2000
Basal area (m² ha⁻¹)	33	18	50	38	48
Mean air temperature (°C)	8.1	7.9	8.1	7.5	6.1
Annual precipitation (mm)	1100	1100	1100	1300	990

systems are given in Table 2. Approximate locations of the three New England carbon flux study sites are shown in Fig. 1, and the three Harvard Forest flux towers are shown in Fig. 2. Data for the Harvard Forest EMS tower used in this chapter

Table 2 Characteristics of the eddy covariance systems used at the five flux tower sites

Site	HF-EMS	HF-LPH	HF-Hemlock	Bartlett	Howland
Sonic anemometer	ATI[a] SAT 211/3K	Campbell[b] CSAT3	Campbell[a] CSAT3	ATI[a] SAT 211/3K	ATI[a] SAT 211/3K
Analyzer manufacturer and model (all closed-path)	Licor[c] LI6262	Licor[c] LI6262	Licor[c] LI6262 LI7000[d]	Licor[c] LI6262	Licor[c] LI6262
Sampling height above canopy (m)	5	4.5	5	3	9
Minimum night turbulence (u*) to accept data (m/s)	0.2	0.35	0.4	0.35	0.25

[a] Applied Technologies Inc. (Boulder, Colorado, USA).
[b] Campbell Scientific, Inc. (Logan, Utah, USA).
[c] Licor Biosciences Inc. (Lincoln, Nebraska, USA).
[d] LI-6262 was used until October 2001, LI7000 was used in 2004 and after.

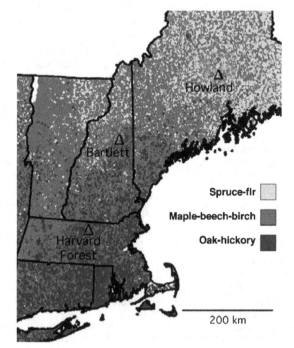

Fig. 1 Locations of study sites, indicated by *black triangles*: Harvard Forest, Massachusetts; Bartlett Experimental Forest (BEF), New Hampshire; and Howland Forest, Maine. Forest type classifications are based on county-level Forest Inventory and Analysis (FIA) data from the USDA Forest Service.

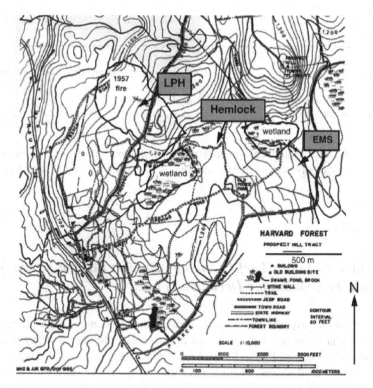

Fig. 2 Locations of Harvard Forest (HF) carbon flux measurement towers: Little Prospect Hill (LPH), Hemlock and Environmental Measurement Site (EMS).

are calculated from data at ftp://ftp.as.harvard.edu/pub/nigec/HU_Wofsy/hf_data in the "Final" folder. Data for the Harvard Forest Hemlock and LPH towers can be found at http://harvardforest.fas.harvard.edu/data/atm.html in data sets HF103 and HF072. Data for the Bartlett Experimental Forest is summarized in Jenkins et al. (2007) and is accessible through the Ameriflux network web site at: http://public. ornl.gov/ameriflux/data-access.shtml, while Howland data is at ftp://epg-ftp. umaine.edu/CDIAC. Throughout this chapter, we use the convention that negative values of net ecosystem exchange (NEE) indicate net carbon uptake by the forest, and positive values indicate carbon flux to the atmosphere. To better portray seasonal changes, in our analyses we use 10-day running means to remove large day-to-day variations in carbon exchange due to synoptic weather events.

In this chapter we use data from this suite of eddy flux towers, along with microclimatic data (Table 3) to show the relationships between detailed observations of canopy phenology, microclimate, and the beginning and end of gross and net carbon uptake at the deciduous Harvard Forest flux tower that presently has a 15-year record. We will also examine effects that differences in slope aspect and wind exposure may

Table 3 Microclimate data used in analysis of phenological change in ecosystem carbon uptake

Microclimate parameter	Measurement location
Air temperature	4–5 m above canopy
Soil temperature	10 cm depth, except where other depths are specified in text and figure captions
Photosynthetically active radiation (PAR)	2–5 m above canopy
Wind direction and speed	At flux measurement height, 4–9 m above canopy

have on seasonal carbon exchange patterns, and differences in the seasonality of net carbon exchange between coniferous, deciduous and mixed forests in New England.

3   Results

3.1   Harvard Forest Leaf Phenology and Net Carbon Exchange of a Predominantly Deciduous Forest

Because most photosynthesis in forests occurs in the leaf canopy, the timing of leaf development in spring and leaf abscission in autumn should show a strong relationship with carbon uptake by the forest. Leaf development and leaf abscission have been observed on 20 species of trees and shrubs at the Harvard Forest since 1991 (Richardson and O'Keefe, in current volume). In Fig. 3a we show average dates of budbreak and the midpoint between budbreak and 75% leaf enlargement for red oak and red maple, which are the two dominant species in the deciduous portions of Harvard Forest. Dates of the first daily net carbon uptake by the forest each year (STARTNCU) and dates at which 20% of maximum daily carbon uptake was reached (20%maxCU) are also plotted. Figure 3b shows the average dates of 50% leaf fall for red oak and red maple in each year, along with dates of the last net carbon uptake (ENDNCU).

Both STARTNCU and 20% maxCU show positive correlations with the average date of budbreak for red oak and red maple (Fig. 4a). There was no statistically significant correlation between the average date for 50% leaf abscission for red oak and red maple and ENDNCU (Fig. 4b), although there is a positive relationship ($r^2 = 0.40$, $p = 0.04$) if the two outliers near the bottom and right-hand end of the graph are removed if the two outliers were removed. The slopes of all regression lines are less than 1, so the time difference between the observed phenological event and the beginning or cessation of net carbon uptake is smaller later in the season. The relatively weak relationship between leaf loss in the fall and ENDNCU is very likely due to the fact that carbon uptake by leaves may stop substantially before leaf fall, and leaves of red oak sometimes do not abscise for weeks or months after they turn brown and are no longer photosynthetically active. An indicator of leaf physiological activity, such as leaf greenness, is likely to correlate better with the end of photosynthetic activity in autumn than leaf abscission.

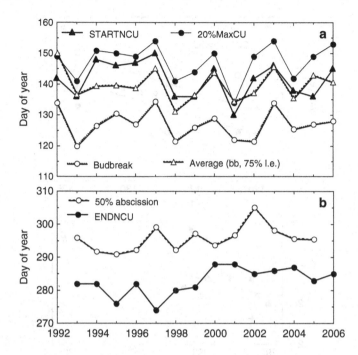

Fig. 3 (a) Average dates of bud break and the midpoint between budbreak and 75% leaf expansion for red oak and red maple at Harvard Forest, plus the dates of first 10-day average daily NEE less than zero (start of net ecosystem carbon uptake or STARTNCU) measured at the EMS flux tower in 1992 through 2006. (b) Dates of 50% leaf abscission for red oak and red maple, and dates of the last 10-day average NEE below zero (ENDCU) measured at the EMS flux tower for the same years. Dates for STARTCU and ENDCU were calculated from data at: ftp://ftp.as.harvard.edu/pub/nigec/HU_Wofsy/hf_data/final.

3.2 Seasonal Patterns of Net Carbon Exchange in Three Forest Types within Harvard Forest

The three Harvard Forest eddy flux measurement towers show a large difference in the seasonality of carbon uptake between the deciduous sites (LPH and EMS) and the coniferous eastern hemlock forest. While net carbon uptake by the deciduous forests occurs almost exclusively between the end of May and mid-October (approximately days 150–285), in the hemlock stand the peak uptake was in April and May, and declined from June through September before a second carbon uptake peak in late October (Fig. 5). Increased net C uptake by hemlock forest in late October must be due to declining ecosystem respiration rather than increasing photosynthesis, as the latter becomes progressively limited by declining daylength and solar angle in autumn.

More subtle differences in the annual pattern of carbon exchange are visible between the two primarily deciduous forests. For instance, the EMS tower (which

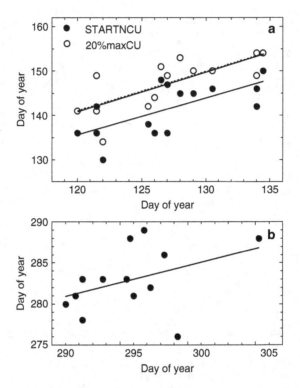

Fig. 4 (a) Average bud-break date for red oak and red maple vs. date of first net carbon uptake (*closed circles*) and date at which daily net ecosystem carbon uptake reached 20% of maximum (*open circles*). Regression line for date of first C uptake; $y = 0.828x + 36.31$, $r^2 = 0.45$, $p = 0.07$, regression for date of 20% of maximum daily C uptake; $y = 0.889x + 34.3$, $r^2 = 0.53$, $p = 0.02$. (b) Average date of 50% leaf abscission for red oak and red maple in 1993–2005 vs. last date of net carbon uptake. Regression line is $y = 0.414x + 160.3$, $r^2 = 0.17$, $p = 0.17$.

contains slightly under 25% coniferous trees by basal area within 500 m of the tower) shows an accelerating negative trend in NEE, indicating an increasing rate of carbon uptake, beginning around the end of April (day 120) and lasting until mid-June (day 165). In contrast the LPH site shows a sharp upturn in NEE in early May (days 120–130) before the phase of increasingly negative NEE that lasts until the end of June (day 180) (Fig. 5). Some of the difference between the two sites is likely due to the presence of conifers at the EMS site, whose high photosynthetic activity during May masks increasing respiration in the deciduous portion of the forest during the part of May when deciduous leaves are still maturing. With a smaller contribution (7%) of conifers in the footprint of the LPH tower, increasing respiration during early May becomes visible in the NEE trajectory.

The onset of net carbon uptake in the forest at EMS occurred on average about 8 days earlier than the forest at LPH, in the 4 years for which we have data for both sites (Fig. 5, Table 4), which may have been the result of differing rates of snow melt and soil warming. The presence of conifers and more southerly aspect of EMS

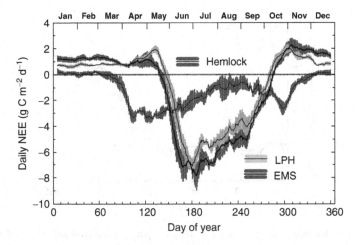

Fig. 5 Mean daily net ecosystem carbon exchange (NEE) for the coniferous eastern hemlock forest in 2001 and 2004–2006 and for the primarily deciduous forests sampled by the LPH and EMS towers in 2003–2006. Negative values indicate net carbon uptake. The lines are plotted from the average running men daily NEE for each day of the year, for 4 years of data at the LPH and EMS towers, and 3–4 years at the Hemlock tower, where a long data gap occurred due to lightning damage in 2005). The gray areas around the lines show standard errors of the mean for values in 3–4 years on a given day of year.

Table 4 First and last day and duration of annual net carbon uptake period at each deciduous forest site, calculated from 10-day running averages

Year	Harvard Forest, Main Tower (EMS)			Harvard Forest, Little Prospect Hill (LPH)			Bartlett Exp. Forest, New Hampshire		
	First	Last	Duration	First	Last	Duration	First	Last	Duration
2003	146	286	140	153	284	131	NA	NA	NA
2004	138	288	150	143	286	143	130	279	149
2005	136	284	148	150	280	130	148	279	131
2006	145	286	141	150	278	128	132	270	138
Average 2003–2006	141	286	145	149	282	133	na	na	na
Average 2004–2006	140	286	146	148	281	133	137	276	139

could have resulted in greater radiation load and lower albedo compared to the north facing purely hardwood stand of LPH. This hypothesis was, indeed, supported by differences in soil temperatures between the two sites (Fig. 6). The W to NW exposure of the LPH site also increases its exposure to NW winds, which in New England tend to be stronger and occur during periods of lower air temperature than other wind directions. For trees at the LPH site, lower soil temperature may slow root growth and any developmental changes requiring elevated metabolism in roots, while the cooling effects of NW winds may also inhibit spring leaf growth. The end of net carbon uptake in October occurred on average 4 days later at the

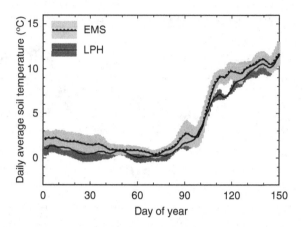

Fig. 6 Seven-day mean soil temperature at 10 cm depth near the HF-EMS and HF-LPH flux towers at Harvard Forest (see Fig. 2) in winter and spring. The *shaded areas* around each line indicate one standard error across 4 years (2003–2006). Soil temperature was measured at 5 m depth at six locations about 250 m SE of the EMS tower, and at 10 cm depth at four locations 200 m SW and 200 m NW of the LPH tower. Soil temperature at 10 cm depth for the EMS tower locations was estimated using the 5 cm measurement plus an adjustment based on a soil temperature profile covering 5–50 cm depths, measured in a location about 4 km away. The mean value at 10 cm for all measurement locations was used for each daily average. (Soil temperatures near the EMS tower and the soil temperature profile courtesy J. Melillo, Ecosystems Center, Marine Biological Laboratory, Woods Hole, MA, USA).

EMS site than at LPH (Fig. 5, Table 4), leading to an average annual duration of net daily carbon uptake which was 12 days shorter at the LPH site (Table 4).

3.3   Phenology of Carbon Exchange in Northern vs. Southern New England

3.3.1   Deciduous Forests

In 2 of the 3 years for which there is simultaneous data, net carbon uptake began earlier at the Bartlett Experimental Forest site in New Hampshire than at Harvard Forest about 200 km further south (Fig. 7). Given the difference in latitude, this is surprising; however, the Bartlett site is more than 100 m lower in elevation than the Harvard Forest sites, and in 2004–2006 the average annual temperature measured at the Bartlett flux tower was only about 0.5°C lower than for the Harvard Forest sites (Table 1). The earlier onset of net carbon uptake at Bartlett compared to Harvard Forest in 2 of 3 years is also consistent with earlier leaf development at Hubbard Brook Experimental Forest which is 30 km W of Bartlett, compared to Harvard Forest (Richardson et al. 2006). In the 3 years of simultaneous data however, the difference

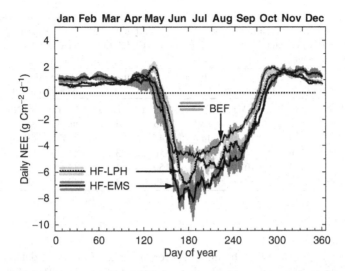

Fig. 7 Average 10-day running means (*lines*) and standard errors (*shading*) of net ecosystem carbon exchange (NEE) for 2004 through 2006 for the predominantly deciduous Harvard Forest sites (HF-LPH and HF-EMS) and the tower at Bartlett Experimental Forest (BEF) in northern New Hampshire. Negative values indicate forest carbon uptake. The *dark* and *light gray bands* for the Bartlett and HF, LPH towers respectively, and the *green band* for the HF main tower (EMS), show standard errors calculated from values on the same day of the year in each of the 3 years.

between the beginning of net C uptake at Bartlett and at the Harvard Forest deciduous sites was very variable. STARTNCU at Bartlett ranged from 13 days earlier to 12 days later than at HF-EMS, and from 18 to 2 days earlier compared to HF-LPH. (The large ranges are due to the year 2005, when net carbon uptake at Bartlett did not begin until May 28, or 16–18 days later than for the other 2 years.) In autumn, the transition from net carbon storage to carbon loss was consistently earlier at Bartlett than at the deciduous Harvard Forest sites (Fig. 7), possibly because the species at Bartlett respond more quickly to decreasing daylength, or because the first hard autumn frosts (below −2°C) came 20–30 days earlier at Bartlett than at Harvard Forest in 2004 through 2006 (day 280–295 at Bartlett, and day 308–315 at Harvard Forest). The average length of the period with net carbon storage at Bartlett was intermediate between the two Harvard Forest sites (Table 4).

3.3.2 Early Photosynthesis in Deciduous Forests Prior to Net Ecosystem Carbon Storage, and Environmental Predictors of Early Photosynthesis and Beginning of Net Carbon Storage

In addition to the date of first net carbon uptake, for each of the deciduous forest sites we calculated the date for the first evidence of strong photosynthesis in spring, as indicated by a daily maximum half-hourly net carbon uptake rate that exceeded the

Table 5 First day of each year in which the maximum 30-min net carbon uptake during day (DAY) exceeded average nighttime carbon efflux, the start of net carbon uptake (24hNET) and the lag between these two dates. Each is derived from a 10-day running average of the respective carbon flux parameter

Year	Harvard Forest, Main Tower			Harvard Forest, Little Prospect Hill			Bartlett Exp. Forest New Hampshire		
	DAY	24hNET	LAG	DAY	24hNET	LAG	DAY	24hNET	LAG
2003	131	146	15	149	153	4	n/a	n/a	n/a
2004	131	138	7	141	143	2	127	130	3
2005	134	136	2	147	150	3	138	148	10
2006	135	145	10	146	150	4	118	132	14
Mean	133	141	8.5	146	149	3.3	128	137	9.0

average nighttime respiration by the forest for that date. From these data, a slightly different picture of the phenology of carbon exchange emerges. Though there was large interannual variation, the date for first strong net carbon uptake preceded the first date of net carbon uptake by about 9 days, on average, at the HF-EMS and BEF sites, but by less than 4 days at the HF-LPH site (Table 5). The shorter lag at the LPH site may be a consequence of later development of photosynthetic capacity at the site, so that it in general leaf development occurred during warmer weather and therefore proceeded faster. The extreme dominance of red oak at the LPH site may also have contributed to the more rapid development of photosynthetic capacity at that site, because at LPH most leaf development occurred synchronously in one species, rather than more slowly in several species with different periods of leaf maturation at the EMS site. The EMS site also has a higher number of understory saplings and shrubs, particularly around the wetland NW of the EMS tower (Fig. 2). Richardson and O'Keefe in current volume report that budburst was slightly earlier in understory species than in canopy trees, although the two groups showed no consistent difference in the date of 75% leaf expansion.

In a search for climatic drivers of the start of net carbon uptake, for each site we calculated daily average soil temperature, cumulative photosynthetically active radiation received after April 1, and "growing degree-days" calculated by summation of daily average temperatures exceeding 4°C. These statistics varied slightly across the years 2003–2006, but also showed marked differences between sites (Table 6). Both for strong daytime net uptake and 24-h net carbon uptake, the HF-LPH site required a larger number of growing degree-days and a higher cumulative PAR than the other sites. The HF-LPH site was less clearly distinguished by the soil temperatures when net carbon uptake became active in spring (Table 6). Thus, HF-LPH site on a northerly slope showed delayed start of net carbon uptake in response to climatic drivers, as well as in respect to calendar date. The Bartlett site tended to require fewer growing degree-days, lower soil temperature and less cumulative PAR prior to the beginning of either strong photosynthesis or daily net

Table 6 Mean and standard error (SE) of cumulative growing degree-days (GDD), daily average soil temperature (T_{soil}; °C), and accumulated PAR (MJ) since April 1 on the date when daily maximum 30-min net carbon uptake exceeded the average nighttime carbon efflux (DAY) or the start of daily net carbon uptake (24h-NET) at the three deciduous forests. Soil temperature was measured at 5 cm depth at HF-EMS and Bartlett, and at 10 cm depth at HF-LPH

Site	Measure of ecosystem carbon uptake	GDD (Mean ± SE)		T_{soil} (Mean ± SE)		PAR (Mean ± SE)	
HF-EMS	DAY	133	1.0	10.5	0.1	1,361	105
HF-LPH	DAY	147	2.4	11.2	0.7	1,778	88
Bartlett	DAY	128	5.8	8.0	1.1	1,209	207
HF-EMS	24h-NET	141	2.5	11.8	1.1	1,657	55
HF-LPH	24h-NET	149	2.1	11.8	0.7	1,860	63
Bartlett	24h-NET	137	5.7	10.1	0.6	1,486	173

carbon uptake compared to other sites. However, a tendency for the forest at Bartlett to require fewer growing degree days, less soil warming, and less incident PAR prior to the initiation of photosynthesis is consistent with the observation by Richardson et al. (2006) that fewer growing degree days accumulated prior leaf development at BEF compared to Harvard Forest. The year-round high hydraulic conductance of diffuse-porous wood in the dominant tree species at Bartlett allows water transport to developing foliage at any time of year, whereas ring-porous red oak, the dominant tree species at Harvard Forest must produce earlywood vessels each year to supply sufficient water to new leaves (see Discussion).

3.3.3 Net Carbon Exchange in Coniferous Forests

Only about 2.5 years of simultaneous data (November 2000 through October 2001, July through December 2004, and 2006) are currently available for the Harvard Forest hemlock forest and the spruce-hemlock dominated forest near Howland, Maine. Daily net carbon uptake at Howland consistently started around day 90. At Harvard Forest net uptake commenced just a few days earlier in 2001, but a full month earlier in 2006 (Fig. 8). Differences in autumn carbon uptake between these sites were observed in all years, with HF-Hemlock showing net carbon uptake for up to two months longer than the Howland site (Fig. 8). The much longer season for carbon uptake at Harvard Forest, especially in 2006, may result from climatic differences. On most days, daily minimum temperatures in October were 2–5°C colder at the Howland site than at Harvard Forest. At Howland in October 2004 and 2006 air temperature fell below 0°C on 8 and 6 days respectively, whereas at

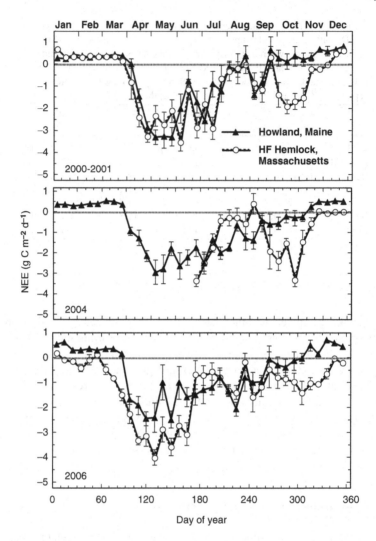

Fig. 8 Daily net carbon exchange averaged over 10-day intervals at the Harvard Forest hemlock site and at Howland, Maine, primarily in 2001, 2004 and 2006. Error bars are standard errors of the ten daily values in each mean; for some winter periods, the error bars are too small to be seen behind the symbols. The last two months in the top panel are in the year 2000, and there were no measurements until late June for the Harvard Forest hemlock site in 2004.

Harvard Forest this happened only once in 2004 and twice in 2006, and then only in late October (Fig. 11). A slightly more open canopy with tree-fall gaps in the relatively old HF hemlock stand may also allow more light to penetrate the canopy at the low sun angles that occur in late autumn, compared to denser canopy in the younger forest at Howland.

Fig. 9 Response of daytime carbon flux to photosynthetically active radiation (PAR) at the Harvard Forest hemlock site and the Howland, Maine forest in October 2001, 2004 and 2006. Fitted curves are from equations of the form C flux = c + $(a \times PAR)/(b + PAR)$, where c is an estimate of daytime ecosystem respiration and a and b describe the shape of the curve.

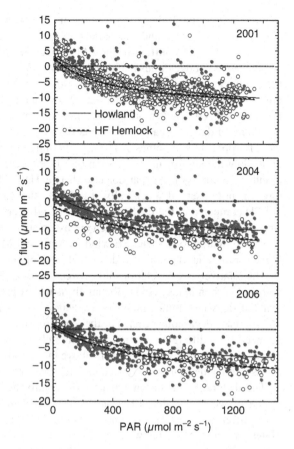

4 Discussion

4.1 Relationship of Leaf Phenology to Net Ecosystem Carbon Exchange and Local Differences Among Harvard Forest Sites

The start of net carbon uptake at HF-EMS was correlated with budbreak and leaf development (Figs. 3a and 4a), but there was also considerable variation among years. For example, the 4-year period, from 1993 through 1996, exhibited delayed start of net carbon uptake in relation to leaf development than did the rest of the time series. While during the other years, the STARTNCU was within 5 days of the mean of budbreak and 75% leaf expansion, in 1993–1996 STARTNCU lagged by nearly 8–11 days. There are at least two possible reasons that the time interval between leaf development and the beginning of net carbon uptake could vary interannually. Firstly, there is variation in ecosystem respiration, the majority of which comes from soil and can be decoupled from canopy development. Secondly, the contribution of conifers in the EMS tower footprint to measured fluxes during the period of deciduous leaf development could vary from year to year, because the conifers occur almost exclusively to the NW of the flux tower, and conifers therefore have a greater influence on the measured fluxes in years with a greater frequency of NW winds more consistent relationship between leaf phenology and the beginning of daily net carbon uptake at the LPH site, which has very few conifers, would support this hypothesis. The 4 years of carbon flux data from LPH in 2003–2006 show no significant correlation of STARTNCU at LPH with observed dates of either budburst or leaf expansion elsewhere at Harvard Forest; however, leaf phenology was not observed specifically at the LPH site. Lower spring soil temperatures at LPH compared to the EMS site (Fig. 6) could delay leaf development, and exposure of foliage to NW winds, which are associated with air temperatures lower by 2–3°C compared to SW and SE wind in April and May (Harvard Forest meteorological station data) could also delay leaf development at the LPH site.

The relatively weak relationship between dates of leaf abscission and the end of daily net carbon uptake (Figs. 3b and 4b) is not surprising when one considers that leaf fall does not immediately follow leaf senescence, especially in red oak, the predominant deciduous species at the Harvard Forest flux tower sites. It points to the need for a different parameter to accurately predict the cessation of net carbon uptake, possibly leaf greenness from above the canopy or decrease in absorbed PAR (measured as the ratio of below-canopy to above-canopy PAR).

The difference in the annual period of daily net carbon uptake between the primarily deciduous sites at Harvard Forest (EMS and LPH) shows that even fairly small differences in forest composition and location may affect the annual period of net carbon uptake. The two flux towers are only 1.1 km apart and differ in elevation by only about 50 m (Table 1), but showed a consistent difference in the dates at which daily net uptake began and ended. A portion of this difference can be attributed to microclimatic differences including lower Tsoil at the LPH than EMS site in spring (Fig. 6). The presence of conifers in the NW sector of the EMS tower footprint (Wofsy et al. 1993; Goulden et al. 1996), where they form about 40% of the basal area (see ftp://ftp.as.harvard.edu/pub/nigec/HU_Wofsy/hf_data/, "Ecological data") clearly contributes to observations of early-season photosynthesis at the site because data from the nearby Hemlock flux tower show substantial carbon uptake in April (Fig. 5), several weeks before budbreak of the dominant deciduous trees (Fig. 3). The residual differences can be attributed primarily to differences in aspect, canopy structure and microclimate, whereas age effects are likely to be minor (Noormets et al. 2007 and in current volume).

The contrast in the phenology of carbon uptake between the predominantly deciduous Harvard Forest sites and the hemlock forest (Fig. 5) shows how differently the conifer forest functions from the deciduous forests. In the hemlock forest, daily net carbon uptake began in late March, daily maximum rates were reached by mid-April, and a decline in daily net carbon uptake began by late June, about the same time when the deciduous forests were achieving maximum daily uptake. Decreasing daily net uptake for the hemlock stand in July and August had a similar pattern to the decrease for the deciduous forests during the same period (actually, the rate of change for the hemlock forest is slightly smaller), but uptake in the hemlock forest then recovered to a second peak in late October, just before time of maximum net carbon loss from the deciduous forests. Average estimated net annual carbon uptake did not differ significantly between the deciduous and hemlock stands (ranging between 3 and 4 Mg C per ha; Hadley et al. 2008; Urbanski et al. 2007; additional data at web addresses given in Methods), but all of the annual uptake occurred during a single large summertime peak for the deciduous forests, and in two smaller spring and fall peaks for the hemlock forest (Fig. 5).

This pattern in net ecosystem carbon exchange can be attributed to high ecosystem respiration in the hemlock forest relative to its maximum photosynthetic capacity. The maximum instantaneous rate of net carbon uptake by the hemlock forest has been measured at about 15 μmol m^{-2} s^{-1} compared to about 25 μmol m^{-2} s^{-1} for deciduous forest, while average ecosystem respiration in mid-to-late summer is typically around 6–8 μmol m^{-2} s^{-1} in both forest types (Hadley et al. 2002, 2008; Wofsy et al. 1993). Ecosystem respiration occurs continuously during day and night throughout the year, increasing with increasing soil and air temperatures. However, there is high carbon uptake only during sunny daytime periods. As a result, with the limited photosynthetic capacity of hemlock foliage even when conditions are optimum for photosynthesis, warm soil and air temperatures can drive total daily respiration in the hemlock forest to levels approaching or exceeding total carbon uptake. The result is in near-zero daily carbon exchange, as seen for the hemlock forest during some late July, August and early September periods, especially in 2001 and 2004 (Fig. 8).

4.2   Differences in Phenology of Ecosystem Carbon Exchange Between Southern and Northern New England

4.2.1   Deciduous Forests

The surprising earlier onset of daily net carbon uptake, as well as the first strong photosynthesis signal, in 2 of 3 years for the Bartlett Experimental forest in northern New Hampshire compared to Harvard Forest sites (Fig. 5, Tables 3–5) could result in part from a difference in the wood anatomy and function of the dominant tree species at the two sites. Red oak, a ring-porous species, forms 46 and 58% of basal area for the Harvard Forest EMS and LPH tower footprints respectively, with diffuse-porous deciduous trees and conifers of secondary importance. In the more diverse forest at Bartlett, red maple and American beech (both diffuse-porous) form the largest fractions of basal area, 28 and 20% respectively, with eastern hemlock third at 17%. The only ring-porous tree in the forest at Bartlett is white ash with just 5% of total basal area. Ring-porous trees depend heavily on the earlywood xylem produced each year for water transport to foliage, and probably in consequence of this, ring-porous trees produce leaves later in spring (Zimmermann and Brown 1971). Although the forest at Bartlett became a net sink for carbon earlier in the year than either of the Harvard Forest deciduous forest sites in 2004 and 2006, it's maximum daily rate of carbon uptake in summer was lower than the Harvard Forest sites, which is consistent with less efficient water transport in diffuse-porous trees (Zimmermann and Brown 1971). Slower water transport may limit stomatal conductance in diffuse porous trees, leading to lower maximum leaf-level photosynthesis by most diffuse-porous species compared to ring-porous oaks (Bassow and Bazzaz 1997) and therefore higher maximum ecosystem-level carbon uptake in forests composed primarily of ring-porous species.

4.2.2   Conifer Forests

The one month delay in the start of net carbon uptake at Howland compared to HF-Hemlock in 2006 is probably mostly a consequence of much earlier spring warming of soil at Harvard Forest in 2006 (Fig. 10) given that Hollinger et al. (1999) showed that the start of the net C uptake season at Howland was coincident with thawing of forest soils. A thaw beginning on March 10, during which the daily minimum air temperature reached 1–7°C for four consecutive nights at Harvard Forest, but reached just 1–2°C for two of these nights at Howland could also have contributed to an earlier start to carbon uptake by hemlocks at Harvard Forest. Leaf conductance to water and carbon exchange in conifers, including eastern hemlock and red spruce, have been shown to exhibit a threshold response to minimum daily air temperature, where air temperatures (and, by inference, foliage temperatures) below freezing can sharply limit stomatal opening and photosynthesis the following day (Fahey et al. 1979; Schwarz et al. 1997; Smith et al. 1984; Hadley 2000). This effect has also been previously seen in ecosystem carbon exchange for the two conifer forests considered here (Hollinger et al. 1999; Hadley and Schedlbauer 2002).

A peak in net carbon uptake in October to early November occurred in the Massachusetts hemlock forest in 2001, 2004 and 2006 but this feature was almost completely absent from the Maine spruce-hemlock forest, aside from a very small increase in C storage in October 2006 (Fig. 8). The autumn carbon uptake at Harvard.

Forest may indicate that the capacity for conifers to store carbon in autumn is expressed in southern but not northern New England, given autumn temperatures typical of the current climate. Previous work at Howland (Hollinger et al. 1999) showed that spruce-hemlock forest C uptake capacity was depressed after air temperatures dropped below a threshold of −2 to −3°C. Similar frost-induced termination of net carbon uptake was observed in a boreal black spruce forest (Goulden et al. 1997). Soil temperature seems unlikely to be limiting to carbon uptake at Howland in October, because the Howland soil temperature at 5 cm depth, although colder than at Harvard Forest, remained above 7°C throughout October in all years (Fig. 10).

However, the fairly small differences in October minimum daily temperatures between Harvard Forest and Howland (Fig. 10) suggest that some other factors may

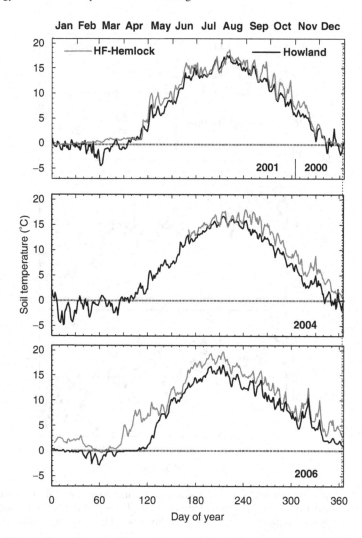

Fig. 10 Measured or estimated soil temperature at 10 cm depth for the HF hemlock and Howland sites in 2000–2001, 2004 and 2006. Values for Howland in 2000, 2001 and 2004 were estimated from measurements at 5 cm, plus a correction based on seasonal temperature gradients between 5 and 20 cm observed in 2006.

also be limiting carbon uptake at Howland in autumn. During most periods, including autumn 2001 and 2004, total daily incident PAR was similar at the two sites, but in the fall of 2006, daily total PAR averaged about 2–5 mol m⁻² or 10–20% lower at Howland than at Harvard Forest. There was also less net carbon uptake during October at a given PAR level in the Howland spruce-hemlock forest than in the Harvard Forest hemlock stand. This occurred particularly at high PAR values in 2004 and 2006, but also at relatively low PAR values in 2001 and 2004 (Fig. 9).

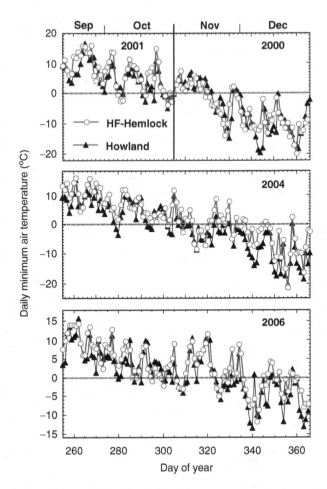

Fig. 11 Mid-September through December daily minimum air temperature for the Harvard Forest hemlock site and the Howland, Maine site. Data for November and December 2000 rather than 2001 are shown in the top panel because the carbon flux data in Fig. 6 are for the earlier year.

In contrast, at high PAR values during summer (July and August) of 2001 and 2006, carbon uptake at Howland was nearly the same as in the Harvard Forest hemlock forest (± 1.5 µmol m^{-2} s^{-1}), and C uptake was higher at Howland by about 3 µmol m^{-2} s^{-1} in 2004. The low October carbon uptake capacity at Howland contributed to very low or negative values of daily carbon uptake for the Howland forest beginning in late September of each year, while net carbon uptake by the Harvard Forest hemlock forest continued into or through November.

Greater carbon uptake by the Harvard Forest hemlock forest, relative to the Maine spruce-hemlock stand at Howland, especially in 2004 and 2006, which had higher spring and fall temperatures than 2000 and 2001, suggest that the carbon balance of the more southerly forest could be more stimulated by climate warming at least in the short term. However, the Harvard Forest hemlock forest and the Howland spruce-hemlock forest also differ in age and structure in ways that may affect canopy photosynthesis, especially in autumn. The Harvard Forest hemlock stand is older and has a higher fraction of large, old trees, some of which have died in the last 10 years, creating a very uneven canopy surface with some large gaps. This may allow more light to penetrate the canopy of the Harvard Forest stand, especially at the low solar angles occurring in autumn, compared to the younger, more even-aged forest at Howland. A comparison between more structurally similar forests may be necessary to establish a climate-related difference in spring and fall carbon uptake.

5 Conclusions

The carbon exchange data we summarize here indicate that in inland areas of New England, forest composition and local topography can have stronger effects than latitude on the springtime start of net carbon uptake by deciduous trees. Although there was not a statistically significant difference between the deciduous BEF site in New Hampshire and the Harvard Forest sites in Massachusetts (not surprising given only 3 years of data from BEF), it is striking that the average date for first net carbon uptake at BEF was earlier than for either of the Harvard Forest deciduous sites (Fig. 7, Table 4). The lower average values of accumulated growing degree-days, soil temperature, and accumulated PAR at BEF during spring transitions to carbon uptake (Table 6), indicate a tree species difference, and not primarily a climatic difference, could responsible for the difference in time of first net carbon uptake between BEF and the other sites. This pattern suggests an important influence of tree species, and perhaps particularly xylem anatomy, on the development of leaves and carbon uptake capacity in forests.

The HF-LPH site was the latest to show net carbon uptake, and showed a more consistent difference from HF-EMS than BEF did from either HF site. Some of the difference between HF-EMS and HF-LPH sites is also attributable to forest composition, particularly the presence of some conifers in the HF-EMS flux tower footprint, but microclimate differences (particularly in soil temperature, see Fig. 6) are likely responsible for some of the lag in onset of net carbon uptake in spring at the HF-LPH site compared to HF-EMS. Soil temperature is an important variable to consider in developing models to predict carbon uptake in spring. In autumn, latitude may significantly influence the cessation of net carbon uptake (Fig. 7, Table 4), given that BEF consistently had the earliest date when there was no net uptake, but a possible effect of tree species cannot be dismissed.

In the contrast between coniferous forests (HF-Hemlock vs. Howland), latitude and associated climate parameters seem to play a significant role. The beginning of net carbon uptake at Howland, which is about 300 km further north, was later in both years for which we have spring data to compare (2001 and 2006) though the difference was only about a week in 2001. More significant in terms of annual carbon storage were autumn periods of net carbon uptake in autumn, which occurred in all 3 years only at the more southerly HF-Hemlock site (Fig. 8). With a general climate warming, autumn carbon uptake by conifer forests may be expected to expand further north.

The correlations between the dates of leaf initiation and expansion over 15 years, and the beginning of net carbon at Harvard Forest are not as strong as might be hoped, if such observations are to be used to predict the beginning of carbon storage in other deciduous forest ecosystems. Variation in ecosystem respiration, not associated with leaf development, is one potential cause of this, but the other problem is that the HF-EMS tower does include over 40% conifers by basal area, as well as a wetland within the NW sector of the tower footprint, and these could influence early spring carbon exchange enough to push the changeover from net carbon loss to carbon uptake earlier, in a manner not predicted by the phenological observations which emphasize deciduous trees, and may not have a consistent relationship to photosynthesis in conifers

Acknowledgements This research was supported by the Office of Science (BER), U.S. Department of Energy, Cooperative Agreement No. DE-FC02-03ER63613, with funding through the Northeast Regional Center of the National Institute for Global Environmental Change, and by the National Institute for Climate Change Research. The National Science Foundation Long-Term Ecological Research (LTER) Program also supported the research at Harvard Forest. ADR and DYH acknowledge funding for the Howland Ameriflux site through the Office of Science (BER), US-DOE, Interagency Agreement No. DE-AI02-07ER64355, and support from the Northeastern Regional Center of the National Institute for Climatic Change Research. Jessica Schedlbauer, and Paul Kuzeja and - assisted in collection and analysis of the data presented for the Harvard Forest Hemlock and Little Prospect Hill sites.

References

Barford, C.C., Wofsy, S.C., Goulden, M.L., Munger, J.W., Pyle, E.H., Urbanski, S.P., Hutyra, L., Saleska, S.R., Fitzjarrald, D. and Moore, K. (2001) Factors controlling long- and short-term sequestration of atmospheric CO_2 in a mid-latitude forest. Science 294, 1688–1691.

Barr, A.G., Black, T.A., Hogg, E.H., Griffis, T.J., Morgenstern, K., Kljun, N., Theede, A. and Nesic, Z. (2007) Climatic controls on the carbon and water balances of a boreal aspen forest, 1994–2003. Global Change Biol. 13, 561–576.

Barr, A.G., Black, T.A., Hogg, E.H., Kljun, N., Morgenstern, K. and Nesic, Z. (2004) Inter-annual variability in the leaf area index of a boreal aspen-hazelnut forest in relation to net ecosystem production. Agric. For. Meteorol. 126, 237–255.

Bassow, S.L. and Bazzaz, F.A. (1997) Intra- and inter-specific variation in canopy photosynthesis in a mixed deciduous forest. Oecologia 109, 507–515.

Fahey, T.J. (1979) Effect of night frost on the transpiration of *Pinus contorta* ssp. latifolia. Oecol. Plant. 14, 483–490.

Goulden, M.J., Daube, B.C., Fan, S.-M. and Sutton, D.J. (1997) Physiological responses of black spruce forest to weather. J. Geophys. Res. (D Atmos.) 102, 28987–28996.

Goulden, M.L., Munger, J.W., Fan, S.M., Daube, B.C. and Wofsy, S.C. (1996) Exchange of carbon dioxide by a deciduous forest: Response to interannual climate variability. Science 271, 1576–1578.

Hadley, J.L. (2000) Effect of daily minimum temperature on photosynthesis in eastern hemlock (*Tsuga canadensis* L.) in autumn and winter. Arct. Antarct. Alp. Res. 32, 368–374.

Hadley, J.L., Kuzeja, P.S., Daley, M.J., Phillips, N.G., Mulcahy, T. and Singh, S. (2008) Water use and carbon exchange of red oak- and eastern hemlock-dominated forests in the northeastern USA: implications for ecosystem-level effects of hemlock woolly adelgid. Tree Physiol. 28, 615–627.

Hadley, J.L. and Schedlbauer, J.L. (2002) Carbon exchange of an old-growth eastern hemlock (*Tsuga canadensis*) forest in central New England. Tree Physiol. 22, 1079–1092.

Hollinger, D.Y., Aber, J.D., Dail, B., Davidson, E.A., Goltz, S.M., Hughes, H., Leclerc, M.Y., Lee, J.T., Richardson, A.D., Rodrigues, C., Scott, N.A., Achuatavarier, D. and Walsh, J. (2004) Spatial and temporal variability in forest-atmosphere CO_2 exchange. Global Change Biol. 10, 1689–1706.

Hollinger, D.Y., Goltz, S.M., Davidson, E.A., Lee, J.T., Tu, K. and Valentine, H.T. (1999) Seasonal patterns and environmental control of carbon dioxide and water vapor exchange in an ecotonal boreal forest. Glob. Change Biol. 5, 891–902.

Jenkins, J.P., Richardson, A.D., Braswell, B.H., Ollinger, S.V., Hollinger, D.Y. and Smith, M.L. (2007) Refining light-use efficiency calculations for a deciduous forest canopy using simultaneous tower-based carbon flux and radiometric measurements. Agric. For. Meteorol. 143, 64–79.

Noormets, A., Chen, J. and Crow, T.R. (2007) Age-dependent changes in ecosystem carbon fluxes in managed forests in northern Wisconsin, USA. Ecosystems 10, 187–203.

Richardson, A.D., Bailey, A.S., Denny, E.G., Martin, C.W. and O'Keefe, J. (2006) Phenology of a northern hardwood forest canopy. Global Change Biol. 12, 1174–1188.

Schwarz, P.A., Fahey, T.J. and Dawson, T.E. (1997) Seasonal air and soil temperature effects on photosynthesis in red spruce (*Picea rubens*) saplings. Tree Physiol. 17, 187–194.

Smith, W.K., Young, D.R., Carter, G.A., Hadley, J.L. and McNaughton, G.M. (1984) Autumn stomatal closure in 6 conifer species of the Central Rocky Mountains. Oecologia 63, 237–242.

Urbanski, S., Barford, C., Wofsy, S., Kucharik, C., Pyle, E., Budney, J., McKain, K., Fitzjarrald, D., Czikowsky, M. and Munger, J.W. (2007) Factors controlling CO_2 exchange on timescales from hourly to decadal at Harvard Forest. J. Geophys. Res. 112, G02020.

Wofsy, S.C., Goulden, M.L., Munger, J.W., Fan, S.M., Bakwin, P.S., Daube, B.C., Bassow, S.L. and Bazzaz, F.A. (1993) Net exchange of CO_2 in a mid-latitudinal forest. Science 260, 1314–1317.

Zimmermann, M.H. and Brown, C.L. (1971) *Trees: Structure and Function*. Springer, New York, pp. 336.

Influence of Phenology and Land Management on Biosphere–Atmosphere Isotopic CO_2 Exchange

Kaycie A. Billmark and Timothy J. Griffis

Abstract Stable isotope and micrometeorological techniques have long been used to study carbon cycle dynamics at a variety of spatial and temporal scales. Combination of these techniques provide a powerful tool for gaining greater process information at the ecosystem and regional scales and can provide a meaningful way to scale processes from leaf to region. In this chapter we review the recent literature and examine the key processes influencing biosphere–atmosphere $^{13}CO_2$ exchange. These processes are examined from the perspective of agricultural land management and rapid seasonal changes in phenology. Novel measurement techniques are introduced that can be used to better quantify the $^{13}CO_2$ exchange between the biosphere and atmosphere to determine how ecosystem processes, land use modifications, and phenology impact the isotopic composition of the atmosphere (i.e. the atmospheric isotopic forcing associated with land surface processes). High temporal resolution isotope mixing ratio and flux measurements, based on tunable diode laser absorption spectroscopy, are presented. The results demonstrate that the isotopic composition of respiration at the ecosystem scale is strongly linked to plant assimilated carbon, which is dependent on plant metabolic physiology and growth phase. We review how this strong isotopic coupling between ecosystem respiration and photosynthesis can impact isotope-based flux partitioning of net ecosystem CO_2 exchange, the variation in the canopy isotopic discrimination parameter, and the resulting isotopic forcing on the atmosphere.

List of symbols

F_N net ecosystem CO_2 exchange (μmol m^{-2} s^{-1})
F_A ecosystem photosynthetic assimilation (μmol m^{-2} s^{-1})
F_R ecosystem respiration (μmol m^{-2} s^{-1})
Δ_{canopy} canopy isotopic discrimination (‰)

K.A. Billmark (✉) and T.J. Griffis
Department of Soil, Water and Climate, University of Minnesota,
Minneapolis and Saint paul, MN, USA
e-mail: kaycie@umn.edu; tgriffis@umn.edu

A. Noormets (ed.), *Phenology of Ecosystem Processes*,
DOI 10.1007/978-1-4419-0026-5_6, © Springer Science+Business Media, LLC 2009

$\mathcal{D}$	canopy isotopic disequilibrium (‰)
C_3	Calvin cycle plant metabolism
C_4	Hatch-Slack cycle plant metabolism
$\delta^{13}C$	carbon isotopic composition (‰)
c_a	canopy air CO_2 mixing ratio (ppm)
c_s	leaf boundary layer CO_2 mixing ratio (ppm)
c_i	stomatal CO_2 mixing ratio (ppm)
c_c	chloroplast CO_2 mixing ratio (ppm)
δ^{13}_N	carbon isotope ratio of the net flux (‰)
δ_A	assimilated carbon isotopic composition (‰)
δ_R	non-foliar respired carbon isotopic composition (‰)
δ_N	carbon isotopic composition of the net CO_2 exchange (‰)
δ_a	atmospheric carbon isotopic composition (‰)
g	total conductance (μmol m^{-2} s^{-1})
g_a	aerodynamic conductance (μmol m^{-2} s^{-1})
g_s	canopy stomatal conductance (μmol m^{-2} s^{-1})
g_m	mesophyll wall conductance (μmol m^{-2} s^{-1})
Δ_b	boundary layer diffusional fractionation (‰)
Δ_s	stomatal diffusional fractionation (‰)
Δ_{diss}	mesophyll dissolution fractionation (‰)
Δ_{aq}	aqueous phase mesophyll transport fractionation (‰)
Δ_f	enzymatic fixation isotopic fractionation (‰)
F_δ	net ecosystem CO_2 isoflux (μmol m^{-2} s^{-1}‰)
K_c	eddy diffusivity of CO_2 (m^2 s^{-1})
$\overline{\rho_a}$	molar density of dry air (mol m^{-3})
M_a	molecular weight of dry air (g mol^{-1})
R_{VPDB}	heavy to light isotopic ratio of NBS-19
w	vertical wind velocity (m s^{-1})
S_c	storage rate of change of CO_2 between ground and measurement height (μmol m^{-2} s^{-1})
C_{wc}	cospectral density of vertical wind velocity and CO_2 mixing ratio (m ppm s^{-1})
R^{13}_N	heavy to light ratio of isotopic fluxes
R_h	heterotrophic component of total ecosystem respiration
R_a	autotrophic component of total ecosystem respiration
F_{Rh}	heterotrophic component of total ecosystem respiration flux (μmol m^{-2} s^{-1})
F_{Ra}	autotrophic component of total ecosystem respiration flux (μmol m^{-2} s^{-1})

1 Introduction

The stable isotope composition of CO_2 in the atmosphere represents an important signal of global change. Interpreting its variation, however, requires a sound understanding of the underlying biophysical processes that govern the isotopic fluxes between the biosphere and atmosphere. The combination of meteorological and stable isotope techniques has significantly advanced our understanding of CO_2 exchange processes in the global carbon cycle. Combined with a mechanistic understanding of $^{13}CO_2$ discrimination it is possible to interpret phenologic and physiologic variability within ecosystems.

The eddy covariance (EC) technique is now used to estimate net ecosystem CO_2 exchange (F_N) in a variety of ecosystems. Long term monitoring networks, such as FLUXNET (Baldocchi 2008), utilize eddy covariance and other micrometeorological approaches to study ecosystem response to environmental variables and global change. Stable isotope techniques have long been in use to study carbon cycling on a variety of spatial and temporal scales, and in particular, provide information about the F_N component fluxes of net photosynthesis (F_A) and ecosystem respiration (F_R). Researchers have combined micrometeorological and stable isotope techniques to partition F_N into the component fluxes (Yakir and Wang 1996; Bowling et al. 2001), and to improve the understanding of isotopic discrimination processes at the ecosystem scale. For example, researchers have observed seasonal and diurnal isotopic trends over forests (Bowling et al. 2002; Baldocchi and Bowling 2003; Knohl et al. 2005) agricultural systems (Zhang et al. 2006; Griffis et al. 2007), and soil (Ekblad and Högberg 2001; Ekblad et al. 2005). Such advances are important to the goal of learning how changes in climate, and subsequent phenologic response will impact F_N and, as well, to extend ecosystem-scale studies to the region using a mechanistic approach (Flanagan and Ehleringer 1998). Given these recent advances, coupled with the development of state-of-the-art measurement technologies, new opportunities are emerging to improve our understanding of land-atmosphere CO_2 exchange processes. For example, high-frequency measurement of the stable isotopologue mixing ratios (Bowling et al. 2003) using laser spectroscopy allows for direct quantification of isotopic fluxes at relatively high temporal resolution (Griffis et al. 2004, 2008). This chapter reviews current state-of-the-art isotopic flux measurement techniques and gives examples of how they can be used to study terrestrial carbon cycle processes. We present high frequency isotopic flux data from an agricultural ecosystem to quantify the impacts of ecosystem phenology, plant physiology and land management on isotopic biosphere-atmosphere exchange and on the contribution of photosynthesis and respiration, to F_N. We also discuss important challenges with using these methodologies, as well as the seasonal and environmental effects on key ecosystem integrated parameters, canopy isotopic discrimination (Δ_{canopy}) and canopy isotopic disequilibrium ($\mathcal{D}$), which in conjunction with the component fluxes provide a better understanding of ecosystem carbon cycling. Continued development of micrometeorological and stable isotope techniques will provide new insights into phenological controls and land-use impacts on biosphere–atmosphere exchange and a greater understanding of the global carbon budget.

2 Carbon Isotopes

2.1 Plant Processes

Mass differences between the stable isotopes of carbon result in isotopic fractionation from the chemical and physical processes associated with carbon cycling. In plants, these processes discriminate against the heavier stable isotope of carbon (^{13}C) so that plants contain relatively less ^{13}C than the atmosphere. Plants most commonly utilize either the Calvin Cycle (C_3) or the Hatch–Slack cycle (C_4) to fix carbon from

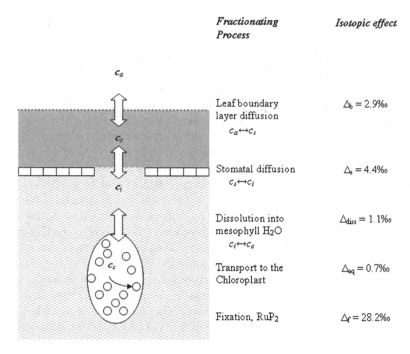

Fig. 1 Isotopic discrimination associated with C_3 plant CO_2 assimilation.

the atmosphere. Key differences in the CO_2 uptake processes associated with C_3 and C_4 plants result in dissimilar carbon isotopic compositions ($\delta^{13}C$) between plants using these photosynthetic pathways.

The process by which CO_2 moves from the atmosphere to fixation within the plant involves multiple steps during which discrimination against $^{13}CO_2$ occurs (Fig. 1). This isotopic fractionation is associated with diffusion through the leaf boundary layer and stomata, dissolution into mesophyll water, transport to the chloroplast, and enzymatic fixation. The nature of this multi-step process creates a concentration gradient from the atmosphere to the plant whereby each reservoir (e.g. canopy air (c_a), leaf boundary layer (c_s), stomata (c_i) and chloroplast (c_c)) has a different CO_2 concentration. Isotopic fractionation occurs as CO_2 moves across a resistance boundary from one reservoir to the next, and during enzymatic fixation. In C_3 plants, isotopic fractionation associated with enzymatic fixation is caused by ribulose bisphosphate (RuP_2) carboxylation. In C_4 plants, discrimination against ^{13}C during fixation is more complex. Here, dissolved CO_2 in the stomata is converted to bicarbonate (HCO_3^-) and fixed by phosphoenolpyruvate (PEP) carboxylase. This primary fixation allows carbon to be relocated to specialized bundle sheath cells where CO_2 is released and refixed by RuP_2 carboxylase as in C_3 plants. Isotopic fractionations are associated with the thermodynamic equilibrium reaction (HCO_3^-) and the enzymatic fixation by PEP carboxylase. There is an additional important isotopic fractionation to consider during this process, owing to the leakiness of the bundle sheath cells. If the bundle sheath reservoir was gas tight, there would be no further fractionation due to a lack of biochemical processes impacting the carbon source. However, leakage from the bundle sheath back to the mesophyll cells allows some fractionation by the RuP_2 carboxylase (Farquhar 1983; Farquhar et al. 1989), resulting in more depleted isotopic signatures than would otherwise be predicted. The various fractionation processes acting during C_3 and C_4 photosynthesis create distinct isotopic signatures between the two vegetation types. C_3 plant leaves have $\delta^{13}C$ signatures that typically range from −25‰ to −28‰, whereas C_4 plant leaves typically range from −11‰ to −14‰ (Farquhar et al. 1989; Cerling et al. 1997).

Although, plant photosynthetic isotopic discrimination is relatively well understood, there is still much to be learned about isotopic fractionations occurring in the plant following carboxylation (Hobbie and Werner 2004; Badeck et al.2005; Gessler et al. 2007). Additional metabolic and transport processes influence isotopic signatures within the plant. As the plant then utilizes stored metabolites, enrichment of respiration products can occur (Duranceau et al. 1999; Ghashghaie et al. 2003; Tcherkez et al. 2003, 2004; Gessler et al. 2007). Phenologic windows in the life cycle of all plants arise when photosynthetic assimilation is suspended and metabolic demand is met through remobilization of stored carbon. It is during these times that consideration must be made when using isotopic signatures of CO_2 to evaluate plant processes and response to environmental variables (Pataki et al. 2003; Badeck et al. 2005; Keitel et al. 2006; Bathellier et al. 2008). One key phenologic window of plant heterotrophy in herbaceous species occurs immediately following germination and continues until leaf emergence. The active vegetative plant growth phases are characterized by the more typically photosynthetic metabolic demand. During these time periods of autotrophy, CO_2 respiration and photosynthetic fluxes are strongly linked (e.g. Scartazza et al. 2004; Griffis et al. 2005a; Knohl et al. 2005; Zobitz et al. 2007) and can be observed to vary with light (e.g. Leavitt and Long 1986; Gessler et al. 2001), vapor pressure deficit (e.g. Bowling et al. 2002; Knohl et al. 2005; Mortazavi et al. 2005; Werner et al. 2006), and water availability (e.g. Flanagan et al. 1996; Ometto et al. 2002; Pataki et al. 2003; Werner et al. 2006).

2.2 Isotope Flux Partitioning Theory

Global scale photosynthesis and ecosystem respiration represent two large opposing fluxes ($\approx$120 Pg C y^{-1}). Small variations in climate, phenology, and land use are expected to impact these processes differently. These variations, therefore, create conditions that process the ratio of heavy to light stable isotopes differently. For example, changing water stress and plant water use efficiency have been shown through theoretical and empirical studies to be correlated with carbon isotope discrimination (Farquhar et al. 1989). Phenologic influence on photosynthesis and ecosystem respiration can be detected in the atmosphere as a result of changing CO_2 concentration and may be traced more precisely to a specific region or underlying process with the additional information and constraints provided by stable isotopes (e.g. Lloyd et al. 1996; Bowling et al. 2001; Randerson et al. 2002; Ogée et al. 2003; Scholze et al. 2003; Aranibar et al. 2006).

At the ecosystem scale, stable isotopes have been used to partition F_N into its component fluxes (Yakir and Wang 1996; Bowling et al. 2001; Zhang et al. 2006; Griffis et al. 2005a; Zobitz et al. 2007). Much of this work relies on two critical assumptions based on detailed observation. First, there exists an isotopic disequilibrium ($\mathcal{D}$) between the isotopic signature of photosynthesis (δ_A) and the isotopic signature of ecosystem respiration (δ_R). This simply states that the difference between the isotope ratio of photosynthesis and that of ecosystem respiration is not zero ($\mathcal{D} = \delta_R - \delta_A \neq 0$). Second, isotopic discrimination associated with ecosystem respiration is negligible when compared to the effects of photosynthesis.

Understanding $\mathcal{D}$ is complex because competing processes exist, which influence the magnitude and sign of disequilibrium. Because the background isotopic composition of the atmosphere is becoming "lighter" from fossil fuel combustion (Suess effect, Francey et al. 1999), recently assimilated CO_2 is expected to be relatively depleted compared to carbon that accumulated in soil organic matter from the pre-industrial era. Additional controls on $\mathcal{D}$, include, ecosystem disturbance (wildfires) and land use change, particularly from natural vegetation to cultivation (Scholze et al. 2008). Therefore, in many studies at the regional, and global scales, it is assumed that $\delta_R > \delta_A$ ($\mathcal{D} > 0$) (Ciais et al. 1995; Fung et al. 1997; Yakir 2004; Suits et al. 2005). However, ecosystem scale studies have estimated that $\mathcal{D}$ can vary and become negative on diurnal and seasonal timescales as a result of plant canopy processes (e.g. Aranibar et al. 2006; Zhang et al. 2006; Zobitz et al. 2007). Negative $\mathcal{D}$ is inconsistent with the Suess effect and other large-scale controls on $\mathcal{D}$, such as land use change, and, therefore, has implications on our understanding of the phenologic influence on terrestrial carbon exchange. Furthermore, since $\mathcal{D}$ is important for inverse models of the global carbon budget, valid parameterization is necessary. This is particularly important for agriculture given that changing climate may alter regional $\mathcal{D}$ through longer fallow seasons and that C_4 production is changing globally.

Recent work has shown that post photosynthetic isotope fractionation occurs resulting in an apparent fractionation of ecosystem respiration (Duranceau et al. 1999; Ghashghaie et al. 2001; Tcherkez et al. 2003; Hobbie and Werner 2004; Xu et al. 2004; Badeck et al. 2005; Gessler et al. 2007; Bathellier et al. 2008). Isotopic fractionation downstream of primary photosynthetic carboxylation, either during assimilate transport or through compartmentalization prior to transport, results in differences in the isotopic signatures of metabolites and in intramolecular distribution of ^{13}C and ^{12}C from fragmentation of the substrate molecule (Tcherkez et al. 2003, 2004). Researchers have observed that dark respiration of C_3 leaves is significantly enriched relative to plant sucrose, the respiration substrate (Duranceau et al. 1999, 2001; Ghashghaie et al. 2001; Tcherkez et al. 2003; Bathellier et al. 2008). Opposing this enrichment are results from Klumpp et al. (2005) and Badeck et al. (2005), which show that root-respired CO_2 from herbaceous species is depleted in ^{13}C. Klumpp et al. (2005) further demonstrated through a mass balance approach that dark respiration in leaves is balanced by root respiration products, such that the overall fractionation in the total respiration flux by plants is low ($\approx 0.7\%o$). It is clear from the above studies that discrimination related to respiration is potentially important, but likely to be less significant than photosynthesis at the ecosystem scale, which can influence isotopic compositions of the atmosphere by as much as $18\%o$ or more.

The analytical framework for isotopic flux partitioning was originally developed by Yakir and Wang (1996) and has since been applied and adapted by numerous investigators (Bowling et al. 2001; Ogée et al. 2003; Suits et al. 2005; Zhang et al. 2006; Zobitz et al. 2007). Based on the isotopic mass balance of $^{13}CO_2$ a system of equations can be solved analytically to determine the component fluxes. F_N is composed of photosynthetic net assimilation (F_A, equal to gross primary production plus foliar respiration) and non-foliar ecosystem respiration (F_R, equal to total ecosystem respiration minus foliar respiration):

$$F_N = F_A + F_R \qquad (1)$$

By convention, fluxes toward the surface are negative and away from the surface are positive, therefore $F_A < 0$ and $F_R > 0$. A similar equation can be written for the isotopic mass balance of CO_2:

$$\delta_N F_N = F_A \left(\delta_a - \Delta_{canopy} \right) + \delta_R F_R \qquad (2)$$

where δ_N is the isotopic composition of F_N, the isotopic signature of the assimilation flux is expressed as the difference between the isotopic signature of canopy air (δ_a) and the canopy isotopic discrimination (Δ_{canopy}). δ_R is the isotopic signature of F_R.

In order to solve the mass balance, we must describe Δ_{canopy} in terms of component fractionations. First, F_A is related to the CO_2 concentration at the site of carboxylation (c_c) via total conductance (g) according to Fick's Law:

$$F_A = g \left(c_c - c_a \right) \qquad (3)$$

where c_a is the CO_2 concentration of canopy air. Conductance (units of mol m^{-2} s^{-1}) consists of aerodynamic conductance (g_a), canopy conductance (g_s), and mesophyll wall conductance (g_m):

$$\frac{1}{g} = \frac{1}{g_a} + \frac{1}{g_s} + \frac{1}{g_m} \qquad (4)$$

The aerodynamic conductance can be calculated according to Blanken and Black (2004). Bulk stomatal conductance, or canopy conductance may be inverted from the Penman-Monteith equation (e.g. Zhang et al. 2006). The mesophyll conductance is typically assumed constant. For example, Pfeffer and Peisker (1998) estimate mesophyll conductance for corn to be 0.87 mol m^{-2} s^{-1} and literature values for soybean are estimated to be approximately 0.4 mol m^{-2} s^{-1} (Bernacchi et al. 2002; Flexas et al. 2008). Given that mesophyll conductance is dependent upon mixing ratio gradients, the assumption of constant mesophyll conductance is not justified.

However, sensitivity in the flux partitioning approach to this parameter is relatively small. Conductance relationships can be expressed for each step:

$$c_s = c_a + \frac{1}{g_a} F_A \tag{5}$$

where c_s is the CO_2 concentration at the leaf surface. Furthermore:

$$c_i = c_s + \frac{1}{g_s} F_A \tag{6}$$

$$c_c = c_i + \frac{1}{g_m} F_A \tag{7}$$

where c_i is the stomatal CO_2 concentration.

By weighting the individual fractionation factors associated with overall canopy discrimination by the gradient of CO_2 from the canopy air through to the site of enzymatic fixation, Δ_{canopy} can be determined:

$$\Delta_{canopy} = \frac{\Delta_b (c_a - c_s)}{c_a} + \frac{\Delta_s (c_s - c_i)}{c_a} + \frac{(\Delta_{diss} + \Delta_{aq})(c_i - c_c)}{c_a} + \frac{\Delta_f c_c}{c_a} \tag{8}$$

where Δ_b is the fractionation by molecular diffusion across the leaf boundary layer (2.9‰), Δ_s is the fractionation by molecular diffusion through stomata (4.4‰), Δ_{diss} and Δ_{aq} are the fractionations associated with dissolution in mesophyll water (1.1‰) and aqueous phase mesophyll transport (0.7‰), respectively, and Δ_f is the fractionation associated with enzymatic fixation (28.2‰ – C_3 plants) (Farquhar et al. 1989).

Finally, Equations (3)–(8) can be rearranged and substituted into Eqn. (2) resulting in a quadratic solution for F_A,

$$F_A^2 = \left(\frac{\Delta_b}{g_a c_a} + \frac{\Delta_s}{g_s c_a} + \frac{\Delta_{diss} + \Delta_{aq}}{g_m c_a} - \frac{\Delta_f}{g c_a} \right) + F_A \left(\delta_a - \Delta_f - \delta_R \right) = \left(\delta_N - \delta_R \right) F_N \tag{9}$$

Here we assume that F_A, ≤ 0 during the daytime.

3 Carbon Isotope Flux Measurements

The partitioning described above requires a direct measurement of the isoflux ($F_\delta = \delta_N F_N$). To date, quantifying $^{13}CO_2$ exchange between the biosphere and atmosphere has presented a formidable challenge. Such measurements have been rare because of the lack of suitable technologies to quantify the isotope ratio of CO_2 or individual isotopologues at the precision and frequency suitable for scalar flux estimation. A number of methods have been used to estimate the $^{13}CO_2$ fluxes including the flux-gradient (Yakir and Wang 1996; Griffis et al. 2004; Lee et al. 2007); relaxed eddy accumulation (Bowling et al. 1999) and the eddy covariance

(EC)/flask isoflux method (Bowling et al. 2001). More recently, direct measurements of isotopic CO_2 exchange have been made using combined eddy covariance-tunable diode laser absorption spectroscopy (EC-TDLAS) (Griffis et al. 2008) and provide a new opportunity to evaluate the isotopic flux partitioning theory. Here we focus on the development of the flux-gradient and EC methods and point out their main limitations.

3.1 Flux-Gradient

The flux-gradient (K-theory) method is based on Monin-Obukhov similarity and has been shown to work well above the roughness sublayer where horizontal flow is homogeneous and the dispersion is considered to be far-field or random (Monin and Obukhov 1954; Raupach 1989; Simpson 1998). In this particular application of measuring stable isotopic fluxes there are two critical assumptions related to similarity: First, the eddy diffusivities (K_c) of the heavy and light isotopes are assumed identical; Second, the sink/source distributions for the CO_2 isotopologues are assumed identical. Provided that the gradient measurement is made above the roughness sublayer, small differences in sink/source distribution of the two isotopologues (i.e. differences in isotope signature between soil and vegetation fluxes) should have little effect on the flux-gradient measurement due to strong mixing. The flux of the two individual isotopologues is obtained via K-theory (Griffis et al. 2004),

$$F_N^x = -K_c \, \overline{\rho_a} \, \frac{\overline{d^x CO_2}}{dz} + \overline{\rho_a} \, \frac{d}{dt} \int_0^z {}^x CO_2(z) dz \qquad (10)$$

where K_c is the eddy diffusivity of CO_2, $\overline{\rho_a}$ is the molar density of dry air, and $\dfrac{\overline{d^x CO_2}}{dz}$ is the time averaged gradient of the ${}^x CO_2$ isotopologue mixing ratio. In the flux ratio approach the eddy diffusivities are assumed identical so that isotopic composition of F_N can be obtained simply from the ratio of the isotopologue gradients in dimensionless units of per mil (‰),

$$\delta_N^{13} = \left(\frac{\dfrac{F_N^{13}}{F_N^{12}}}{R_{VPDB}} - 1 \right) \qquad (11)$$

where δ_N^{13} is the carbon isotope ratio of the net flux, R_{VPDB} is the heavy to light isotope ratio of the National Institute of Science and Technology's (formerly the National Bureau of Standards) standard NBS-19 reported on the Vienna PeeDee Belemnite scale (VPDB). Although laser spectroscopy techniques are based on measurement of individual isotopologues of CO_2, not their ratios, isotope molar mixing ratios are calibrated against international standard values based on the VPDB scale.

The development of TDLAS technology and its application to isotope studies has greatly increased the capacity for making continuous flux measurements using the gradient approach. Yet, it remains a challenge to resolve small isotopic gradients, especially above rough surfaces such as forests. Given other potential problems related to mismatch in footprint for each inlet, counter-gradient transport, and decoupling of the gradient from the local vertical flux density within the roughness sublayer and non-uniform sink/source distribution (Raupach 1989; Kaimal and Finnigan 1994) there has been significant interest in combining EC and TDLAS.

3.2 Eddy-Covariance

Improved TDLAS frequency response and air sampling has provided a new capacity to measure isotopic fluxes directly with the EC methodology (Griffis et al. 2008). Following Reynolds decomposition and averaging, the isotopic flux (F_N^x) can be determined from,

$$F_N^x = \overline{\rho_a w' c^{x'}} + S_c^x = \overline{\rho_a} \int C_{wc^x}(f) df + S_c^x \tag{12}$$

where, $\overline{w' c^{x'}}$ is the covariance of the vertical wind velocity (w) and the heavy ($^x CO_2$) molar mixing ratio, and S_c^x is the rate of change in $^x CO_2$ storage between the ground and the eddy covariance measurement height. C_{wc^x} is the cospectral density of the fluctuations in vertical wind velocity and isotopic CO_2 mixing ratio. The isotopic signature of the flux can then be obtained as in Eqn. (11).

One of the main advantages of the flux-based approach is that the source footprint is more easily defined than that of the classic mixing model (i.e. Keeling plots) (Griffis et al. 2007). Defining the footprint function of a Keeling plot is challenging because of the influence of air mass (back) trajectory. Further, the footprint function of a concentration measurement is considerably larger than that of a flux so that, as is often the case, surface heterogeneity can complicate the interpretation of the Keeling plot. Finally, unlike mixing models, the EC calculation is not sensitive to the background isotope ratio of the atmosphere.

4 Applications and Results

4.1 High Temporal Resolution of Isotopic CO₂ Exchange

Since 2003, high-frequency (seconds to minutes) isotope measurements have been made within an agricultural ecosystem near St. Paul Minnesota. The site is located on a glacial outwash plain, with well-drained silt loam soil. The area was originally a mixed prairie oak savanna that was brought into cultivation shortly after the Civil War, and, therefore, has a long history of C_3/C_4 vegetation.

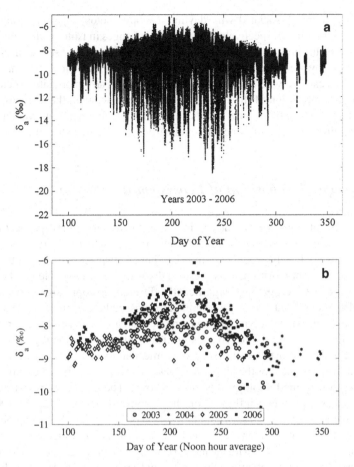

Fig. 2 Isotope measurements at the University of Minnesota field research station measured 3 m above the surface: (**a**) High frequency measurements, (**b**) Noon hour averages.

The primary crops were initially C_3 grains, specifically wheat, oats, and barley, giving way to corn in the early decades of the twentieth century. The area became an agricultural experiment station after World War II and is currently in a *Zea mays* (corn)/*Glycine max* (soybean) rotation. Isotope measurements at the site are averaged over 2–6 min intervals, depending on the calibration scheme, and can be used to obtain a variety of information, such as daily midday average isotopic ratios (Fig. 2), daytime averages, 24-h daily averages, etc. These time series illustrate a number of important features. First, at shorter time-scales (minutes to hours) there is significant variation in the isotopic composition of the surface layer. Much of this variation is driven by boundary layer dynamics, such as the transition from the stable nocturnal boundary layer to a convective well-mixed boundary layer. Second, there is a very strong seasonal influence, including an enrichment in the isotopic signature of the daytime atmosphere and subsequent strong depletion in nighttime isotopic signatures associated with daytime canopy photosynthesis and nighttime ecosystem respiration, respectively. The seasonal changes in isotopic discrimination clearly indicate the timing of canopy development, peak growth and senescence. Third, there exists a physiological influence on the isotopic composition of the atmosphere based on plant canopy type. The high data density provides an opportunity to determine the isotopic composition of the fluxes and to better understand the controls on F_N. Further, these data provide new inputs for model development and validation to help constrain physiological processes and carbon sink/source strength.

4.2   Isotopic Composition of Ecosystem Respiration

Roughly 12% of Earth's land surface is in vegetative agricultural production; an area that is nearly the size of South America (Leff et al. 2004). Considering that these systems are highly productive, they can have a disproportionate influence on the atmosphere's composition. This is especially true from an isotopic point of view because C_3 and C_4 crops have significantly different isotopic signatures. In the Upper Midwest United States the dominate crops include corn (a C_4 plant) and soybean (a C_3 plant) and a typical management strategy is a corn-soybean rotation. Such practices are expected to have a significant influence on the regional carbon balance. Figure 3 illustrates the very dynamic nature of δ_R for an agricultural ecosystem that is typical of the Upper Midwest. In this case, δ_R was measured using the flux-gradient methodology described above. These data demonstrate that phenology has a profound influence on the isotopic signature of F_R. In 2005, C_4 corn was planted on Day of Year 123. For a period of 12–15 days following planting, δ_R was maintained at between −24‰ and −28‰. This isotopic signature is typical of C_3 plants and is associated with the incorporation of the previous year's crop residue upon tillage just prior to planting. Soon after leaf emergence, these values became enriched in ^{13}C relative to the bulk soil organic carbon ($\delta^{13}C° \approx -18‰$) indicating that δ_R is strongly controlled by autotrophic respiration. As the seeds germinated and leaves emerged, δ_R showed a steady $^{13}CO_2$ enrichment, reaching values as high as −10‰ during peak growth. Maximum leaf area index (LAI) of approximately 5.4 was achieved for this corn canopy on day of year 224. A similar trend in the opposite direction was observed during years in C_3 crop rotation, when δ_R becomes increasingly depleted down to approximately −23‰ during peak canopy development. Soybean (C_3) planting occurred on Day of Year 145 in 2006. Leaf emergence soon after, resulted in decreasing isotopic signatures, reflective of C_3 physiology. Maximum LAI of approximately 5.3 was observed for this soybean canopy on day of year 218. This indicates that land-use choices impact the carbon chemistry of the atmosphere in profound ways due to the plant metabolic pathway and the phenologic developmental timing.

The seasonal shift from respiration of soil organic matter during the fallow season, to respiration coupled with plant photosynthesis is important to understanding source or sink strength of the agricultural landscape. Griffis et al. (2005b) combined the isotopic mass balance principle with a numerical optimization

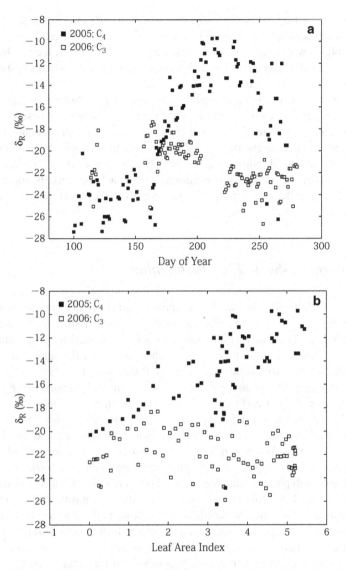

Fig. 3 Isotopic composition of nighttime total ecosystem respiration (**a**) as a function of Day of Year, and (**b**) as a function of LAI. Short gaps in the time series data occurred in 2006 due to instrumental failure. 2006 plant canopy is *Glycine max*; 2005 plant canopy is *Zea mays*.

method to determine the relative importance of autotrophic and heterotrophic respiration during the C_4 phase of the rotation by estimating that R_a accounted for nearly 44% of F_R on average and reached a maximum of 59% during peak growth. These observations were in excellent agreement with experiments conducted by Högberg et al. (2001) and Rochette et al. (1991). Further, this optimization scheme was used to constrain the isotopic signatures of R_a and R_h and demonstrated that R_h showed strong seasonal variation. Thus, it is clear that phenologic development of different plants within agricultural rotation may significantly influence the timing and magnitude of δ_R directly through F_{Ra} and indirectly as F_{Rh} adjusts to the supply of recently fixed root exudates.

The implications of the rapid equilibration between the isotopic composition of F_{Ra} and F_{Rh} are important on a number of fronts. First, it is difficult to separate these contributions using a simple two-end member mixing model. Second, bulk soil organic matter is not necessarily a good indicator of the isotopic composition of F_R. Third, the rapid equilibration suggests that phenology may in fact undermine our ability to employ the isotopic flux partitioning approach, because the methodology will fail when the differences between δ_A and δ_R are below the isotopic flux measurement resolution (Zhang et al. 2006).

4.3 Ecosystem-Scale Flux Partitioning

Zhang et al. (2006) demonstrated that isotopic partitioning, using a flux gradient approach, during a C_4 crop phase worked best during the early and late growing season when D was large. Here, we revisit isotopic partitioning considering a C_3 growing phase using two different methodological approaches: improved flux-gradient sampling system and direct measurement of the isotopic fluxes using the EC technique.

Figure 4 highlights phenologic patterns in the diurnal response of F_N. As the canopy developed from early growing season through peak canopy development, daytime F_N became more negative, which indicates a strong CO_2 flux into the canopy. The onset of senescence was typified by a return to less negative F_N, as the photosynthetic flux decreased. When trends in F_N are reported as a function of LAI, one can clearly observe the strong coupling of plant development and ecosystem exchange. In the early growth phase, F_N during midday hours (10–15 h) showed a nearly linear relationship with LAI. Additionally, there was little variability in measured flux, indicating that environmental variables within a normal range are less important during this growth stage. It is likely that the nature of the monoculture contributes to this effect, in that plants have maximum light availability at this time prior to canopy closure and are not under strict competition for other resources, such as nutrients and water. In regions dominated by agriculture, this effect of timing specificity and strength of early plant development is significant. As the canopy continues to mature, midday F_N continues to increase, although the increase is smaller with respect to LAI. At peak canopy development, F_N reached its maximum uptake, but was highly variable. As would be expected, F_N decreased with the onset of senescence, owing to reduced photosynthetic capacity.

Trends in F_N were mimicked by that of the isoflux, but in the opposite direction, since isoflux is a product of a negative flux and a negative isotopic signature (Fig. 5). One can observe the timing of morning photosynthetic onset throughout the season by the sharp rise in the isoflux, which occurred earlier in the day during the earlier

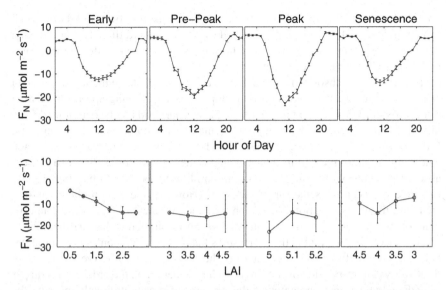

Fig. 4 Ensemble diurnal variations of net ecosystem exchange during critical plant development stages; Early development stage (LAI≤3), Day of Year 176–190, 2006, Pre-Peak development stage (3 < LAI < 5), Day of Year 190–202, 2006, Peak development stage (LAI ≥ 5), Day of Year 221–235, 2006, and Onset of Senescence (5 > LAI > 3), Day of Year 235–250, 2006. Error bars are standard deviation of the ensembled time period mean (*top row*) and LAI from 10–15 h (*bottom row*).

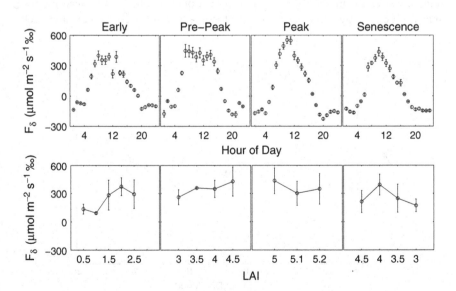

Fig. 5 Ensemble diurnal variations of canopy isoflux during critical plant development stages using the gradient approach. *Top row* presented as a time series and bottom row as a function of LAI.

growth periods, owing to earlier sunrise. The largest isoflux occurred during peak canopy growth and was typically achieved by noon standard time. It is at this time that the canopy exerts a strong isotopic forcing, which influences the isotopic composition of CO_2 in the atmosphere.

To estimate the absolute magnitude of the daytime photosynthetic flux, we can evaluate F_N and the isoflux using the isotopic flux partitioning approach. Here, a critical assumption is made that daytime δ_R is constant and can be determined from nighttime measurements. By doing so, F_N may be partitioned into its component fluxes of F_A and F_R according to Eqns. (1) and (9). We used a flux gradient approach where δ_R is determined from adjacent nighttime CO_2 isotopic fluxes (δ_N in the absence of photosynthesis). This assumption allowed us to both partition F_N and calculate a key canopy-scale parameter, Δ_{canopy}. However, there are problems associated with this δ_R assumption. First, nighttime measurement of δ_R represent an estimate of the isotopic signature of total ecosystem respiration (foliar and nonfoliar), whereas, isotope partitioning theory requires that F_R represent nonfoliar respiration. Second, through direct measurement of nighttime values, we observe variability in δ_R of ecosystem respiration over a single nighttime period on the order of several per mil depending on environmental conditions. Since we cannot directly measure the variability in δ_R at the canopy scale during the day, we must assume a constant daytime δ_R value from nighttime measurements. Given that there is observed variability in δ_R (Fig. 6), it is clear that the assumption of constant daytime δ_R is violated. Third, as was previously discussed, several researchers have shown post-photosynthetic isotope fractionation occurs (Duranceau et al. 1999; Ghashghaie et al. 2001; Tcherkez et al. 2003; Badeck et al. 2005; Bathellier et al. 2008). Owing to our lack of understanding of these processes and their specific timing, post-photosynthetic fractionations pose significant challenges to the combined isotope and micrometeorological approach.

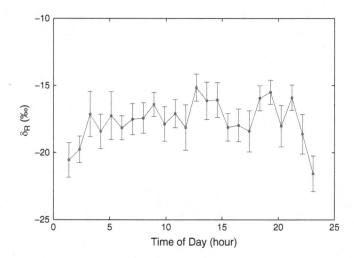

Fig. 6 Isotopic diurnal variability of δ_R measured on bare soil using automated chambers coupled to a TDLAS system. Data were ensemble averaged over a 9 day period.

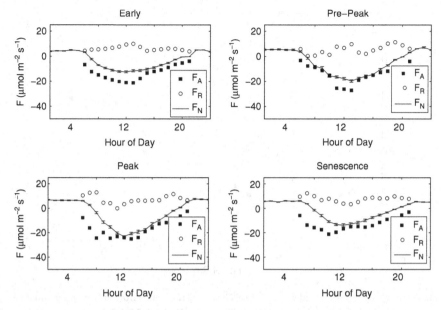

Fig. 7 Ensemble diurnal variations of photosynthesis and respiration from isotopic partitioning using the gradient method.

Improvements in technology, such as automated chambers that measure ecosystem respiration 24 h per day and EC-TDLAS where isoflux is directly measured, are making advances to better understand controls on δ_R.

Despite concerns regarding assumptions inherent in the isotope partitioning theory, this approach may yield robust estimates of F_A and F_R (Fig. 7). This method is most robust when looking at ensemble averages of phenologic growth stages; ensemble averaging reduces noise and improves signal detection. During some growth stages, we estimated that F_A was large in the morning hours relative to F_N.

Further, in some cases midday and/or early afternoon hours showed a decrease in F_A relative to F_N. This phenomenon, or midday anomaly, has been observed by others in similar ecosystem scale studies (e.g. Bowling et al. 2003; Zhang et al. 2006; Zobitz et al. 2007), and is often attributed to failure in the isotopic approach. As turbulent intensity increases toward midday, the isotopic gradient is reduced, and can become too small to be resolved by the measurement technique. This issue, however, may be overcome by new EC-TDLAS. Figure 8 demonstrates isotopic partitioning during the peak growth stage in 2006 using EC-TDLAS. Here we observed a relatively constant offset in the estimated F_A from F_N measurements, with little evidence of a midday anomaly. The F_A estimates from the gradient method and EC-TDLAS (Fig. 9) showed that the flux-gradient method progressively underestimated F_A compared to EC-TDLAS. This underestimation may be an artifact of small gradients relative to the instrument noise, or may be associated with footprint differences. Further evaluation of the two methods is currently underway to better understand these factors.

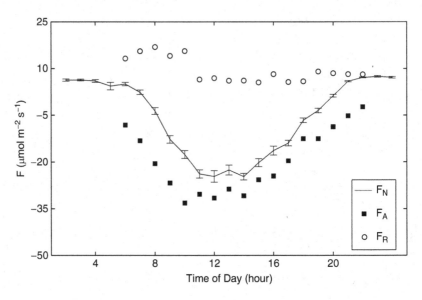

Fig. 8 Ensemble diurnal variations of photosynthesis and respiration from the isotopic partitioning of a C_3 canopy using the EC-TDLAS method during the Peak growth stage, Day of Year 221–235, 2006.

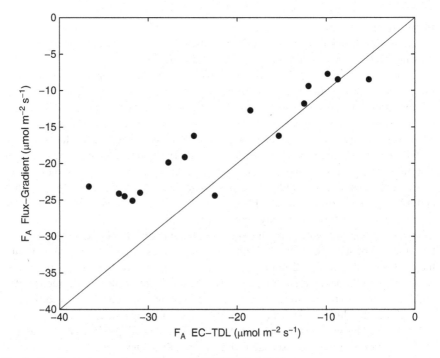

Fig. 9 F_A estimated using the gradient and EC-TDLAS methods.

Another difficulty associated with the isotope flux partitioning approach can partly be attributed to similarity between the δ_A and δ_R. $\mathcal{D}$ values near zero can result in failure of the isotope partitioning method. For our study, the conversion of land from a C_3 to a C_4 ecosystem imposes a step change in canopy isotopic discrimination and results in a progressive change in δ_R related to the decomposition of recently assimilated carbon and accumulated soil carbon pool, which then influences $\mathcal{D}$ estimation. Griffis et al. (2005b) found that $\mathcal{D}$ was greatest during the early growing phase and diminished over the course of the growing season. Zhang et al. (2006) and Zobitz et al. (2007) noted that the sign of $\mathcal{D}$ was often negative (contrary to expectation), indicating that δ_A was more enriched than δ_R. This trend has been observed by others (Bowling et al. 2003; Miller et al. 2003; Aranibar et al. 2006; Zhang et al. 2006; Zobitz et al. 2007), and is attributed to the enrichment of canopy respiration compared to soil respiration (Bowling et al. 2005), respiration of enriched phloem sugars contributing to the assimilation signature (Scartazza et al. 2004), and the possibility of unknown isotopic effects associated with respiration as discussed earlier. Although, the latter two uncertainties are difficult to address, we speculate that continued measurements using automated chambers and EC-TDLAS may be able to shed light on these issues.

A key parameter derived from the isotopic partitioning approach is canopy isotopic discrimination (Δ_{canopy}). This parameter is often applied in regional and global inverse modeling studies, where understanding plant photosynthetic processing of carbon is critical, for the purpose of closing the carbon budget (e.g. Ciais et al. 1995; Fung et al. 1997). As such, robust estimation of Δ_{canopy} is important, because small changes to this parameter on the order of a few per mil can influence, for example, the inferred global terrestrial carbon sink by as much as 25% (Randerson et al. 2002; Fung et al. 1997). During the peak growing season, Δ_{canopy} shows a morning rise and a late afternoon drop in isotopic discrimination (Fig. 10). The morning increase in Δ_{canopy} is attributed to the changing conductance associated with environmental factors and the change in CO_2 concentration gradient from canopy air to chloroplast as nighttime CO_2 rich boundary layer conditions give way to daytime turbulence. Similarly, the late afternoon decrease in Δ_{canopy} results from environmentally induced increased resistance of CO_2 transport to the site of fixation. However, variability associated with methodological differences make accurate estimation of this parameter difficult.

The diurnal range in Δ_{canopy} is large and can span several per mil during the peak growing season, when Δ_{canopy} reaches its maximum of approximately 18‰ (Fig. 10). These daytime values are in agreement with similar studies over natural C_3 vegetation (e.g. Bowling et al. 2005; Zobitz et al. 2007). Daily average Δ_{canopy} values indicate that as the canopy reaches maximum development, isotopic discrimination increases and then remains constant (data not shown). The timing of this daily average maximum coincides with maximum leaf area at the site (approximately Day of Year 224). At this point, the plant canopy has completed its physical development such that CO_2 can be most efficiently utilized for additional metabolic needs (vegetative growth, flowering and seed production). Interestingly, isotopic discrimination does not decrease as the canopy begins to senesce, suggesting that the structures involved in delivering CO_2 to the chloroplast are not degraded until photosynthesis ceases.

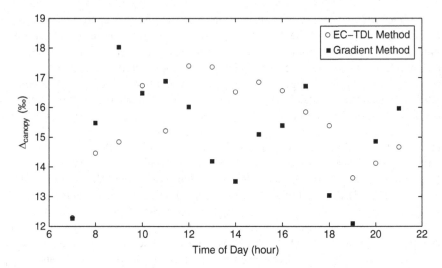

Fig. 10 Canopy discrimination (Δ_{canopy}) during peak growth stage using the gradient and EC-TDLAS methods.

Through continued research and refining of ecosystem-scale flux partitioning, we may begin to address key questions specifically related to how carbon is cycled during different phenologic growth stages and how we can use this information to better manage our agricultural systems. One concern that exists is how the carbon source/sink dynamic will change in agricultural systems given a warmer climate and changing land use. Under traditional management, agricultural systems are likely to source CO_2 to the atmosphere as active soil respiration fallow seasons become longer. Employing a combined stable isotope and micrometeorological approach to study systems under different management strategies, such as reduced tillage and cover cropping, will allow researchers to find solutions to management questions. For example, will covercrops change carbon residence time in agricultural ecosystems (as evidenced by $\mathcal{D}$ or δ_R), and/or will management decisions affect the flux of respired carbon to the atmosphere.

5   Conclusions

Stable isotope and micrometeorological techniques are important tools for studying biosphere–atmosphere exchange processes. Through use of these techniques, we can better evaluate the impact of climate variation, changing land surface properties, and phenology on the isotopic composition of the atmosphere. Such forensic tools are essential for detecting and understanding changes in our atmosphere. From an agricultural and land use change perspective these tools are particularly powerful because of the large isotopic differences observed between C_3 and C_4 crops and the large fluxes associated with these highly productive ecosystems. Our increasing capacity to measure isotopic fluxes on a routine basis should continue to provide new insights regarding the impacts of natural and anthropogenic change on the atmosphere.

In this chapter we demonstrated using novel techniques that respiration during plant development is strongly coupled to canopy assimilated carbon, specific to plant metabolic pathway. Additionally, plant development influenced the strength and change of net biosphere–atmosphere exchange as well as the isotopic signature of CO_2 in the atmosphere. Further, we demonstrated that using a combined isotope and micrometeorological approach allows the net exchange to be partitioned into the gross fluxes of photosynthesis and respiration. However, isotopic partitioning is limited by measurement uncertainty in key parameters and, in particular, by the estimation of δ_R. We believe that further methodological and instrumental development, such as EC-TDLAS, as well as other established and emerging techniques, such as isotopic analysis coupled with automated soil respiration chambers, will significantly advance the efficacy of this technique. Through continuous high-resolution CO_2 stable isotope observations we will have the capacity to better detect and understand phenologic response to change.

Acknowledgements We express our sincere thanks to John Baker who has had a significant impact on the development of this work. We also thank numerous technicians and students who have provided field assistance including Matt Erickson, Bill Breiter, Jim Brozowski, Travis Bavin, Jennifer Corcoran, Jeremy Smith, Kyounghee Kim and Lisa Welp. We thank Dr. T. A. Black for helping us with the integration of the automated chamber system with the TDL system. Financial support for this project was provided by the National Science Foundation, ATM-0546476 and the Office of Science (BER), U.S. Department of Energy, Grant No. DE-FG02-03ER63684.

References

Aranibar, J.N., Berry, J.A., Riley, W.J., Pataki, D.E., Law, B.E. and Ehleringer, J.R. (2006) Combining meteorology, eddy fluxes, isotope measurements, and modeling to understand environmental controls of carbon isotope discrimination at the canopy scale. Global Change Biol. 12, 710–730.

Badeck, F.-W., Tcherkez, G., Nogués, S., Piel, C. and Ghashghaie, J. (2005) Post-photosynthetic fractionation of stable carbon isotopes between plant organs - a widespread phenomenon. Rapid Commun. Mass Sp. 19, 1381–1391.

Baldocchi, D.D. (2008) Breathing of the terrestrial biosphere: lessons learned from a global network of carbon dioxide flux measurement systems. Aust. J. Bot. 56, 1–26.

Baldocchi, D.D. and Bowling, D.R. (2003) Modeling discrimination of $^{13}CO_2$ above and within a temperate broad-leaved forest canopy on hourly to seasonal time scales. Plant Cell Environ. 26, 231–244.

Bathellier, C., Badeck, F.-W., Couzi, P., Harscoët, S., Mauve, C. and Ghashghaie, J. (2008) Divergence in $\delta^{13}C$ of dark respired CO_2 and bulk organic matter occurs during the transition between heterotrophy and autotrophy in *Phaseolus vulgaris* plants. New Phytol. 177, 406–418.

Bernacchi, C.J., Portis, A.R., Nakano, H., von Caemmerer, S. and Long, S.P. (2002) Temperature response of mesophyll conductance. implications for the determination of Rubisco enzyme kinetics and for limitations to photosynthesis *in vivo*. Plant Physiol. 130, 1992–1998.

Blanken, P.D. and Black, T.A. (2004) The canopy conductance of a boreal aspen forest, Prince Albert National Park, Canada. Hydrol. Proc. 18, 1561–1578.

Bowling, D.R., Baldocchi, D.D. and Monson, R.K. (1999) Dynamics of isotopic exchange of carbon dioxide in a Tennessee deciduous forest. Global Biogeochem. Cycles 13, 903–922.

Bowling, D.R., Burns, S.P., Conway, T.J., Monson, R.K. and White, J.W.C. (2005) Extensive observations of CO_2 carbon isotope content in and above a high-elevation subalpine forest. Global Biogeochem. Cycles 19 15, 83.

Bowling, D.R., McDowell, N.G., Bond, B.J., Law, B. and Ehleringer, J.R. (2002) ^{13}C content of ecosystem respiration is linked to precipitation and vapor pressure deficit. Oecologia 131, 113–124.

Bowling, D.R., Sargent, S., Tanner, B. and Ehleringer, J.R. (2003) Tunable diode laser absorption spectroscopy for stable isotope studies of ecosystem-atmosphere CO_2 exchange. Agric. For. Meteorol. 118, 1–19.

Bowling, D.R., Tans, P.P. and Monson, R.K. (2001) Partitioning net ecosystem carbon exchange with isotopic fluxes of CO_2. Global Change Biol. 7, 127–145.

Cerling, T.E., Harris, J.M., MacFadden, B.J., Leakey, M.G., Quade, J., Eisenmann, V. and Ehleringer, J.R. (1997) Global vegetation change through the Miocene/Pliocene boundary. Nature 389, 153–158.

Ciais, P., Tans, P.P., White, J.W.C., Trolier, M., Francey, R.J., Berry, J.A., Randall, D.R., Sellers, P.J., Collatz, J.G. and Schimel, D.S. (1995) Partitioning of ocean and land uptake of CO_2 as inferred by $\delta^{13}C$ measurements from the NOAA Climate Monitoring and Diagnostics Laboratory Global Air Sampling network. J. Geophys. Res. 100, 5051–5070.

Duranceau, M., Ghashghaie, J., Badeck, F., Deleens, E. and Cornic, G. (1999) $\delta^{13}C$ of CO_2 respired in the dark in relation to $\delta^{13}C$ of leaf carbohydrates in Phaseolus vulgaris L. under progressive drought. Plant Cell Environ. 22, 515–523.

Duranceau, M., Ghashghaie, J. and Brugnoli, E. (2001) Carbon isotope discrimination during photosynthesis and dark respiration in intact leaves of Nicotiana sylvestris: comparison between wild type and mitochondrial mutant plants. Aust. J. Plant Physiol. 28, 65–71.

Ekblad, A., Bostrom, B., Holm, A. and Comstedt, D. (2005) Forest soil respiration rate and $\delta^{13}C$ is regulated by recent above ground weather conditions. Oecologia 143, 136–142.

Ekblad, A. and Högberg, P. (2001) Natural abundance of ^{13}C in CO_2 respired from forest soils reveals speed of link between tree photosynthesis and root respiration. Oecologia 127, 305–308.

Farquhar, G.D. (1983) On the nature of carbon isotope discrimination in C_4 species. Aust. J. Plant Physiol. 10, 205–226.

Farquhar, G.D., Ehleringer, J.R. and Hubick, K.T. (1989) Carbon isotope discrimination and photosynthesis. Annu. Rev. Plant Physiol. Mol. Biol. 40, 503–537.

Flanagan, L.B., Brooks, J.R., Varney, G.T., Berry, S.C. and Ehleringer, J.R. (1996) Carbon isotope discrimination during photosynthesis and the isotope ratio of respired CO_2 in boreal forest ecosystems. Global Biogeochem. Cycles 10, 629–640.

Flanagan, L.B. and Ehleringer, J.R. (1998) Ecosystem-atmosphere CO_2 exchange: interpreting signals of change using stable isotope ratios. Trends Ecol. Evol. 13, 10–14.

Flexas, J., Ribas-Carbó, M., Diaz-Espejo, A., Galmés, J. and Medrano, H. (2008) Mesophyll conductance to CO_2: current knowledge and future prospects. Plant Cell Environ. 31, 602–621.

Francey, R.J., Allison, C.E., Etheridge, D.M., Trudinger, C.M., Enting, I.G., Leuenberger, M., Langenfelds, R.L., Michel, E. and Steele, L.P. (1999) A 1000-year high precision record of $\delta^{13}C$ in atmospheric CO_2. Tellus B 51, 170–193.

Fung, I., Field, C.B., Berry, J.A., Thompson, M.V., Randerson, J.T., Malmstrom, C.M., Vitousek, P.M., Collatz, G.J., Sellers, P.J., Randall, D.A., Denning, A.S., Badeck, F. and John, J. (1997) Carbon 13 exchanges between the atmosphere and biosphere. Global Biogeochem. Cycles 11, 507–533.

Gessler, A., Keitel, C., Kodama, N., Weston, C., Winters, A.J., Keith, H., Grice, K., Leuning, R. and Farquhar, G.D. (2007) $\delta^{13}C$ of organic matter transported from leaves to the roots in Eucalyptus delegatensis: short-term variations and relation to respired CO_2. Funct. Plant Biol. 34, 692–706.

Gessler, A., Schrempp, S., Matzarakis, A., Mayer, H., Rennenberg, H. and Adams, M.A. (2001) Radiation modifies the effect of water availability on the carbon isotope composition of beach (Fagus sylvatica). New Phytol. 150, 653–664.

Ghashghaie, J., Badeck, F.-W., Lanigan, G., Nogués, S., Tcherkez, G., Deleens, E., Cornic, G. and Griffiths, H. (2003) Carbon isotope fractionation during dark respiration and photorespiration in C_3 plants. Phytochem. Rev. 2, 145–161.

Ghashghaie, J., Duranceau, M., Badeck, F.W., Cornic, G., Addeline, M.T. and Deleens, E. (2001) $\delta^{13}C$ of CO_2 respired in the dark in relation to $\delta^{13}C$ of leaf metabolites: comparison between *Nicotiana sylvestris* and *Helianthus annuus* under drought. Plant Cell Environ. 24, 505–515.

Griffis, T.J., Baker, J.M., Sargent, S.D., Tanner, B.D. and Zhang, J. (2004) Measuring field-scale isotopic CO_2 fluxes with tunable diode laser absorption spectroscopy and micrometeorological techniques. Agric. For. Meteorol. 124, 15–29, 38.

Griffis, T.J., Baker, J.M. and Zhang, J. (2005a) Seasonal dynamics and partitioning of isotopic CO_2 exchange in C_3/C_4 managed ecosystem. Agric. For. Meteorol. 132, 1–19.

Griffis, T.J., Lee, X., Baker, J.M., Sargent, S.D. and King, J.Y. (2005b) Feasibility of quantifying ecosystem-atmosphere $C^{18}O^{16}O$ exchange using laser spectroscopy and the flux-gradient method. Agric. For. Meteorol. 135, 44–60.

Griffis, T.J., Sargent, S.D., Baker, J.M., Lee, X., Tanner, B.D., Greene, J., Swiatek, E. and Billmark, K. (2008) Direct measurement of biosphere–atmosphere isotopic CO_2 exchange using the eddy covariance technique. J. Geophys. Res. 113, D08304.

Griffis, T.J., Zhang, J., Baker, J.M., Kljun, N. and Billmark, K. (2007) Determining carbon isotope signatures from micrometeorological measurements: Implications for studying biosphere-atmosphere exchange processes. Bound.-Lay. Meteorol. 123, 295–316.

Hobbie, E.A. and Werner, R.A. (2004) Intramolecular, compound-specific, and bulk carbon isotope patterns in C_3 and C_4 plants: a review and synthesis. New Phytol. 161, 371–385.

Högberg, P., Nordgren, A., Buchmann, N., Taylor, A.F.S., Ekblad, A., Högberg, M.N., Nyberg, G., Ottosson-Lofvenius, M. and Read, D.J. (2001) Large-scale forest girdling shows that current photosynthesis drives soil respiration. Nature 411, 789–792.

Kaimal, J. and Finnigan, J. (1994) *Atmospheric Boundary Layer Flows: Their Structure and Measurement.* Oxford University Press, Oxford, pp. 289.

Keitel, C., Matzarakis, A., Rennenberg, H. and Gessler, A. (2006) Carbon isotopic composition and oxygen isotopic enrichment in phloem and total leaf organic matter of European beech (Fagus sylvatica L.) along a climate gradient. Plant Cell Environ. 29, 1492–1507.

Klumpp, K., Schäufele, R., Lötscher, M., Lattanzi, F.A., Feneis, W. and Schnyder, H. (2005) C-isotope composition of CO_2 respired by shoots and roots: fractionation during dark respiration? Plant Cell and Environ. 28, 241–250.

Knohl, A., Werner, R.A., Brand, W.A. and Buchmann, N. (2005) Shortterm variations in $\delta^{13}C$ of ecosystem respiration reveals link between assimilation and respiration in a deciduous forest. Oecologia 142, 70–82.

Leavitt, S.W. and Long, A. (1986) Stable carbon isotope variability in tree foliage and wood. Ecology 67, 1002–1010.

Lee, X.H., Kim, K. and Smith, R. (2007) Temporal variations of the $^{18}O/^{16}O$ signal of the whole-canopy transpiration in a temperate forest. Global Biogeochem. Cycles 21, GB3013.

Leff, B., Ramankutty, N. and Foley, J.A. (2004) Geographic distribution of major crops across the world. Global Biogeochem. Cycles 18, GB1009.

Lloyd, J., Kruijt, B., Hollinger, D.Y., Grace, J., Francey, R.J., Wong, S.C., Kelliher, F.M., Miranda, A.C., Farquhar, G.D., Gash, J.H.C., Vygodskaya, N.N., Wright, I.R., Miranda, H.S. and Schulze, E.D. (1996) Vegetation effects on the isotopic composition of atmospheric CO_2 at local and regional scales: Theoretical aspects and a comparison between rain forest in amazonia and a boreal forest in Siberia. Aust. J. Plant Physiol. 23, 371–399.

Miller, J.B., Tans, P.P., White, J.W.C., Conway, T.J. and Vaughn, B.W. (2003) The atmospheric signal of terrestrial carbon isotopic discrimination and its implication for partitioning carbon fluxes. Tellus 55B, 197–206.

Monin, A.S. and Obukhov, A.M. (1954) Basic laws of turbulent mixing near the ground. Trudy Geofiz. Ins. Akad. Nauk SSSR 24, 163–187. (In Russian)

Mortazavi, B., Chanton, J.P., Prater, J.L., Oishi, A.C., Oren, R. and Katul, G.G. (2005) Temporal variability in ^{13}C of respired CO_2 in a pine and a hardwood forest subject to similar climatic conditions. Oecologia 142, 57–69.

Ogée, J., Peylin, P., Ciais, P., Bariac, T., Brunet, Y., Berbigier, P., Roche, C., Richard, P., Bardoux, G. and Bonnefond, J.M. (2003) Partitioning net ecosystem carbon exchange into net assimilation and respiration using ($^{13}CO_2$) measurements: A cost-effective sampling strategy. Global Biogeochem. Cycles 17, GB1070.

Ometto, J.P.H., Flanagan, L.B., Martinelli, L.A., Moreira, M.Z., Higuchi, N. and Ehleringer, J.R. (2002) Carbon isotope discrimination in forest and pasture ecosystems of the Amazon basin, Brazil. Global Biogeochem. Cycles 16, GB1109.

Pataki, D.E., Ehleringer, J.R., Flanagan, L.B., Yakir, D., Bowling, D.R., Still, C.J., Buchmann, N., Kaplan, J.O. and Berry, J.A. (2003) The application and interpretation of Keeling plots in terrestrial carbon cycle research. Global Biogeochem. Cycles 17, 1022.

Pfeffer, M. and Peisker, M. (1998) CO_2 gas exchange and phosphoenolpyruvate carboxylase activity in leaves of Zea mays L. Photosynth. Res. 58, 281–291.

Randerson, J.T., Chapin, F.S., Harden, J.W., Neff, J.C. and Harmon, M.E. (2002) Net ecosystem production: A comprehensive measure of net carbon accumulation by ecosystems. Ecol. Appl. 12, 937–947.

Raupach, M.R. (1989) Applying lagrangian fluid-mechanics to infer scalar source distributions from concentration profiles in plant canopies. Agric. For. Meteorol. 47, 85–108.

Rochette, P., Pattey, E., Desjardins, R.L., Dwyer, L.M., Stewart, D.W. and Dube, P.A. (1991) Estimation of maize Zea Mays L. canopy conductance by scaling up leaf stomatal conductance. Agric. For. Meteorol. 54, 241–261.

Scartazza, A., Mata, C., Matteucci, G., Yakir, D., Moscatello, S. and Brugnoli, E. (2004) Comparisons of $\delta^{13}C$ of photosynthetic products and ecosystem respiratory CO_2 and their responses to seasonal climate variability. Oecologia 140, 340–351.

Scholze, M., Ciais, P. and Heimann, M. (2008) Modeling terrestrial ^{13}C cycling: Climate, land use and fire. Global Biogeochem. Cycles 22, GB1009.

Scholze, M., Kaplan, J.O., Knorr, W. and Heimann, M. (2003) Climate and interannual variability of the atmosphere-biosphere $^{13}CO_2$ flux. Geophys. Res. Lett. 30, 1097.

Simpson, T.J. (1998) Application of isotopic methods to secondary metabolic pathways. Biosynthesis 195, 1–48.

Suits, N.S., Denning, A.S., Berry, J.A., Still, C.J., Kaduk, J., Miller, J.B. and Baker, I.T. (2005) Simulation of carbon isotope discrimination of the terrestrial biosphere. Global Biogeochem. Cycles 19, GB1017.

Tcherkez, G., Farquhar, G.D., Badeck, F. and Ghashghaie, J. (2004) Theoretical considerations about carbon isotope distribution in glucose of C3 plants. Funct. Plant Biol. 31, 857–877.

Tcherkez, G., Nogués, S., Bleton, J., Cornic, G., Badeck, F.W. and Ghashghaie, J. (2003) Metabolic origin of carbon isotope composition of leaf dark-respired CO_2 in French bean. Plant Physiol. 131, 237–244.

Werner, R.A., Unger, S., Pereira, J.S., Maia, R., David, T.S., Kurz-Besson, C., David, J.S. and Maguas, C. (2006) Importance of short-term dynamics in carbon isotope ratios of ecosystem respiration ($\delta^{13}C$) in a Mediterranean oak woodland and linkage to environmental factors. New Phytol. 172, 330–346.

Xu, C.Y., Lin, G.H., Griffin, K.L. and Sambrotto, R.N. (2004) Leaf respiratory CO_2 is ^{13}C-enriched relative to leaf organic components in five species of C_3 plants. New Phytol. 163, 499–505.

Yakir, D. (2004) The stable isotopic composition of atmospheric CO_2, In: The Atmosphere: Treatise on Geochemistry. Vol.4. Elsevier, Amsterdam, pp. 175–212.

Yakir, D. and Wang, X.F. (1996) Fluxes of CO_2 and water between terrestrial vegetation and the atmosphere estimated from isotope measurements. Nature 380(6574), 515–517.

Zhang, J., Griffis, T.J. and Baker, J.M. (2006) Using continuous stable isotope measurements to partition net ecosystem CO_2 exchange. Plant Cell Environ. 29, 483–496.

Zobitz, J.M., Burns, S.P., Ogée, J., Reichstein, M. and Bowling, R. (2007) Partitioning net ecosystem exchange of CO_2: A comparison of a Bayesian/isotope approach to environmental regression methods. J. Geophys. Res. 112, G03013.11.1

Part II
Biological Feedbacks

Phenology of Plant Production in the Northwestern Great Plains: Relationships with Carbon Isotope Discrimination, Net Ecosystem Productivity and Ecosystem Respiration

Lawrence B. Flanagan

Abstract This chapter represents a case study of seasonal and annual variation in above-ground biomass production in a northern temperate grassland during a 9 year period (1998–2006). I describe the relationship between variation in biomass production and the major environmental factor controlling this variation, precipitation inputs. Annual peak biomass production and leaf $\delta^{13}C$ values were negatively correlated, and this relationship was consistent with lower biomass production being predominantly controlled by reduced water availability. Two patterns in the relationship between peak above-ground biomass and annual net ecosystem productivity (NEP) were observed. First, there was a strong linear relationship between biomass and NEP in 4 years when peak leaf area production was held significantly below the maximum by low water availability. Second, among 4 years with similar maximum LAI values, there was wide variation in NEP largely due to differences in the length of time leaf tissue was photosynthetically active. Water availability also had a significant effect on ecosystem respiration because it controlled grassland phenology. The product of plant biomass and soil water content was a good proxy for estimating ecosystem respiration because of relationships with both autotrophic and heterotrophic activities. Seasonal variation in grassland respiration was more strongly linked to carbon substrate availability and soil moisture than to shifts in temperature.

1 Introduction

Phenology is the study of the timing of biological events and the abiotic and biotic factors that regulate the timing (Jolly et al. 2005). In grassland ecosystems plant growth and development can vary tremendously within and among growing seasons. This fluctuation in plant growth in grasslands is largely controlled by the amount

L.B. Flanagan (✉)
Department of Biological Sciences, University of Lethbridge, Lethbridge, AB, Canada
e-mail: larry.flanagan@uleth.ca

A. Noormets (ed.), *Phenology of Ecosystem Processes*,
DOI 10.1007/978-1-4419-0026-5_7, © Springer Science+Business Media, LLC 2009

of precipitation received (Knapp and Smith 2001; Knapp et al. 2006). In addition, grasslands have very large asymmetric responses to variation in precipitation. Plant productivity increases in wet years are much more pronounced than reductions in productivity during dry years (Knapp and Smith 2001). This seasonal and annual variation in the phenology of plant growth and development will have significant effects on ecosystem carbon and water fluxes (Churkina et al. 2005). For example, maximum leaf area production is one major factor controlling photosynthesis, transpiration and other components of ecosystem carbon and water budgets (Chapin et al. 2002; van Dijk et al. 2005). The timing of leaf-out and leaf-senescence regulates the length of the photosynthetic season in ecosystems dominated by deciduous plants. Ecosystem respiration is also affected by the phenology of plant growth and development (Xu et al. 2004; Flanagan and Johnson 2005). Photosynthetically-produced carbohydrates that are exuded from plant roots are a major source of carbon that affects the respiration rate of soil microbes. The rate and timing of the release of these respiratory substrates and the growth of plant roots strongly control the total ecosystem respiration rate (Xu et al. 2004).

While precipitation may be the most important abiotic factor controlling biological activity in grasslands, other environmental factors like temperature sums (e.g. growing degree days) and changes in photoperiod and light quality can also strongly influence plant phenology (Jolly et al. 2005). Short-term episodic events, like temperature extremes (spring snowfalls, frost events, summer heat waves) or pulse inputs of rain or floods, also have the potential to modify the dynamics of plant growth and ecosystem physiology (Xu et al. 2004; Ciais et al. 2005). The complete effects of unusual or episodic events also may not appear immediately, but only become apparent as "carry-over effects" from one growing season to another.

This chapter represents a case study of seasonal and annual variation in above-ground biomass production in a northern temperate grassland during a 9 year period (1998–2006). I describe the relationship between variation in biomass production and the major environmental factor controlling this variation, precipitation inputs. Measurements of the carbon isotope composition of plant tissue illustrate the close links between water-use and carbon gain during photosynthesis in this semi-arid ecosystem. In addition, I demonstrate the consequences of variation in the amount and timing of plant growth for net ecosystem production and ecosystem respiration. Some examples of "carry-over effects" from one growing season to another will also be presented.

2   Methods

The Great Plains refers to grasslands at the center of the North American continent that extend from Alberta, Canada to Texas, USA, covering approximately 2.6 million square kilometers or 14% of the total continental land area (Ostlie et al. 1997; Savage 2004). This large region has been mapped and sub-classified in many different ways by a variety of organizations and agencies.

Savage (2004) describes the ecological variety of the region as a mosaic of 15 eco-regions based on maps published jointly in 1999 by the World Wildlife Fund Canada and its associate in the USA. The dominant two eco-regions of the Great Plains are the northwestern short/mixed grasslands and the southern short grasslands. These eco-region names reflect the influence of temperature and moisture interactions in this semi-arid, continental area on plant height and species composition. The focus of this case study is an Ameriflux and Fluxnet-Canada research site located just west of the city limits of Lethbridge, Alberta, Canada (Lat. N:49.43°; Long. W:112.56°, 951 m above sea level) in the northwestern short/mixed grassland. The plant community consisted of the dominant grasses *Agropyron dasystachyum* [(Hook.) Scrib.] and *A. smithii* (Rydb.). Other major plant species represented to a lesser extent include: *Vicia americana* (Nutt.), *Artemesia frigida* (Willd.), *Carex filifolia* (Nutt.), *Stipa comata* (Trin. And Rupr.), *Stipa viridula* (Trin.) and *Bouteloua gracilis* [(H.B.K.) Lag.] (Flanagan and Johnson 2005). Average (±SD) canopy height during 2 years of very contrasting weather and plant productivity were 18.5 ± 3.4 cm (2001) and 34.4 ± 9.5 cm (2002). The following statistics illustrate the continental, semi-arid conditions apparent in this region of the Great Plains: mean daily temperatures (1908–1999) for January and July are $-8.6°C$ and $18.0°C$, respectively; mean annual precipitation (1908–1999) is 401.5 mm, with 32% falling in May and June; average pan evaporation (Class A) exceeds the average precipitation by at least 200% and often 300% during the summer months (Wever et al. 2002).

A weather station was established to provide meteorological data as described in detail previously (Flanagan et al. 2002; Wever et al. 2002). Measurements and equipment relevant to this study are briefly described below. Copper-constantan thermocouples were used to measure soil temperatures at soil depths of 2, 4, 8 and 16 cm. Volumetric soil moisture content was measured (over a 0–15 cm depth) using four replicate soil water reflectometers (CS-615, Campbell Scientific Ltd., Edmonton, Alberta). The measurements made by CS-615 probes had been previously calibrated relative to manual soil volumetric measurements. To do this soil samples (0–15 cm depth) were collected on a weekly basis using a soil corer. The known volume of soil was weighed, dried at 105°C for at least 48 h, and then re-weighed. The gravimetric moisture content was converted into volumetric measurement using the bulk density of the soil. Data are presented as available soil moisture, which was defined as the ratio of actual extractable water (difference between a given volumetric measurement and the minimum volumetric soil water content) to maximum extractable water (difference between maximum and minimum volumetric soil water content). The maximum (0.452 m^3 m^{-3}, on June 19, 2002) and minimum (0.100 m^3 m^{-3}, on October 22, 2001) volumetric soil water contents were those recorded during the period January 2001–December 2002. Total precipitation was recorded in 15-min intervals by a tipping bucket rain gauge (TE525, Texas Electronics, Inc., Dallas, Texas). With the exception of the rain gauge, all data were recorded as half-hourly averages on dataloggers (CR10, CR10X, CR23X Campbell Scientific Ltd., Edmonton, Alberta). Data from the Lethbridge airport or the Lethbridge Agriculture and Agri-Food Canada Research

Center were used in instances where meteorological data from the site were missing and for precipitation during the winter months as the tipping bucket rain gauge is not appropriate for measuring precipitation that falls as snow.

Replicate samples (n = 6) for total aboveground live biomass were collected by clipping vegetation within a 20 cm by 50 cm quadrat. The quadrats were placed in randomly selected 1 × 1.5 m sub-plots located within two larger 20 × 20 m plots, one northeast and the other southeast of the instrument hut. The harvested plant samples were dried in an oven at 60°C for at least 24 h and then weighed (Mettler PJ400, Greifensee, Switzerland) for above-ground biomass. A biomass index was defined as the ratio of the observed live biomass on any given date to the maximum live biomass measured during the period January 2001–December 2002 (253.9 g biomass m^{-2}, on July 16, 2002).

The dried above-ground plant samples were ground to a fine powder with a tissue grinder or a mortar and pestle. The ground samples were prepared for measurements of carbon isotopic composition by combustion. A 1–2 mg sub-sample of organic material was sealed in a tin capsule and loaded into an elemental analyzer for combustion (Carlo Erba). The carbon dioxide generated from the combustion was purified in a gas chromatographic column and passed directly to the inlet of a gas isotope ratio mass spectrometer (Delta Plus, Finnigan Mat, San Jose, CA, USA). The carbon isotope ratios were expressed as $\delta^{13}C$:

$$\delta^{13}C = \left[\frac{R_{sample}}{R_{std}} - 1 \right] \tag{1}$$

where R is the molar ratio of heavy to light isotope and where the subscript std refers to the international standard Vienna Pee Dee Belemnite (V-PDB). The $\delta^{13}C$ values are conveniently expressed in parts per thousand or per mil (‰).

The eddy covariance technique was used to measure net ecosystem CO_2 exchange and the fluxes of water vapor and sensible heat on a continuous basis. A detailed description of our methods and data processing procedures was provided by Flanagan et al. (2002) and Wever et al. (2002). The Fluxnet-Canada standard protocol was used for gap-filling data and calculating annual net ecosystem productivity (NEP) (Barr et al. 2004). The annual integrated NEP values have an uncertainty of approximately ± 30 g C m^{-2} year^{-1}. The calculation of the NEP uncertainty was described in detail by Flanagan and Johnson (2005).

Total ecosystem respiration (ER) was measured with a portable gas exchange system (LI-6200, LI-COR, Lincoln, Nebraska) and a dynamic, closed chamber (LI-6000-09 Soil respiration chamber, LI-COR, Lincoln, Nebraska) that was vented to the atmosphere to maintain pressure equilibrium. The chamber was attached to polyvinyl chloride collars (15 cm tall) that were inserted into the soil approximately 6 cm depth. The ground area enclosed by a collar was 71.6 cm^2. The height above ground of the inside of the chamber plus the collar was approximately 20.5 cm. The live vegetation in the collar was left intact so that the measurement represented total ecosystem respiration (soil plus above-ground vegetation). Measurements were made at six different collar locations during several intervals throughout a day in order to calculate a temperature dependence (to soil temperature at 4 cm depth) of respiration during that day. We fitted the following equation to our data in order to calculate a standardized rate of respiration at a reference temperature of 10°C (R_{10}) (Lloyd and Taylor 1994):

$$ER = R_{10} \cdot Q_{10}^{\left(\frac{T_{soil} - T_{ref}}{10}\right)} \tag{2}$$

where Q_{10} is the temperature sensitivity coefficient (dimensionless) that describes the magnitude of change in respiration rate for a 10°C change in temperature, T_{soil} is the soil temperature (°C), and T_{ref} is the reference temperature of 10°C. The R_{10} value can be thought of as a measure of the capacity of an ecosystem for respiration that will depend on factors influencing both autotrophic and heterotrophic respiration. Further details of the calculations involving Eqn. (2) are described in Flanagan and Johnson (2005).

3 Results and Discussion

3.1 Phenology of Plant Production in the Great Plains

Several aspects of phenology varied dramatically over the 9-year period at the Lethbridge research site (Fig. 1a–c). For example, we observed significant differences among years in the timing of the start of plant growth, the date of maximum biomass production, the length of time leaf area remained at or near its maximum value, and the date at which the plant canopy was completely senescent (Fig. 1a–c). In addition, there was greater than threefold variation in the maximum above-ground biomass production among the study years (Fig. 2). In general plant phenology responds to changes in environmental conditions, although the magnitude of response and the mechanisms responsible differ among ecosystems. For example, the start of leaf development can often be predicted by thermal conditions as indicated by growing degree days or other similar parameters (Barr et al. 2004; van Dijk et al. 2005), although soil thawing rather than air temperature may be the primary control in arctic ecosystems (van Wijk et al. 2003). Alternatively, in a water-limited desert ecosystem the timing of the first significant rain may be the factor that initiates plant growth (Beatley 1974). Jolly et al. (2005) proposed a general index (the growing season index (GSI)) that combined the effects of temperature, radiation (or photoperiod) and moisture availability (as indicated by vapor pressure deficit) to predict leaf phenology of a global basis with significant success.

The variation in phenology and productivity observed at the Lethbridge grassland is typical of semi-arid ecosystems, and was largely controlled by changes in the amount and timing of precipitation inputs that in turn control primary productivity by affecting plant water status and soil nutrient availability (Flanagan et al. 2002; Wever et al. 2002; Flanagan and Johnson 2005).

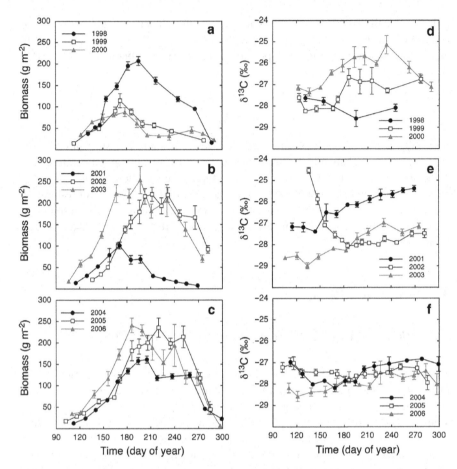

Fig. 1 Comparison of seasonal variation in the amount of above-ground green plant biomass in a grassland near Lethbridge, Alberta, Canada during 1998–2006 (panels **a–c**). Also shown are the associated measurements of the carbon isotope composition ($\delta^{13}C$, ‰) of the biomass samples (panel **d–f**). Values represent averages ±SE, n = 6.

During our studies, the amount of precipitation received during the main growing season (April–August) varied about fourfold, with 4 years having precipitation equal to or above the long-term average (average recorded during 1971–2005), and 4 years having growing precipitation amounts below the long-term mean (Fig. 3). In general, there was a strong linear relationship between the amount of growing season precipitation and peak above-ground biomass production in 6 of 8 study years (Fig. 4). In both 2003 and 2006, relatively high biomass production occurred in years with only moderate amounts of growing season precipitation, but in both of these years there was significant soil moisture "carry-over" from the previous year. The growing seasons of both 2002 and 2005 had precipitation that was significantly above average (Fig. 3), and soil water content remained high in the fall and

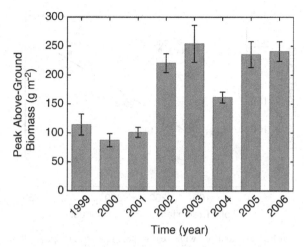

Fig. 2 Comparison of the seasonal peak value for the above-ground green plant biomass in a grassland near Lethbridge, Alberta, Canada during 1999–2006. Values represent averages ±SE, n = 6.

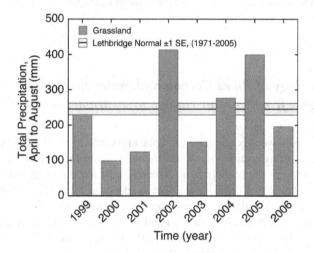

Fig. 3 Annual variation in total growing season (April through August) precipitation in relation to average precipitation (±SE) recorded during 1971–2005 at Lethbridge, Alberta, Canada.

winter periods preceding the 2003 and 2006 growing seasons. As a consequence, only moderate amounts of precipitation distributed at appropriate intervals during the growing season maintained soil moisture and soil nutrient availability at levels favorable for relatively high plant productivity in 2003 and 2006, despite the moderate, total amount of growing season precipitation. It is also interesting to note that 2003 and 2006 were the 2 years with the earliest start to the growing season, perhaps because of a reduced soil moisture limitation associated with the carry-over of moisture between years.

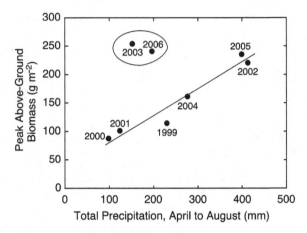

Fig. 4 Relationship between total growing season precipitation and peak above-ground green plant biomass in a grassland near Lethbridge, Alberta, Canada during 1999–2006. Values for above-ground biomass represent averages, n = 6. In 2003 and 2006 there was high peak above-ground biomass production despite only moderate input of growing season precipitation. Both 2003 and 2006 followed years with very high precipitation inputs during the growing season.

3.2 Phenology of Plant Carbon and Water Relations: Insights from Carbon Isotope Measurements

The relationship between precipitation and plant productivity in grassland ecosystems results from moisture availability simultaneously controlling plant carbon uptake and water balance. Both photosynthetic carbon acquisition and water use are strongly influenced by plant guard cells that are able to adjust the rates of net photosynthesis and transpiration by regulating stomatal aperture size and, therefore, the diffusion of carbon dioxide and water vapor into and out of leaves (Farquhar and Sharkey 1982). During photosynthetic gas exchange the stable carbon isotope composition of plant tissue is altered by isotope effects that reflect the nature of the tradeoff between increasing stomatal conductance to improve the rate of CO_2 diffusion in to a leaf, and minimizing the loss of water vapor by diffusion out of a leaf during transpiration (Farquhar et al. 1982, 1989). As a plant increases stomatal conductance and the supply of CO_2 to photosynthetic biochemistry in the chloroplast, discrimina-tion against CO_2 molecules containing ^{13}C occurs to a greater extent and the carbon isotope composition of plant tissue becomes more depleted in ^{13}C (lower $\delta^{13}C$ values). In contrast, the carbon isotope composition of plant tissue becomes more enriched in ^{13}C (higher $\delta^{13}C$ values) as stomatal conductance is reduced and the degree of stomatal limitation to CO_2 uptake and water vapor loss is increased. Farquhar et al. (1982, 1989) developed theory and reviewed empirical evidence for a relationship between the carbon isotope composition of plant tissue and the ratio of leaf intercellular

Fig. 5 Relationship between peak above-ground green plant biomass and carbon isotope composition ($\delta^{13}C$, ‰) of the biomass samples in a grassland near Lethbridge, Alberta, Canada during 1999–2006. Values represent averages, n = 6.

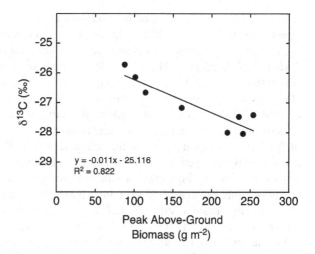

CO$_2$ (C$_i$) to ambient CO$_2$ concentration (C$_a$). In addition, it has been proposed that the C$_i$/C$_a$ ratio reflects the tradeoff between CO$_2$ uptake and water loss, and carbon isotope measurements have been used to demonstrate variation in this physiological characteristic in response to environmental changes and variation among plants with different life histories or functional groups (Ehleringer 1993a, b). For example, leaf C$_i$/C$_a$ ratio tends to be higher in grasses than forbs, higher in annuals than perennials, higher in broad-leaf deciduous trees than conifer trees (Smedley et al. 1991; Brooks et al. 1997; Ponton et al. 2006). Measurement of the carbon isotope composition of plant tissue is a very useful tool, therefore, to examine phenological variation in physiological response to aridland conditions.

We observed a strong negative correlation between peak above-ground biomass and leaf $\delta^{13}C$ values during 1999–2006 (Fig. 5). This relationship was consistent with lower biomass production being primarily caused by reduced water availability, which also resulted in reduction of stomatal conductance, increased stomatal limitation of photosynthesis and higher $\delta^{13}C$ values. In addition to the general relationship shown in Fig. 5, seasonal patterns of variation in the carbon isotope composition of biomass provided more detailed insights into the control of phenology in this grassland of the Great Plains (Fig. 1 d-f). The two rows in Fig. 1 represent biomass production and $\delta^{13}C$ values in groups of 3 years with contrasting patterns of change in environmental conditions. Figure 1a, d present data from 1998 to 2000, and during this time period environmental conditions changed from abundant growing season precipitation in 1998 (337.3 mm precipitation April–August) and high plant growth through to a year (1999) with normal growing season precipitation and moderate-to-low plant production, to a very dry year (2000) with limited plant growth. Abundant rains in late May and June of 1998 stimulated high biomass production and an associated decline in $\delta^{13}C$. The progressively lower precipitation in 1999 and 2000 that reduced biomass production also resulted in increased $\delta^{13}C$

values. A late season pulse of precipitation stimulated a second phase of plant growth during September 2000, with an associated reduction in $\delta^{13}C$ values.

A contrasting sequence of change in environmental conditions was illustrated during 2001–2003 (Fig. 1b, e). The low precipitation and plant growth in 2001 was followed by above-normal precipitation and abundant plant production in 2002. There was a progressive increase in $\delta^{13}C$ values during the growing season of 2001, and some of the carbohydrate fixed during the dry, late growing season of 2001 appears to have been stored and used to construct the leaf tissue at the start of 2002. This was illustrated by the initial very high $\delta^{13}C$ values in 2002, which subsequently dropped as soil moisture and other environmental conditions improved, with very high plant productivity occurring in response to the favorable conditions that developed later in June 2002. The abundant precipitation in the 2002 growing season resulted in high soil moisture being carried over through the fall and winter and into the start of the 2003 growing season. This contributed to high plant productivity early in 2003, and the moderate precipitation in 2003 was sufficient to sustain high biomass production throughout the growing season, although $\delta^{13}C$ values increased somewhat in July and August 2003 relative to 2002, illustrating the influence of reduced moisture inputs on photosynthetic gas exchange in 2003.

Finally, the years 2004–2006 all showed relatively high plant productivity with much more subtle seasonal patterns of variation in green biomass and $\delta^{13}C$ values (Fig. 1c, f). The growing season precipitation in 2004 was just slightly above normal (Fig. 3) and so peak biomass production was intermediate between that observed in 1999–2001 and the four high productivity years (2002, 2003, 2005, 2006; Fig. 2). The relatively dry conditions, that started in the later part of July and extended through September, resulted in higher $\delta^{13}C$ values being recorded at this time in 2004 than in 2005 or 2006. High summer precipitation supported high productivity and relatively constant $\delta^{13}C$ values during 2005. The abundant precipitation in the 2005 growing season again resulted in high soil moisture being carried over through the fall and winter and into the start of the 2006 growing season. An early start to plant growth was observed in 2006 with relatively low $\delta^{13}C$ values. The moderate growing season precipitation in 2006 was again sufficient to maintain high plant productivity with only a minor increase in $\delta^{13}C$ values during July and August compared to those measured at the start of the 2006 season.

3.3   Consequences of Variation in Phenology and Leaf Area for Net Ecosystem Productivity

Differences among ecosystems in carbon acquisition are strongly determined by the amount of leaf area produced and the length of time that the leaf area is photosynthetically active (Chapin et al. 2002; Churkina et al. 2005; Gu et al. this volume). Consistent with this statement we observed two patterns between peak above-ground

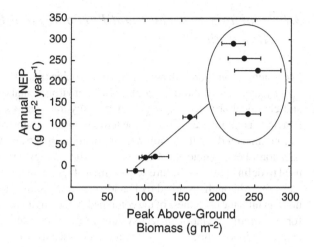

Fig. 6 Relationship between peak above-ground green plant biomass and net ecosystem production (NEP) measured by eddy covariance in a grassland near Lethbridge, Alberta, Canada during 1999–2006. Values for the biomass represent averages ±SE, n = 6. The NEP values are annual integrated values with an uncertainty of approximately ±30 g C m^{-2} year^{-1}.

biomass and annual net ecosystem productivity at the Lethbridge research site (Fig. 6). First, for the 4 years with peak biomass below the maximum for this site (approximately 240–250 g m^{-2}), there was a strong linear relationship between biomass and NEP. This linear relationship reflects the major control of maximum leaf area index (LAI) on NEP, as peak biomass was strongly correlated with LAI (Flanagan et al. 2002). The low moisture inputs, which occurred particularly during 1999–2001 (Fig. 3), limited LAI development with strong effects on annual ecosystem photosynthesis and NEP (Flanagan et al. 2002). Second, among the 4 years that had very similar peak biomass averaging approximately 240–250 g m^{-2}, and, therefore approximately equal LAI of about 1.0, there was wide variation in NEP (Fig. 6). This variation in NEP under approximately equal LAI was largely due to differences in the length of time leaf tissue was green and photosynthetically active (canopy duration), with minor influences of environmental factors on stomatal limitation of photosynthetic gas exchange (Fig. 1). The observation that maximum LAI and canopy duration are major controls on NEP has important implications for the use of simple remote sensing reflectance indices like normalized difference vegetation index (NDVI) and enhanced vegetation index (EVI) to estimate the NEP over large areas. Both NDVI and EVI are good proxies for maximum LAI and the length of the photosynthetic period, and Churkina et al. (2005) have shown that these indices could be used successfully to estimate NEP at 28 different ecosystem flux sites across the globe. In addition, microwave backscatter measurements may provide an estimate of grassland biomass/productivity that is independent of visible and near infrared reflectance measurements (Frolking et al. 2005). These simple approaches are complementary to other uses of surface reflectance measurements in conjunction with ecosystem models to calculate NEP based on separate determination of ecosystem photosynthesis and several component respiration fluxes (Running et al. 2000).

3.4   Linkages Between Plant Phenology and Ecosystem Respiration

Ecosystem respiration is dependent on autotrophic (plant) and heterotrophic (microbe) activity, and both of these are controlled by environmental conditions and the quantity and chemical structure (quality) of carbohydrate and other respiratory substrates (Davidson and Janssens 2006). The temperature and substrate controls on respiration can be analyzed by the combined application of Arrhenius and Michalis–Menten equations. For example, the Arrhenius activation energy term (E_a) (or Q_{10} value) is often used to define the temperature sensitivities of the parameters in the Michalis–Menten equation. Although, as noted by Davidson and Janssens (2006), temperature acclimation can limit the usefulness of the combined Arrhenius and Michalis-Menten approaches for explaining variation in ecosystem respiration rates. Temperature acclimation can result in a number of significant effects such as (a) alterations to the abundance of enzymes involved in respiration; (b) variation in the type of enzymes involved (e.g. isozymes produced under different environmental conditions by one organism, or changes to the species composition of microbial organisms active under different environmental conditions); and (c) other modifications to cell structure (e.g. changes to membrane lipid composition) that influence enzyme activity.

Plant phenology is linked to changes in ecosystem respiration because, as shown by Högberg et al. (2001), plant respiration contributes a major fraction of the carbon dioxide released to the soil and atmosphere during ecosystem respiration, and because plant tissue or its chemical products form the majority of respiratory substrates used by microbial respiration (Davidson and Janssens 2006). Plant growth and development involves many changes in cell and enzyme characteristics, and the magnitude of plant respiratory activity should vary in proportion to plant growth. Soil microbial growth and development is closely tied to plant activity because plant exudates and senescent plant tissue provide respiratory substrates for microbes, but also because both plant and microbe activities are strongly influenced by the same changes in the environmental conditions. While temperature plays an important role in controlling respiration rates in grassland ecosystems, water availability has a more significant role because it limits both the amount and timing of plant and microbe growth and development. In other words, water availability controls grassland plant phenology which influences autotrophic respiration and its acclimation processes, and also controls substrate availability to, and the activity of, soil microbes.

A comparison between dry (2001) and wet (2002) years can illustrate the influence of plant phenology and its interaction with environmental conditions for effects on grassland ecosystem respiration (Flanagan and Johnson 2005). The temperature response curves for ecosystem respiration at Lethbridge changed dramatically during the growing season in 2001 (Fig. 7a). In June 2001 the temperature response curve was quite similar to that recorded during June 2002. However, the slope of relationship between respiration and temperature declined (Fig. 7a) as soil moisture was reduced through the growing season in 2001 (Fig. 8). By contrast, there was little change in the slope of the respiration–temperature response curve

Fig. 7 Seasonal variation in the temperature dependence of ecosystem respiration rate measured using dynamic, closed-system (darkened) chambers in a grassland near Lethbridge, Alberta, Canada during (**a**) 2001 and (**b**) 2002. The different symbols represent measurements made throughout one day as soil temperature varied. Soil temperature was measured at a depth of 4 cm. Sample dates are indicated for the different symbols. Respiration values represent averages, n = 6. The graph is reproduced from Flanagan and Johnson (2005).

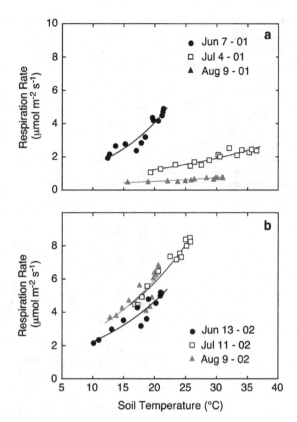

during 2002, although the maximum rates of respiration increased from June through to August (Fig. 7b). The lower maximum respiration rates and the decline in temperature response of respiration observed in July and August of 2001 were the result of lower soil moisture and the associated reduced biomass production relative to that measured in 2002 (Fig. 8). A large fraction (94%) of the variation in R_{10} values calculated from data in both 2001 and 2002 could be explained by seasonal variation in the product of the above-ground biomass index and the available soil moisture index (Fig. 9). This observation suggests that the amount of plant above-ground biomass is a good proxy for estimating total autotrophic respiration. In addition, soil moisture appears to be the dominant environmental factor in this semi-arid grassland that influences changes in both autotrophic and heterotrophic respiration during the growing season. Craine et al. (1999) have also shown that grassland soil respiration was more strongly linked to changes in carbon availability and soil moisture than to shifts in temperature. The interaction between soil moisture and plant activity was the dominant control on the magnitude of ecosystem respiration in a California grassland (Xu et al. 2004). The highest rates of respiration observed by Xu et al. (2004) generally occurred under cool temperatures (approximately 10°C) in the winter when soil moisture and plant activity were highest. Only very

Fig. 8 Comparison of seasonal changes in total above-ground green plant biomass (**a**), and available soil moisture (**b**) in a grassland near Lethbridge, Alberta, Canada during 2001 and 2002. Biomass and soil moisture are both expressed on a relative scale (0–1). The data are reproduced from Flanagan and Johnson (2005).

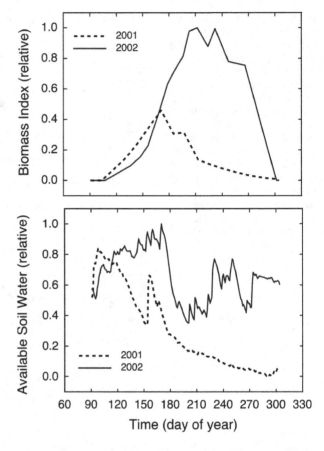

Fig. 9 Relationship between the product of changes in above-ground biomass and soil moisture and variation in the standardized rate of respiration, normalized to 10°C (R_{10}). Biomass and soil moisture are both expressed on a relative scale (0–1). These data are based on measurements made in a grassland near Lethbridge, Alberta, Canada during 2001 and 2002. The graph is reproduced from Flanagan and Johnson (2005).

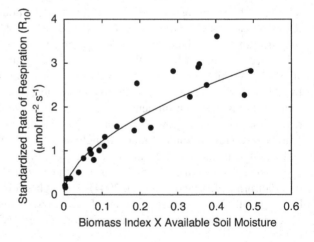

low respiration rates (normally less than 1 μmol m^{-2} s^{-1}) were measured in summer when soil temperature was high (approximately 30°C) but soil moisture was very low and the grass community was dormant.

4 Conclusions

Several aspects of grassland phenology and productivity varied during a 9-year study period, with much of the variation controlled by the amount and timing of precipitation inputs during the growing season. We observed a negative relationship between biomass production and leaf δ^{13}C values, and this relationship was consistent with lower biomass production being predominantly controlled by reduced water availability. Seasonal patterns of variation in leaf δ^{13}C provided more detailed insights into the control of phenology of plant production in this Great Plains grassland. We also observed two patterns in the relationship between peak above-ground biomass and annual net ecosystem productivity (NEP). First, there was a strong linear relationship between biomass and NEP in 4 years when peak leaf area production was held significantly below the maximum by low water availability. Second, among 4 years with similar maximum LAI values, there was wide variation in NEP largely due to differences in the length of time leaf tissue was photosynthetically active. Water availability also had a significant effect on ecosystem respiration because it controlled grassland phenology. The product of plant biomass and soil water content was a good proxy for estimating ecosystem respiration because of relationships with both autotrophic and heterotrophic activities. Seasonal variation in grassland respiration was more strongly linked to substrate availability and soil moisture than to shifts in temperature.

Acknowledgements Financial support was provided by grants from the Natural Sciences and Engineering Research Council of Canada and the University of Lethbridge. Peter Carlson, Chera Emrick, Nicole Geske, Bruce Johnson, Francine May, Ngaire Nix and Linda Wever provided excellent technical assistance. Stewart Rood kindly allowed my associates and I to work on his land.

References

Barr, A.G., Black, T.A., Hogg, E.H., Kljun, N., Morgenstern, K. and Nesic, Z. (2004) Inter-annual variability in the leaf area index of a boreal aspen-hazelnut forest in relation to net ecosystem production. Agric. Forest Meteorol. 126, 237–255.

Beatley, J.C. (1974) Phenological events and their environmental triggers in Mojave desert ecosystems. Ecology 55, 856–863.

Brooks, J.R., Flanagan, L.B., Buchmann, N. and Ehleringer, J.R. (1997) Carbon isotope composition of boreal plants: functional grouping of life forms. Oecologia 110, 301–311.

Chapin, F.S., Matson, P.A. and Mooney, H.A. (2002) *Principles of Terrestrial Ecosystem Ecology.* Springer, New York.

Churkina, G., Schimel, D., Braswell, B.H. and Xiao, X. (2005) Spatial analysis of growing season length control over net ecosystem exchange. Global Change Biol. 11, 1777–1787.

Ciais, P., Reichstein, M., Viovy, N., Granier, A., Ogee, J., Allard, V., Aubinet, M., Buch-mann, N., Berhofer, Chr., Carrara, A., Chevallier, F., De Noblet, N., Friend, A.D., Friedlingstein, P., Grunwald, T., Heinesch, B., Keronen, P., Knohl, A., Krinner, G., Loustau, D., Manca, G., Matteucci, G., Miglietta, F., Ourcival, J.M., Papale, D., Pile-gaard K., Rambal, S., Seufert, G., Soussana, J.F., Sanz, M.J., Schulze, E-D., Vesala, T. and Valentini, R. (2005) Europe-wide reduction in primary productivity caused by the heat and drought in 2003. Nature 437, 529–533.

Craine, J.W., Wedin, D.A. and Chapin, F.S. (1999) Predominance of ecophysiological controls on soil CO_2 flux in a Minnesota grassland. Plant Soil 207, 77–86.

Davidson, E.A. and Janssens, I.A. (2006) Temperature sensitivity of soil carbon decomposition and feedbacks to climate change. Nature 440, 165–173.

Ehleringer, J.R. (1993a) Carbon and water relations in desert plants: an isotopic perspective. In: J.R. Ehleringer, A.E. Hall, G.D. Farquhar (Eds.), *Stable Isotopes and Plant Carbon-Water Relations*. Academic, San Diego, CA, pp. 155–172.

Ehleringer, J.R. (1993b) Gas-exchange implications of isotopic variation in arid-land plants. In: J.A.C. Smith, H. Griffiths (Eds.), *Water Deficits: Plant Responses from Cells to Community*. BIOS Scientific Publishers Ltd, Oxford, pp. 265–284.

Farquhar, G.D., Ehleringer, J.R. and Hubick, K.T. (1989) Carbon isotope discrimination and photosynthesis. Annu. Rev. Plant Physiol. Mol. Biol. 40, 503–537.

Farquhar, G.D., O'Leary, M.H. and Berry, J.A. (1982) On the relationship between carbon isotope discrimination and the intercellular carbon dioxide concentration in leaves. Aust. J. Plant Physiol. 9, 121–137.

Farquhar, G.D. and Sharkey, T.D. (1982) Stomatal conductance and photosynthesis. Annu. Rev. Plant Physiol. 33, 317–345.

Flanagan, L.B., Wever, L.A. and Carlson, P.J. (2002) Seasonal and interannual variation in carbon dioxide exchange and carbon balance in a northern temperate grassland. Global Change Biol. 8, 599–615.

Flanagan, L.B. and Johnson, B.G. (2005) Interacting effects of temperature, soil moisture and plant biomass production on ecosystem respiration in a northern temperate grassland. Agric. Forest Meteorol. 130, 237–253.

Frolking, S., Fahnestock, M., Milliman, T., McDonald, K. and Kimball, J. (2005) Interannual variability in North American grassland biomass/productivity detected by SeaWinds scatterometer backscatter. Geophys. Res. Lett. 32, L21409.

Högberg, P., Nordgren, A., Buchmann, N., Taylor, A.F.S., Ekblad, A., Högberg, M.N., Nyberg, G., Ottosson-Löfvenius, M. and Read, D.J. (2001) Large-scale forest girdling shows that current photosynthesis drives soil respiration. Nature 411, 789–792.

Jolly, W.M., Nemani, R. and Running, S.W. (2005) A generalized, bioclimatic index to predict foliar phenology in response to climate. Global Change Biol. 11, 619–632.

Knapp, A.K. and Smith, M.D. (2001) Variation among biomes in temporal dynamics of aboveground primary production. Science 291, 481–484.

Knapp, A.K., Bruns, C.E., Fynn, R.W.S., Kirkman, K.P., Morris, C.D. and Smith, M.D. (2006) Convergence and contingency in production-precipitation relationships in North American and South African C4 grasslands. Oecologia 149, 456–464.

Lloyd, J. and Taylor, J.A. (1994) On the temperature dependence of soil respiration. Funct. Ecol. 8, 315–323.

Ostlie, W.R., Schneider, R.E., Aldrich, J.M., Faust, T.M., McKim, R.L.B. and Chaplin, S.J. (1997) *The Status of Biodiversity in the Great Plains*. The Nature Conservancy, Arlington, VA.

Ponton, S., Flanagan, L.B., Alstad, K.P., Johnson, B.G., Morgenstern, K., Kljun, N., Black, T.A. and Barr, A.G. (2006) Comparison of ecosystem water-use efficiency among Douglas-fir forest, aspen forest and grassland using eddy covariance and carbon isotope techniques. Global Change Biol. 12, 294–310.

Running, S.W., Thornton, P.E., Nemani, R.R. and Glassy, J.M. (2000) Global terrestrial gross and net primary productivity from the earth observing system. In: O.E. Sala, R.B. Jackson, H.A. Mooney, R.W. Howarth (Eds.), *Methods in Ecosystem Science*. Springer, New York, pp. 44–57.

Savage, C. (2004) *Prairie: a Natural History*. Greystone Books, Douglas & McIntyre Publishing Group, Vancouver.

Smedley, M.P., Dawson, T.E., Comstock, J.P., Donovan, L.A., Sherrill, D.E., Cook, C.S. and Ehleringer, J.R. (1991) Seasonal carbon isotope discrimination in a grassland community. Oecologia 85, 314–320.

van Dijk, A.I.J.M., Dolman, A.J. and Schulze, E.-D. (2005) Radiation, temperature, and leaf are explain ecosystem carbon fluxes in boreal and temperate European forests. Global Biogeochem. Cycles 19, GB2029.

van Wijk, M.T., Williams, M., Laundre, J.A. and Shaver, G.R. (2003) Interannual variability of plant phenology in tussock tundra: modeling interactions of plant productivity, plant phenology, snowmelt and soil thaw. Global Change Biol. 9, 743–748.

Wever, L.A., Flanagan, L.B. and Carlson, P.J. (2002) Seasonal and interannual variation in evapotranspiration, energy balance and surface conductance in a northern temperate grassland. Agric. Forest Meteorol. 112, 31–49.

Xu, L., Baldocchi, D.D. and Tang, J. (2004) How soil moisture, rain pulses, and growth alter the response of ecosystem respiration to temperature. Global Biogeochem. Cycles 18, doi: 10.1029/2004GB002281.

Is Temporal Variation of Soil Respiration Linked to the Phenology of Photosynthesis?

Eric A. Davidson and N. Michele Holbrook

Abstract If recent photosynthate is the primary source of carbohydrates for root respiration and possibly for much of soil microbial respiration through root exudates, then temporal variation in soil respiration (SR) may be linked to plant phenological patterns at hourly to seasonal time scales. Here we review the evidence for this linkage and identify the research needs for improving our understanding of the physiological and ecological linkages between photosynthesis and respiration in ecosystems. The linkage is clearest at the season time scale, where the importance of substrate supply to belowground carbon processes follows a seasonal pattern in temperate and boreal ecosystems. Correlations of SR with canopy light, temperature, and vapor pressure deficit also suggest a link between root respiration and canopy photosynthesis on times scales of a few hours to about 3 weeks. Temporal correlations between photosynthetic activity and SR make it tempting to view patterns in SR as the direct outcome of variation in substrate delivery rates, but these analyses provide only inferential evidence of a physiological linkage. Isotopic labeling studies indicate that the lag between fixation by foliage and respiration of the label in the rhizosphere is usually on the order of a day or more in forest ecosystems. More rapid transmission of pressure/concentration waves through the phloem is theoretically possible, and current understanding of phloem physiology and the regulation of growth suggests that the linkage between canopy and root processes is based on more than the mass transport of substrate from sources to sinks. However, improved understanding of assimilate transport and partitioning is needed before variation in SR patterns can be linked mechanistically to the physiology and the phenology of the plants fueling belowground metabolism.

E.A. Davidson (✉)
The Woods Hole Research Center, Woods Hole, MA, USA
e-mail: edavidson@whrc.org

N.M. Holbrook
Department of Organismic and Evolutionary Biology,
Harvard University, Cambridge, MA, USA
e-mail: holbrook@fas.harvard.edu

A. Noormets (ed.), *Phenology of Ecosystem Processes*,
DOI 10.1007/978-1-4419-0026-5_8, © Springer Science+Business Media, LLC 2009

1   Introduction

The efflux of CO_2 from the soil surface, known as soil respiration (SR), is primarily
a combination of microbial respiration and root respiration, mediated by gas transport
within the soil. Hence, SR is affected by a suite of environmental factors that control
the biological metabolism of a variety of soil-dwelling organisms and that
control physical processes of gaseous diffusion and convection. Among these
environmental factors, temperature is the most obvious and the one most commonly
correlated with SR (Davidson and Janssens 2006; Hibbard et al. 2005). However,
the role of substrate supply to roots and to soil microbes is gaining increasing
recognition as an important determinant of variation in SR (Davidson et al. 2006a;
Ryan and Law 2005). Rapid changes in substrate availability that accompany wetting
of dry soil (Birch 1958; Borken et al. 2003; Bottner 1985; Kieft et al. 1987), girdling
of trees (Högberg et al. 2001), and shading and clipping of grasses (Craine et al.
1999; Wan and Luo 2003) clearly affect soil respiration independently of temperature
at time scales from hours to months. The ultimate source of carbohydrates for root
and soil microbial respiration is primarily plant photosynthate, which suggests a
potentially important link between plant phenology and SR, as mediated by the
supply of photosynthate for respiration of roots, mycorrhizae, and heterotrophs
utilizing root exudates within the rhizosphere. The objectives of this chapter are to
review the evidence for this linkage from diel to seasonal scales and to identify the
research needs for improving our understanding of these physiological and ecological
linkages between photosynthesis and respiration in ecosystems.

2   The Evidence for Linkages Between Plant Phenology and Soil Respiration

Variation among ecosystems in annual rates of SR is due primarily to differences
in site productivity (Hibbard et al. 2005; Janssens et al. 2001; Reichstein et al.
2003; Ryan and Law 2005; Sampson et al. 2007). Leaf area index (LAI), used as a
crude surrogate for site productivity, has been correlated with annual respiration
among forest and grassland study sites, presumably because the greater the site
LAI, the more substrates were produced for respiration (Hibbard et al. 2005;
Reichstein et al. 2003).

2.1   Seasonal Variation

Seasonal variation in SR in temperate and boreal ecosystems usually covaries with
temperature, but substrate supply also varies seasonally, and its effects may be
confounded, in part, by variation attributed to temperature variation. Variation in

SR due to seasonal patterns of plant phenology can, in turn, affect the apparent temperature sensitivities of respiration. Curiel-Yuste et al. (2004) argued that greater apparent Q_{10} values (the factor by which observed respiration increases for each 10-degree increase in temperature) for SR measured across seasons in a Belgian hardwood forest compared to an adjacent conifer forest reflected greater seasonality of belowground C allocation by the hardwoods. When Q_{10}'s were calculated for only 2-month intervals, soil respiration in the hardwood and conifer stands had nearly identical temperature sensitivities, indicating similar responses to diel and synoptic scale variation of temperature. Only when winter and summer observations were combined, did the hardwoods appear to have greater temperature sensitivity for soil respiration, presumably due to the greater seasonality of photosynthesis and subsequent supply of substrate belowground in the hardwood stand. Hence, the greater seasonal Q_{10} in the hardwood site may have been mostly a phenological response rather than different temperature sensitivity, *per se*.

Seasonal variation in C allocation can affect both maintenance respiration and growth of roots, mycorrhizae, and rhizosphere microorganisms. When a pulse of root growth occurs in the spring, then the amount of respiring tissue increases simultaneously with temperature-dependent increases in specific root respiration (i.e., CO_2 production per gram of tissue or per unit of enzyme capacity). In this case, the apparent Q_{10} of soil respiration (i.e., observed CO_2 efflux as a function of temperature) across seasons reflects a combination of seasonal variation in root growth and the temperature responses of specific root respiration rates, both of which correlate positively with temperature. Hanson et al. (2003) reported a lower Q_{10} for soil respiration in an oak forest in Tennessee, USA, when dates associated with root growth (observed in minirhizotrons) were excluded. Similarly, in trenched plots without roots and control plots with roots in temperate forests, Boone et al. (1998) reported Q_{10}'s of 2.5 and 3.5, respectively, and Epron et al. (1999) reported 2.3 and 3.9, respectively. Boone et al. (1998) calculated a Q_{10} of 4.6 for the root respiration inferred from the difference between the control and trenched plots. However, as Boone et al. (1998) also pointed out, this root respiration Q_{10} includes the effects of both seasonal changes in root biomass (i.e., root growth) and direct responses of existing root biomass to changing temperature.

Root elongation has been shown to peak several weeks after foliar expansion in the hardwood forest of Oak Ridge and at several other northern mixed forests (Joslin et al. 2001), thus suggesting that the autotrophic component of soil respiration probably lags aboveground autotrophic respiration in the spring. Cardon et al. (2002) found that SR was inversely correlated with shoot flush in potted oaks, which is consistent with allocation to roots after leaves have expanded and become mature, reflecting a higher prioritization within the plant for springtime shoot growth compared to root growth (Marcelis 1996; Wardlaw 1990). Contrary to earlier studies that related root elongation only to temperature, Joslin et al. (2001) found that a phenology index and soil water content were the most important correlates with root elongation, and that temperature was not significant. There is good evidence that at least some of the root elongation in these tree species is fueled by recently fixed photosynthate (Gaudinski et al. 2001; Joslin et al. 2001).

The carbon source for root respiration in a boreal forest has been shown to come from stored carbohydrates in early spring and from more recent photosynthate in late spring (Cisneros-Dozal et al. 2006). Thus, in addition to the temperature responses of extant enzymes, it is also important to understand phenological changes in the abundance of reactive enzymes present in temporally varying stocks of roots, fungi, leaves, and non-structural carbohydrate.

The components of total ecosystem respiration do not respond to temperature and phenology in complete synchrony. At both Howland (Maine) and Harvard (Massachusetts) forests, aboveground respiration accelerates first in the early spring, coincident with bud break and foliar expansion, which results in soil respiration being only 30–40% of total ecosystem respiration in the early spring Fig. 1. Soil respiration then gradually increases to about 60–70% of total ecosystem respiration during the summer and 90–100% in the autumn and winter.

This asynchrony of aboveground and belowground phenological patterns may contribute to seasonal hysteresis of apparent temperature sensitivities. Seasonal hysteresis of apparent Q_{10}'s for total ecosystem respiration and soil respiration at a spruce-dominated study site in Howland, Maine, USA, provides an example of multiple processes interacting to produce highly variable apparent temperature sensitivities. Soil respiration always exhibited a higher Q_{10} in the spring than in the

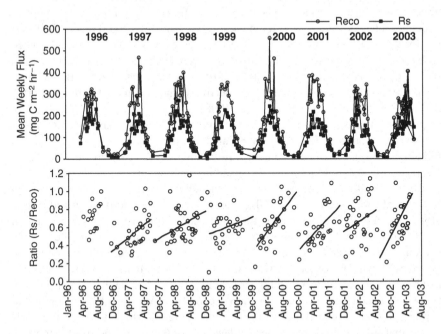

Fig. 1 Total ecosystem respiration (Reco), soil respiration (Rs) at the Howland Forest, Maine, USA (*upper panel*); and the ratio of soil respiration:ecosystem respiration (*lower panel*). Note that the ratio falls below 0.4 every spring and increases to near 1.0 by the autumn. From Davidson et al. (2006b).

Table 1 Ratio of ecosystem respiration (ER) and soil respiration (SR) at 15°C to respiration at 5°C (Q_{10}) at the Howland forest, Maine, USA (from Davidson et al. 2006b)

Season	Flux	1997	1998	1999	2000	2001	2002	All years
Spring	ER	2.8	2.7	3.2	3.9	3.5	2.9	3.1
	SR	4.5	3.0	3.3	5.0	3.9	3.2	3.5
Fall	ER	4.1	3.5	3.2	4.1	3.1	3.1	3.4
	SR	2.8	2.6	3.2	3.0	2.3	1.9	2.5

autumn Table 1, perhaps because of springtime root growth, as discussed above. Soils warm from the top downward in the spring, and they cool from the top downward in the autumn, so hysteresis based on temperature measured at a fixed depth could also be influenced by varying soil depths of CO_2 production (Reichstein et al. 2005). During the spring, soil respiration Q_{10} was always higher than total ecosystem respiration Q_{10}, but the reverse was observed in the autumn (Table 1). In the autumn, both nighttime respiration and daytime net ecosystem exchange of CO_2 dropped sharply immediately after the first frost and remained low for the rest of the autumn and winter (Hollinger et al. 1999). If both photosynthesis and respiration drop sharply in response to the first autumn frost, the apparent temperature sensitivity of ecosystem respiration may be elevated in the autumn because of this physiological threshold effect that induces relatively abrupt dormancy.

2.2 Synoptic Scale Variation

Synoptic scale variation in weather patterns are known to affect canopy processes, but the link to belowground processes is only beginning to be recognized. Variation in rates of soil respiration have been correlated with air temperature and vapor pressure deficit (VPD) that occurred 1–4 days earlier in a Norway spruce stand (Ekblad et al. 2005) and in a mixed coniferous boreal forest stand (Ekblad and Högberg 2001), inferring a lag time of several days between the effect of weather on photosynthesis and the subsequent effect on root respiration and exudation. Analysis of $^{13}CO_2$ in soil respiration in the Norway spruce sites also revealed a 3–7 day lag between drought-induced fractionation of photosynthate and the isotopic signal of the soil CO_2 efflux (Ekblad et al. 2005). In situ radiocarbon labeling in a black spruce forest of Manitoba, Canada, revealed that the peak of the ^{14}C appearance in roots and in rhizosphere respiration occurred 4 days after labeling (Carbone et al. 2007), which is consistent with previously inferred lags from ^{13}C analyses in boreal forests (Eklad et al. 2005), but longer than the 2-day lag measured with ^{14}C in young poplar trees (Horwath et al. 1994). A lag of 5–10 days was observed between variation in VPD and the ^{13}C signature of total ecosystem respiration in several temperate forests of Oregon, USA (Bowling et al. 2002). A 4–5 day lag was observed between ^{13}C in total ecosystem respiration and the ratio of VPD to photosynthetically active radiation in a deciduous forest in Germany (Knohl et al. 2005).

Using automated soil CO_2 profile measurements and a diffusivity model to estimate SR, Baldocchi et al. (2006) found significant correlations between SR and eddy-covariance-based measurements of canopy photosynthesis in an oak savanna, with lags of zero and 14 days. They interpreted the zero lag to demonstrate diel effects and the 14-day lag to be the impact of a similar 14-day lag observed between canopy photosynthesis and VPD. Gaumont-Guay et al. (2008) found a 2–3 week lag between mean daytime photosynthesis measured by eddy covariance and mean nighttime root respiration calculated from root exclusion plots in a boreal black spruce forest. Hence the evidence from treed ecosystems indicates a lag of several days to 3 weeks between canopy processes and SR.

Using radiocarbon labeling in California grasses and shrubs, Carbone and Trumbore (2007) reported that about half of the label transported belowground was respired within 24 h of labeling, about another quarter during the next 5 days, and the remainder during the subsequent month. Wan and Luo (2003) demonstrated that clipping of a tallgrass prairie significantly reduced soil respiration within 2 days. Although the number of studies remains small, it is tempting to draw the conclusion that, as might be expected, the lag time between canopy processes and SR is shorter for grasses and shrubs than for trees. This response presumably reflects the time required for substrate to travel from the canopy to the roots. In any case, evidence is accumulating that meteorological conditions affecting photosynthesis in a variety of ecosystems may also affect SR with lag times of 1 day to 2 weeks.

2.3 Diel Variation

Diel variation in SR has been correlated to diel variation in soil temperature (Janssens and Pilegaard 2003; Xu and Ye 2001), but the diel pattern of SR sometimes lags that of soil temperature. Hence, several studies have been designed to decompose the diel variation of SR into two components – one that is sensitive to soil temperature and one that is not. Tang et al. (2005) showed that SR under oak canopies appears to be decoupled from soil temperature, whereas SR in grassy areas showed the expected correlation with soil temperature. The SR under oak canopies was, instead, correlated with photosynthesis, but with a lag of 7–12 h. The authors inferred that the lack of correlation with soil temperature under the tree canopy and the lagged correlation with photosynthesis was the result of oak rhizosphere respiration that was linked to lagged diel patterns of substrate availability. Gaumont-Guay et al. (2008) showed a similar 12-h lag between photosynthesis and root respiration in a boreal black spruce forest. Liu et al. (2006) also demonstrated that part of the diel cycle of SR in a mixed deciduous forest of Tennessee, USA, was independent of soil temperature and was correlated with photosynthetically active radiation (PAR) with a 1-h lag. No correlation with PAR was observed during the dormant season, so this diel response appears to interact with the phenology of photosynthesis.

Several oral presentations at a workshop on automated measurements of soil respiration, held in Durham, New Hampshire, USA, in September 2007, presented similar observations of diel variation in SR being out of phase with soil temperature and related in some way to light, temperature, or VPD of the aboveground environment (Carbone and Vargas 2007). Although this body of work is not yet available in the literature, it appears that the evidence for an effect of supply of recent photosynthate on diel patterns of SR is rapidly growing. However, the published data are too few to begin investigation of what controls variation in the observed lag times among sites or the relative importance of temperature-dependent and substrate-dependent processes. Moreover, correlations between components of SR diel patterns with canopy light, temperature, and VPD provide only inferential evidence of a physiological linkage. Care must be taken to rule out confounding factors, such as diel variation in wind speed, which could change soil CO_2 concentration profiles and fluxes. At present, we cannot explain the physiological mechanism of how canopy processes can affect rhizosphere respiration within hours.

3 Physiological Links of Phloem Transport

Although isotopic labeling studies indicate that the lag between fixation by foliage and respiration of the label in the rhizosphere is usually on the order of a day or more (Carbone and Trumbore 2007; Carbone et al. 2007; Ekblad et al. 2005; Horwath et al. 1994; Högberg et al. 2008), more rapid transmission of pressure/concentration waves through the phloem, which could provide roots with information on shoot level processes, is theoretically possible (Thompson and Holbrook, 2004). Because root respiration and microbial respiration of root exudates comprise a large fraction of total SR, substrate supply to roots is increasingly recognized as an important controlling factor (Davidson et al. 2006a; Ryan and Law 2005), but we know little about controls on phloem transport to roots and communication between canopy and root processes. In this section, we explore the physiological basis for transport of substrate and signals between the canopy and roots, nearly all of which comes from the study of relatively small, non-woody plants.

What is the nature of the linkage between shoot activity and root growth? Temporal correlations between photosynthetic activity and SR make it tempting to view patterns in SR as the direct outcome of variation in substrate delivery rates. Indeed, the lag times are in approximate agreement with our best understanding of phloem transport rates (Fisher 2002), and because much of the organic substrates for SR are carried there via the phloem, it is hard to imagine how phloem transport could not, at some level, provide some degree of control. Nevertheless, the literature examining what is often referred to as "sink strength," which describes the ability of growing regions to acquire or compete for phloem-transported materials, makes it clear that treating the phloem as nothing more than a conveyor belt is overly simplistic, with the opportunities for physiological control over the distribution of photoassimilates occurring at multiple levels (Fisher 2002). In addition to failing to take into account the active role played by sink tissues, this perspective ignores the buffering capacity of the phloem itself resulting from the leakage and retrieval of solutes along the transport pathway (Thorpe et al. 2005).

A major paradox in understanding the coordination of above and belowground processes is that although increased shoot activity appears to lead to increased root activity (e.g., Muller et al. 1998), there is little evidence that the growth of roots or other sinks is dictated by substrate availability (Farrar 1996; Pritchard et al. 2005). Manipulating substrate availability in barley by removing part of the root system or exogenous sugar application had no effect on growth or respiration within an hour (Bingham and Farrar 1988; Farrar and Minchin 1991). However, excised roots responded immediately (Williams and Farrar 1990), suggesting that if root growth was substrate limited, then increasing substrate availability should have elicited a response. Manipulations of pH to increase elongation rates in maize roots had no effect on either turgor or osmotic pressure (Winch and Pritchard 1999), indicating that solute delivery must have also increased to maintain turgor and demonstrating a close relationship between solute import and extension growth. Of course, it is possible to create a situation in which growth is substrate limited. Extension growth in *Arabidopsis* roots grown on agar at low light responded to exogenous sugars (Freixes et al. 2002). However, when the leaves were provided with adequate light levels, root extension rates were insensitive to the presence of sugars in the media. The take-home message from these studies is that the unloading of solutes from the phloem and the growth that they support are highly coordinated. Treatments that speed up or retard elongation do so without significantly perturbing turgor or osmotic pressures, indicating substantial homeostasis of water relations within the growing zone (Pritchard et al. 2005).

The fact that root growth is not necessarily the passive outcome of mass transport through the phloem does not mean that belowground activity is uncoupled from aboveground metabolism. Indeed, the ability of sinks to control over their own activity may allow them to respond more effectively to changes in resource availability at the whole plant level. Sieve tubes experience substantial turgor pressures as a consequence of their high solute concentrations (van Bel and Hafke 2005). Although pressure drops must accompany the movement of phloem sap from source to sink regions, these are likely to be small relative to the pressure drops that exist between the phloem and surrounding tissues. Thus, the phloem can be thought of as forming a "high pressure manifold system" (Fisher 2002) in which the positional effects on phloem delivery are small relative to the control exerted by the sink tissues. In the case of developing roots, phloem unloading into the apical region of increasing relative elemental growth rate occurs via plasmodesmata (symplasmic unloading) (Oparka et al. 1994). More recent work (Pritchard et al. 2005) suggests that unloading across the plasma membrane into the cell wall (apoplasmic unloading) may dominate in more proximal regions of the root zone. Both provide opportunities for regulating solute delivery. Symplasmic unloading takes place through plasmodesmata and thus is sensitive to changes in plasmodesmatal resistance (Baluska et al. 2001; Fisher and Oparka 1996; Schulz 1995), while apoplasmic unloading rates can be modified by changes in the expression of enzymes that influence both diffusion gradients within the cell wall and the uptake of solutes into living cells (including resorption back into the phloem). The point here is that there are substantial opportunities for physiological control of unloading rates and that this may dictate rates of solute delivery, resulting in transport that is at least as much sink-controlled as source-controlled. Because it takes time for sinks to alter their ability to utilize more photosynthate (Minchin and Lacointe 2005; Minchin et al. 1997), linkages between activities above and belowground will reflect the physiology of both sources and sinks.

There is no doubt that coordination between belowground growth and metabolism and the output and activity of aboveground meristems is critical for all plants (Bloom 2005). Coordination between shoot and root activity requires some means of communicating changes in the fortunes of one sphere to the other. The existence of many sucrose sensitive genes (e.g., Koch et al. 1996) demonstrates that, in addition to being both a product and a substrate, sucrose can have a regulatory role (Farrar 1996). From the point of view of linking above and belowground phenology, however, this distinction could be viewed as moot as both depend on mass transfer of sucrose. Vascular tissues can transport information faster than they convey materials via mass transport, both by electrical signals, thought to be important in communicating herbivore damage, and by pressure pulses that propagate at much higher speeds. Mathematical analysis of phloem transport indicates that changes in loading rates results in pressure-concentration waves that have the potential to transmit information rapidly to distant sinks (Thompson and Holbrook 2004). Although research on whether such pressure-concentration waves actually elicit changes in sink physiology is lacking, it raises the possibility of a more rapid and potentially nuanced form of communication between sources and sinks. This mathematical analysis of the potential for pressure-concentration waves within the phloem has been cited as support for a physiological basis for rapid (≤ 12 h) linkage between photosynthesis and root respiration (Gaumont-Guay et al. 2008; Tang et al. 2005), but the empirical evidence from studies of phloem physiology can neither confirm nor refute the importance of this mechanism for trees at the ecosystem scale.

4 Conclusions

At the seasonal time scale, ample evidence regarding the importance of substrate supply to belowground carbon processes suggests that the answer to the question posted in the title of the chapter must be yes, that seasonal variation of soil respiration is linked to the phenology of photosynthesis. The evidence at shorter time scales is also compelling, but less complete, and confounding factors still need to be disentangled. Studies that combine automated SR and eddy covariance measurements are producing growing evidence for a link between root respiration and canopy photosynthesis on times scales of hours to weeks, but we lack sufficient knowledge of the processes of phloem transport to explain this linkage through physiological mechanisms. It is important to underscore the tremendous knowledge gaps regarding control of assimilate partitioning and growth. Not only are the tools available for studying phloem transport limited, but most studies have been conducted on relatively small, typically agronomic, plants. Whether insights derived from such studies apply equally well to trees and other woody species remains to be seen. Nevertheless, coordination between belowground growth and metabolism and the output and activity of aboveground meristems is critical for all plants. Current understanding of phloem physiology and the regulation of growth suggests that this linkage is based on more than the mass transport of substrates from sources to sinks. Thus, the often-observed pattern of root growth occurring only after spring shoot growth has ceased reflects a lower prioritization within the plant, rather than their distant location from the leaves. Similarly, while increases in photosynthesis typically result in increased growth rates (but not altered partitioning), these should not be interpreted as indicating a causal role for mass action. Instead, we are forced to accept a more physiological perspective on the linkage between the phenology of photosynthesis and the temporal patterns of soil respiration. While this does not provide an easy (or a single) answer to the question of the time scale at which such linkages may exist, it does offer the promise that as we understand better the nature of assimilate transport and partitioning we will be able to connect variation in SR patterns with the biology of the plants fueling belowground metabolism.

Acknowledgements This research was supported by the U.S. Department of Energy's Office of Science (BER) through Grant Nos. DE-FG02–00ER63002 and 07-DG-11242300–091 and through the Northeastern Regional Center of the National Institute for Climatic Change Research, Grant No. DE-FC02–06ER64157.

References

Baldocchi, D., Tang, J. and Xu, L. (2006) How switches and lags in biophysical regulators affect spatial-temporal variation of soil respiration in an oak-grass savanna. J. Geophys. Res. 111, 1–13.

Baluska, F., Cvrckova, F., Kendrick-Jones, J. and Volkmann, D. (2001) Sink plasmodesmata as gateways for phloem unloading. Myosin VIII and calreticulin as molecular determinants of sink strength? Plant Physiol. 126, 39–46.

Bingham, I.J. and Farrar, J.F. (1988) Regulation of respiration in roots of barley. Physiol. Plantarum 70, 491–498.

Birch, H.F. (1958) The effect of soil drying on humus decomposition and nitrogen availability. Plant Soil 10, 9–32.

Bloom, A. (2005) Coordination between roots and shoots. In: N.M. Holbrook and M.A. Zwieniecki (Eds.), Vascular Transport in Plants. Academic, New York, pp. 241–256.

Boone, R.D., Nadelhoffer, K.J., Canary, J.D. and Kaye, J.P. (1998) Roots exert a strong influence on the temperature sensitivity of soil respiration. Nature 396, 570–572.

Borken, W., Davidson, E.A., Savage, K., Gaudinski, J. and Trumbore, S.E. (2003) Drying and wetting effects on carbon dioxide release from organic horizons. Soil Sci. Soc. Am. J. 67, 1888–1896.

Bottner, P. (1985) Response of microbial biomass to alternate moist and dry conditions in a soil incubated with ^{14}C- and ^{15}N-labelled plant material. Soil Biol. Biochem. 17, 329–337.

Bowling, D.R., McDowell, N.G. and Bond, B. (2002) ^{13}C content of ecosystem respiration is linded to precipitation and vapor pressure deficit. Oecologia 131, 113–124.

Carbone, M.S. and Vargas, R. (2007) Automated soil respiration measurements: new information, opportunities and challenges. New Phytol. 177, 295–297.

Carbone, M.S. and Trumbore, S.E. (2007) Contribution of new photosynthetic assimilates to respiration by perennial grasses and shrubs: residence times and allocation patterns. New Phytol. 176, 1–12.

Carbone, M.S., Czimczik, C.I., McDuffee, K.E. and Trumbore, S.E. (2007) Allocation and residence time of photosynthetic products in a boreal forest using a low-level ^{14}C pulse-chase labeling technique. Global Change Biol. 13, 466–477.

Cardon, Z.G., Czaja, A.D., Funk, J.L. and Vitt, P.L. (2002) Periodic carbon flushing to roots of *Quercus rubra* saplings affects soil respiration and rhizosphere microbial biomass. Oecologia 133, 215–223.

Cisneros-Dozal, L.M., Trumbore, S.E. and Hanson, P.J. (2006) Partitioning sources of soil-repired CO_2 and their seasonal variation using a uniqe radiocarbon tracer. Global Change Biol. 12, 194–204.

Craine, J.M., Wedin, D.A. and Chapin, F.S. (1999) Predominance of ecophysiological controls on soil CO_2 flux in a Minnesota grassland. Plant Soil 207, 77–86.

Curiel Yuste, J., Janssens, I.A., Carrara, A. and Ceulemans, R. (2004) Annual Q_{10} of soil respiration reflects plant phenological patterns as well as temperature sensitivity. Global Change Biol. 10, 161–169.

Davidson, E.A. and Janssens, I. (2006) Temperature sensitivity of soil carbon decomposition and feedbacks to climate change. Nature 440, 165–173.

Davidson, E.A., Janssens, I.A. and Luo, Y. (2006a) On the variability of respiration in terrestrial ecosystems: moving beyond Q_{10}. Global Change Biol. 12, 154–164.

Davidson, E.A., Richardson, A.D., Savage, K.E. and Hollinger, D.Y. (2006b) A distinct seasonal pattern of the ratio of soil respiration to total ecosystem respiration in a spruce-dominated forest. Global Change Biol. 12, 230–239.

Ekblad, A. and Hogberg, P. (2001) Natural abundance of ^{13}C in CO_2 respired from forest soils reveals speed of link between tree photosynthesis and root respiration. Oecologia 127, 305–308.

Ekblad, A., Bostro, B., Holm, A. and Comstedt, D. (2005) Forest soil respiration rate and $d^{13}C$ is regulated by recent above ground weather conditions. Oecologia 143, 136–142.

Epron, D., Farque, L., Lucot, E. and Badot, P.M. (1999) Soil CO_2 efflux in a beech forest: the contribution of root respiration. Ann. Forest Sci. 56, 289–295.

Farrar, J.F. (1996) Sinks - integral parts of a whole plant. J. Exp. Bot. 47, 1273–1280.

Farrar, J.F. and Minchin, P.E.H. (1991) Carbon partitioning in split root systems of barley: relation to metabolism. J. Exp. Bot. 42, 1261–1269.

Fisher, D.B. (2002). Long-distance transport. In: B. Buchanan, W. Gruissem and R.L. Jones (Eds.), Biochemistry and Molecular Biology of Plants. Wiley, New York, pp. 730–785.

Fisher, D.B. and Oparka, K.J. (1996) Post-phloem transport: principles and problems J. Exp. Bot. 47, 1141–1154.

Freixes, S., Thibaud, M., Tardieu, F. and Muller, B. (2002) Root elongation and branching is related to local hexose concentration in Arabidopsis thaliana seedlings. Plant Cell Environ. 25, 1357–1366.

Gaudinski, J.B., Trumbore, S.E., Davidson, E.A., Cook, A.C., Markewitz, D. and Richter, D.D. (2001) The age of fine-root carbon in three forests of the eastern United States measured by radiocarbon. Oecologia 129, 420–429.

Gaumont-Guay, D., Black, T.A., Barr, A.G., Jassal, R.S. and Nesic, Z. (2008) Biophysical controls on rhizospheric and heterotrophic components of soil respiration in a boreal black spruce stand. Tree Physiol. 28, 161–171.

Hanson, P.J., O'Neill, E.G., Chambers, M.L.S., Riggs, J.S., Joslin, J.D. and Wolfe, M.H. (2003). Soil respiration and litter decomposition. In: H. P.J. and W. S.D. (Eds.), North American Temperate Deciduous Forest Responses to Changing Precipitation Regimes Springer, New York, pp. 163–189.

Hibbard, K.A., Law, B.E., Reichstein, M. and Sulzman, J. (2005) An analysis of soil respiration across northern hemisphere temperate ecosystems. Biogeochemistry 73, 29–70.

Högberg, P., Högberg, M.N., Göttlicher, S.G., Betson, N.R., Keel, S.G., Metcalfe, D.B., Campbell, C., Schindlbacher, A., Hurry, V., Lundmark, T., Linder, S. and Näsholm, T. (2008) High temporal resolution tracing of photosynthate carbon from the tree canopy to forest soil micro-organisms. New Phytol. 177, 220–228.

Högberg, P., Nordgren, A., Buchmann, N., Taylor, A.F., Ekblad, A., Högberg, M.N., Nyberg, G., Ottosson-Lafvenius, M. and Read, D.J. (2001) Large-scale forest girdling shows that current photosynthesis drives soil respiration. Nature 411, 789–792.

Hollinger, D.Y., Goltz, S.M., Davidson, E.A., Lee, J.T., Tu, K. and Valentine, H.T. (1999) Seasonal patterns and environmental control of carbon dioxide and water vapor exchange in an ecotonal boreal forest. Global Change Biol. 5, 891–902.

Horwath, W.R., Pregitzer, K.S. and Paul, E.A. (1994) ^{14}C Allocation in tree-soil systems. Tree Physiol. 14, 1163–1176.

Janssens, I.A. and Pilegaard, K. (2003) Large seasonal changes in Q_{10} of soil respiration in a beech forest. Global Change Biol. 9, 911–918.

Janssens, I.A., Lankreijer, H. and Matteucci, G. (2001) Productivity overshadows temperature in determining soil and ecosystem respiration across European forests. Global Change Biol. 7, 269–278.

Joslin, J.D., Wolfe, M.H. and Hanson, P.J. (2001) Factors controlling the timing of root elongation intensity in a mature upland oak forest. Plant Soil 228, 201–212.

Kieft, T.L., Soroker, E. and Firestone, M.K. (1987) Microbial biomass response to a rapid increase in water potential when dry soil is wetted. Soil Biol. Biochem. 19, 119–126.

Knohl, A., Werner, R.A., Brand, W.A. and Buchmann, N. (2005) Short-term variations in δ ^{13}C of ecosystem respiration reveals link between assimilation and respiration in a deciduous forest. Oecologia 142, 70–82.

Koch, K., Wu, Y. and Xu, J. (1996) Sugar and metabolic regulation of genes for sucrose metabolism: potential influence of maize sucrose synthase and soluble invertase responses on carbon partitioning and sugar sensing. J. Exp. Bot. 47, 1179–1186.

Liu, Q., Edwards, N.T., Post, W.M., Gu, L., Ledford, J. and Lenhart, S. (2006) Temperature-independent diel variation in soil repiration observed from a temperate deciduous forest. Global Change Biol. 12, 2136–2145.

Marcelis, L.F.M. (1996) Sink strength as a determinant of dry matter partitioning in the whole plant. J. Exp. Bot. 47, 1281–1292.

Minchin, P.E.H. and Lacointe, A. (2005) New understanding on phloem physiology and possible consequences for modeling long-distance carbon transport. New Phytol. 166, 771–779.

Minchin, P.E.H., Thorpe, M.R., Wunsche, J., Palmer, J. and Picton, R.F. (1997) Carbon partitioning in apple trees: short and long term adaptation of fruits to change in available photosynthate. J. Exp. Bot. 48, 1401–1406.

Muller, B., Stosser, M. and Tardieu, F. (1998) Spatial distributions of tissue expansion and cell division rates are related to irradiance and to sugar content in the growing zone of maize roots. Plant Cell Environ. 21, 149–158.

Oparka, K.J., Duckett, C.M., Prior, D.A.M. and Fisher, D.B. (1994) Real-time imaging of phloem unloading in the root tip of Arabidopsis. Plant J. 6, 756–766.

Pritchard, J., Ford-Lloyd, B. and Newbury, H.J. (2005). Roots as an integrated part of the translocation pathway. In: N.M. Holbrook and M.A. Zwieniecki (Eds.), Vascular Transport in Plants. Academic, New York, pp. 157–179.

Reichstein, M., Rey, A., Freibauer, A., Tenhunen, J., Valentini, R., Banza, J., Casals, P., Cheng, Y., Grunzweig, J.M., Irvine, J., Joffre, R., Law, B.E., Loustau, D., Migietta, F., Oechel, W., Ourcival, J.M., Pereira, J.S., Peressotti, A., Ponti, F., Qi, Y., Rambal, S., Rayment, M., Romanya, J., Rosi, F., Tedeschi, V., Tirone, G., Xu, M. and Yakir, D. (2003) Modeling temporal and large-scale spatial variability of soil respiration from soil water availability, temperature and vegetation productivity indices. Global Biogeochem. Cycles 17, 1–15.

Reichstein, M., Subke, J.-A., Angeli, A.C. and Tenhunen, J.D. (2005) Does the temperature sensitivity of decomposition of soil organic matter depend upon water content, soil horizon, or incubation time? Global Change Biol. 11, 1754–1767.

Ryan, M.G. and Law, B.E. (2005) Interpreting, measuring, and modeling soil respiration. Biogeochemistry 73, 3–27.

Sampson, D.A., Janssens, I.A., Curiel Yuste, J. and Ceulemans, R. (2007) Basal rates of soil respiration are correlated with photosynthesis in a mixed temperater forest. Global Change Biol. 13, 2008–2017.

Schulz, A. (1995) Plasmodesmatal widening accompanies the short term increase in symplastic unloading in pea root tips under osmotic stress. Protoplasma 188, 22–37.

Tang, J., Baldocchi, D. and Xu, L. (2005) Tree photosynthesis modulates soil respiration on a diurnal time scale. Global Change Biol. 11, 1298–1304.

Thompson, M. and Holbrook, N. (2004) Scaling phloem transport: information transmission. Plant Cell Environ. 27, 509–519.

Thorpe, M., Minchin, P., Gould, N. and McQueen, J. (2005). The stem apoplast: a potential communication channel in plant growth regulation. In: N.M. Holbrook and M.A. Zwienieck (Eds.), Vascular Transport in Plants. Academic, New York, pp. 203–221.

van Bel, A.J.E. and Hafke, J.B. (2005). Physiochemical determinants of phloem transport. In: N.M. Holbrook and M.A. Zwienieck (Eds.), Vascular Transport in Plants. Academic, New York, pp. 19–44.

Wan, S. and Luo, Y. (2003) Substrate regulation of soil respiration in a tallgrass prairie: Results of a clipping and shading experiment. Global Biogeochem. Cycles 17, 1054.

Wardlaw, I.F. (1990) The control of carbon partitioning in plants. New Phytol. 116, 341–381.

Williams, J.H.H. and Farrar, J.F. (1990) Control of barley root respiration. Physiol. Plantarum 79, 259–266.

Winch, S.K. and Pritchard, J. (1999) Acid-induced wall loosening is confined to the accelerating region of the root growing zone. J. Exp. Bot. 50, 1481–1487.

Xu, M. and Ye, Q. (2001) Spatial and seasonal variations of Q_{10} determined by soil respiration measurements at a Sierra Nevadan forest. Global Biogeochem. Cycles 15, 687–696.

The Annual Cycle of Development of Trees and Process-Based Modelling of Growth to Scale Up From the Tree to the Stand

Koen Kramer and Heikki Hänninen

Abstract Climate change affects both the annual cycle of tree development and the processes related to tree growth. The annual cycle of development manifests as observable phenological events such as leaf unfolding, flowering and leaf fall, but also includes less apparent traits, such as changes in frost hardiness and photosynthetic capacity. Seasonality in these traits can be due either to a fixed sequence of events that take place even in a constant environment, or to fluctuations in environmental factors. Thus, in a constant environment, the latter mode of development displays no seasonality. In addition, and depending on the trait considered, the internal state of development affects the tree's capacity to respond to environmental factors. Given that the effects of climate change on the seasonality of a particular phenological trait may depend on interactions between fixed and fluctuating development traits, in order to explore these effects the entire annual cycle of development must be modelled. The processes related to tree growth include photosynthesis, respiration and allocation at the level of the individual tree; at stand level they include resource availability and biotic interactions. In this chapter we present the general theory of the annual cycle of development of trees, with examples of climate change effects on phenological traits with different mode of development for tree species in the boreal, temperate and Mediterranean zone of Europe. A process-based model on tree growth is outlined, with focus on scaling up from the tree to the stand level in time and space. Examples of climate change are presented, based on a model that couples the annual cycle of development and the growth of trees. Phenological events are characterized by responses to temperature that are under strong selective pressure. Future lines of development in this field of research include an assessment of the adaptive potential of phenological events to climate change. An example of this genetic approach is also presented.

K. Kramer (✉)
Centre of Ecosystem Studies, Wageningen University and Research Centre,
Wageningen, Gelderland, The Netherlands
e-mail: koen.kramer@wur.nl

H. Hänninen
Department of Biological and Environmental Sciences, University of Helsinki,
Helsinki, Finland
e-mail: heikki.hanninen@helsinki.fi

A. Noormets (ed.), *Phenology of Ecosystem Processes*,
DOI 10.1007/978-1-4419-0026-5_9, © Springer Science+Business Media, LLC 2009

1   Introduction

The zones where boreal and temperate trees grow have a very pronounced seasonal climate. In the southern temperate and the Mediterranean regions of the northern hemisphere it is the availability of water that varies seasonally, whereas in the more northern conditions the seasonality is caused by the variation in incoming solar radiation and in air temperature – especially the latter. Boreal and temperate trees are adapted to the seasonality of air temperature: in their annual cycle of development the frost-hardy dormant state and susceptible active growth state are synchronized with the annual climatic cycle (Fuchigami et al. 1982; Koski and Sievänen 1985; Sarvas 1972, 1974).

The annual cycle has to be explicitly addressed when building process-based models for the growth of tree stands, as otherwise the stand models may significantly overestimate growth. Bergh et al. (1998), for instance, found that the gross primary production of a northern Swedish stand of Norway spruce (*Picea abies* (L.) Karst.) was overestimated by 40% when the simulation addressed only the instantaneous effects of climatic factors, without considering the long-term restrictions on the inherent photosynthetic capacity caused by low air temperature and the limiting effects caused by frozen soil. Fortunately, there is a long tradition in modelling the annual cycle of boreal and temperate trees (Cannell 1990; Hari 1972; Pelkonen and Hari 1980), so models for the different aspects of the annual cycle are available for stand modellers. Recently we reviewed the modelling of the annual cycle at some length, including in our review the equations for a selection of published models (Hänninen and Kramer 2007). The following discussion of the theoretical fundamentals of phenological modelling and the classification of the models of the annual cycle is largely based on that review.

The remainder of this chapter is structured as follows. In Sect. 2 we present general aspects on modelling the tree's annual cycle of development. In Sect. 3 we give examples on how the phenological models have been used to assess the effects of climatic change on tree species. In Sect. 4 process-based models of tree growth and their use for scaling up in space and time are discussed. In Sect. 5 we provide a synthesis of the two previous sections, i.e. we present cases where phenological models have been coupled with process-based growth models and show how these integrated models have been used for assessing the effects of climatic change on broadleaved deciduous and needle leaved evergreen trees species of the boreal, temperate and Mediterranean zone in Europe. Finally, in Sect. 6 we discuss future developments.

2   Principles for Modelling the Annual Cycle of Trees

In the models of the annual cycle (Fig. 1), the key state variable is *the state of development*, *S(t)*. It quantifies the phase of a given attribute of the annual cycle. The state of development can either describe the physiological attributes of the annual cycle or quantify the annual ontogenetic cycle of trees. The physiological

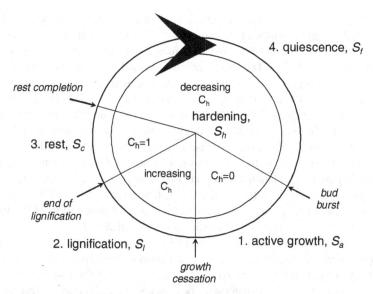

Fig. 1 The annual cycle of trees of boreal and temperate zones. The *outer circle* describes fixed sequence development and phenological events. The *inner circle* describes frost hardiness as example of a trait with some fluctuating development. The capacity for hardening depends on the state of ontogenetic development of the fixed sequence cycle. See the Appendix for details on the models of the annual cycle of trees for the 4 phases shown on the Figure.

attributes include frost hardiness (Leinonen 1996a; Repo et al. 1990a) and photosynthetic capacity of needles (Mäkelä et al. 2004; Pelkonen and Hari 1980) of evergreen conifers. In this case, the models of changes in the state of development can be tested with empirical data, since the physiological properties (e.g. frost hardiness) are directly measurable (Leinonen 1996a; Repo 1992). If the state of development represents ontogenetic development, the empirical part of the studies is more complicated, since it is impossible to continuously measure the changing value of the state of development (Fuchigami et al. 1982; Sarvas 1972, 1974). The best examples of this approach are various temperature sum and day degree models that predict the timing of bud burst of deciduous trees during spring (Chuine 1998; Chuine et al. 1998; Hänninen 1990; Hänninen 1995; Kramer 1994a, b). These models predict the phenological event on the basis of the cumulative sum of temperature; the event's threshold is specific to the event in question, and also to the species and provenance. As empirical observations are only available for the date of the occurrence of the event, strictly speaking the only phenomenon that is being modelled is the timing of the event. In physiological terms, however, the phenological event is the final outcome of a microscopic ontogenetic development that has been going on within the bud for a long time. The progress of this ontogenetic development before the phenological event becomes visible is simulated by the increasing value of the temperature sum, i.e. with the value of the $S(t)$ state of development.

As in any other dynamic model, the course of the state of development over time is simulated by first calculating the value of its first time derivative, i.e. the value of *the rate of development, R(t)*, and then integrating *R(t)* over time (Hari 1972; Hari and Rasanen 1970). The effect of environmental factors on phenology is modelled via the responses of the *R(t)* to the environment. The best known example of such a model is the classic temperature sum (Reaumur 1735) or thermal time model (Cannell and Smith 1983). In this case, the value of *R(t)* equals zero if the ambient temperature is zero below a given threshold and increases linearly above that threshold. The crucial phase of modelling is the formulation of the environmental responses of the rate of development, since the following integration over time is pure mathematical deduction. Thus, once the environmental responses of rate of development are determined, then the time course of state of development, *S(t)*, is also determined for any environment for which the time course of the driving environmental factors is known (Hänninen and Lundell 2007).

As the morphological and physiological phenomena of the annual cycle of development are very diverse, various assumptions must be made when modelling the annual cycle. Hänninen and Kramer (2007) classified the phenomena of the annual cycle into two main categories. The concept of *fixed sequence development* covers phenomena belonging to the irreversible ontogenetic development, for instance those leading to bud burst during spring (Kramer 1994a, b; Sarvas 1972, 1974). In this category, the value of the state of development, *S(t)*, either increases or stays constant, but it never decreases. In other words, the rate of development, *R(t)*, is either positive or zero, but never negative. On the other hand, the concept of *fluctuating development* characterizes various phenomena of physiological acclimation which are at least partly reversible. For instance, this category includes the springtime recovery of the photosynthetic capacity (Pelkonen 1980; Pelkonen and Hari 1980).. In this category, the value of the state of development, *S(t)*, can increase as well as decrease, and the value of the rate of development, *R(t)*, can be either positive or negative. In *integrated models* both types of development are modelled simultaneously, so that the fixed sequence phenomena affect the fluctuating phenomena. This is the case, for instance, in the more comprehensive models of frost hardiness, where the phase of the annual growth cycle restricts the hardening of the trees. During the dormant phase the frost hardiness is fully reversible, i.e. when the temperature drops again after an intermittent mild spell during winter, rehardening follows dehardening, whereas during the phase of active growth there is very little, if any, rehardening (Kellomäki et al. 1992, 1995; Leinonen 1996b).

We further classify the models based on the assumptions concerning the environmental responses of the rate of development (Hänninen and Kramer 2007). In *E-models* the environmental responses stay constant, as is the case for instance in the classic temperature sum models. This simplifying assumption of constancy is justifiable when addressing only a limited part of the annual cycle, but when addressing the whole cycle, however, more complicated *ES-models* are needed. In these models the environmental responses change during the annual cycle. A given air temperature, for instance, may bring about either hardening or dehardening, depending on the phase of the annual cycle. In some models in the fluctuating development category, the prevailing state of development of a given attribute of the cycle affects the response of that attribute's rate of development to air temperature. Models incorporating feedback in this way have been developed for photosynthetic capacity (Pelkonen and Hari 1980) and for frost hardiness (Leinonen et al. 1995; Repo et al. 1990a). Furthermore, in some integrated models, the state of development of a given attribute of the cycle affects the environmental response of the rate of development of another attribute. This is the case in integrated models of frost hardiness, where the state of the ontogenetic development affects the environmental responses of hardening/dehardening (Kellomäki et al. 1992, 1995; Leinonen 1996b).

3   Effects of Climate Change on the Phenology of Trees

The impact climate change has on a tree's functioning depends on the category of development the affected trait belongs to: a given change in temperature affects traits with fixed sequence development differently than a trait with fluctuating development. Furthermore, climate change affects state-dependent changes differently than state-independent changes. In this section we present examples illustrating the general theory of modelling the annual cycle of trees as outlined above. See the Appendix for the equations and parameter values used in the various phenological models.

3.1   Bud Burst

Tree bud burst is a classic example of a fixed sequence development. It occurs when the cumulative chilling (Eqn. 4) and forcing (Eqn. 5) temperatures exceed a certain threshold. Climate change influences how rapidly chilling temperature is accumulated during the rest phase; it also influences forcing during quiescence. In this way it influences the date in spring on which bud burst occurs. A change in the date of bud burst leads to a change in the probability of frost damage and in the amount of radiation that is available for growth during the growing season. However, different tree species differ in their response to climate change. Table 1 presents the average date of leaf unfolding and leaf fall over the period 1894–1959, and the change in these phenological events in relation to average winter and summer temperature (Kramer et al. 1996).

Using these responses of leaf unfolding and leaf fall to temperature, the consequences of global warming on both the probability of frost damage and on the duration of the growing season can be assessed. Consider the tree species in Table 1: given the simulated advance of the date of leaf unfolding (Fig. 2a), in NW Europe the probability of frost damage to these species is likely to decrease. As the rates of advancement for leaf unfolding and leaf fall are not

Table 1 Phenological characteristics of *Betula, Fagus* and Quercus in the Netherlands. U – average date of leaf unfolding, P_0 – probability of sub-zero temperature in a symmetrical 11-d period around U, dU/dT_w – change in U with change in winter temperature (d C^{-1}), F – average date of leaf fall, dF/dT_s – change in F with change in summer temperature (d C^{-1}). Negative values indicate advancement of date of leaf unfolding or of leaf fall

Species	U	P_0	dU/dT_w	F	dF/dT_s
Betula pubescens	April 22	0.58	−5	October 4	−3
Fagus sylvatica	May 2	0.37	−4	October 16	0
Quercus robur	May 6	0.18	−5	October 20	−5

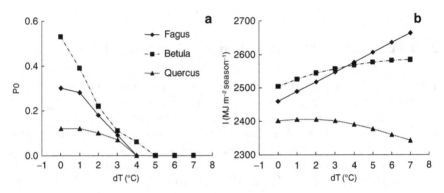

Fig. 2 Effect of an increase of temperature on: (**a**) probability of frost damage, P_0, and (**b**) available radiation during the growing season, I. Reproduced with permission from Kramer (1996). © Inter-Research.

the same, climate change is projected to affect the length of the growing season and thus the amount of radiation that is available for growth (Fig. 2b). This analysis suggests that *Fagus* is likely to gain the most from the warming climate, whereas if the temperature increases by more than 3°C, the amount of light available would decrease for *Quercus*.

3.2   Frost Hardiness

Frost hardiness is an example of a trait whose development fluctuates with changes in temperature. It can be operationally defined as the lowest temperature at which there is no visual damage to tissues within the bud. However more advanced methods are now available to determine the degree of frost hardiness (Repo 1993; Repo and Lappi 1989; Repo et al. 1990b); it can be simulated by assuming a stationary level of frost hardiness that depends on temperature and day length ((17); Leinonen 1996b). The stationary level is attained only if the environmental conditions are constant for a certain period of time. How quickly the actual level of frost hardiness

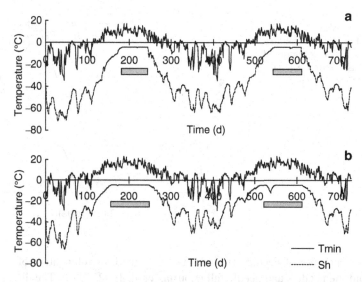

Fig. 3 Implications of climate change for *Pinus sylvestris* in Finland on the duration of the growing season (*grey horizontal bars*), and the level of frost hardiness, S_h. A. current climate, B. future Finnish climate based on an increase of 6°C in daily temperature. Frost damage occurs if the minimum daily temperature, T_{min}, is less than S_h. Data of minimum temperature of the current climate are observations from Joensuu, Finland.

attains the stationary level then depends on an ecophysiological time constant and the difference between the actual and stationary levels of frost hardiness (Eqn. 6).

An example of frost hardiness and the duration of the growing season of *Pinus sylvestris* in the current climate and in a hypothetically warmer climate (Fig. 3) shows that in the projected future Finnish climate (temperature increase by 6°C), the difference between the minimum daily temperature (T_{min}) and the level of frost hardiness (S_h) decreases. Thus, under these conditions, there is a greater probability of the minimum daily ambient temperature falling below the level frost hardiness of the plant, thus resulting in frost damage. Also the variability in minimum temperature may increase in a warmer climate. This could lead to an increased probability of frost damage even in the absence of warming if the state of frost hardiness of the plant can not track the increased variability in temperature (depending on the value of the time constant in (6)). We did not include that aspect of climate change in this example.

3.3   Water-Driven Phenology

Changes in water availability affect the phenology of trees via different mechanisms than the impacts of temperature. As an example, the development of leaf area index (LAI) of *Pinus pinaster* Aiton is presented. In coniferous species, water availability

Fig. 4 Development of leaf
area index (LAI) of *Pinus
pinaster* and *Molinea caerulea* in
south-east France. Reproduced
with permission from Kramer
et al. (2000). © Springer.

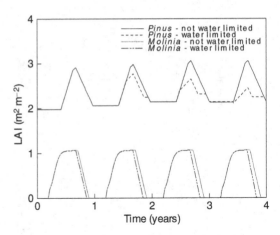

affects the number of needle primordia being formed, the elongation of both the
shoot and the needles and needle fall (Loustau et al. 1992, 1996). The direct effect
of summer drought on coniferous trees is a reduction in leaf area index (LAI)
because shoot- and needle elongation is reduced and needle fall increased. The
indirect effect of summer drought is a reduction in needle primordia which affects
the LAI of the next year. This is because in coniferous trees, a segment of the shoot
is associated with a bundle of needles (Loustau et al. 1992, 1996). Thus, shoot
elongation stops as soon as all primordia are developed into needles even if water
is in ample supply. Consecutive dry summers have a more pronounced impact on
the LAI of *P. pinaster* than a single dry year (Fig. 4). This mechanism affects both
the development of LAI over the growing season and the date on which the LAI
attains its maximum value.

In contrast to conifers, the impact of limiting water availability on the LAI of a
deciduous grass *Molinia* is instantaneous, and the stress from consecutive dry years
does not compound (Fig. 4). For deciduous or evergreen broadleaved Mediterranean
trees, there is also no such indirect effect of water limitation on the next years leaf
area as elongation of both leaves and shoot depend on instantaneous water supply.

4   Process-Based Tree Growth Models

Tree phenology is an important aspect of a tree's responses to climate change, as it
determines the onset and cessation of the functioning of the tree, as well as the
tree's frost hardiness and photosynthetic capacity. As was made clear in the previous
section, it is necessary to simulate the entire annual cycle, rather than parts of it
(e.g. a temperature sum leading to bud burst) because there are interactions between
physiological attributes and ontogenetic development. Hence, to evaluate climate
change impacts on tree growth, the entire annual cycle as described above needs to
be integrated in process-based models on tree growth. In this section we present one

such model of tree growth, FORGRO, to illustrate the scaling up of processes in space and time. In Sect. 5 we present the importance on phenology for climate change responses at these larger scales.

FORGRO describes stand properties, expressed per hectare, including: tree density, stem volume, tree height, stem diameter at breast height (DBH), canopy dimensions, biomass of foliage, branches, stem (heartwood and sapwood) coarse roots and fine roots. The model describes the physical environment (light, temperature, vapour pressure deficit) in the vegetation and in the soil in detail. The interception and attenuation of light in the canopy are described in detail because of the non-linear relationship between photosynthesis and available light (Spitters et al. 1986). FORGRO includes a leaf energy balance (Goudriaan and Van Laar 1994), thus a vertical temperature gradient, as photosynthesis is strongly affected by temperature. Maintenance respiration is proportional to the amount of respiring biomass and increases exponentially with temperature (Cannell and Thornley 2000; Penning de Vries et al. 1974; Ryan 1995; Ryan et al. 1996). Growth respiration is proportional to total growth but is not temperature-dependent (Penning de Vries et al. 1974). For the allocation of photosynthates among tree components it is assumed that the internal development of partitioning rations depends solely on the tree's dimensions (Grote 1998). This assumption can be derived from pipe model theory and the principle of a functional balance between tree components (Mäkelä and Hari 1986; Valentine 1988). Allocation is especially important for the long-term dynamics of forest growth; however, it is not important for comparing the predicted and observed exchange of CO_2 and H_2O. The hydrological aspect of FORGRO includes interception and evaporation of rain by the canopy, and transpiration of water taken up from the soil by the vegetation (Kropff 1993). The links between carbon and water cycles in the soil and the vegetation are through the effects of soil moisture and air vapour pressure deficit on stomatal conductance (Leuning 1995; Leuning et al. 1995).

4.1 Scaling Up in Space

The scaling up in space entails moving from leaf level to canopy level, which is primarily dependent on accurately characterizing the vertical light profile through the canopy and on assessing the light climate for individual leaves (Leuning et al. 1995). The amount of radiation absorbed by a given leaf layer affects the leaf temperature and stomatal conductance and thus the rates of photosynthesis and transpiration per layer and for the whole canopy. In the FORGRO model the integration of assimilation and transpiration over the foliage layers is performed using a nested Gaussian integration technique (Goudriaan and Van Laar 1994). This is an efficient technique for the numerical integration of a known, smooth function. The basic idea is to evaluate that function at representative positions on the total integration interval and to assign weights to each of these function evaluations. In the case of the light profile in a canopy, the amount of radiation absorbed at different depths in the canopy depends on the incoming direct and diffuse shortwave radiation, and the downward attenuation and scattering of light results in a declining proportion of sunlit foliage and an increasing proportion of shaded foliage (Spitters et al. 1986). Very good estimates of the integral, i.e. the whole canopy value of absorbed radiation, photosynthesis and transpiration, are obtained if, for the shaded foliage, three positions, i.e. foliage layers within the canopy, are considered; for the sunlit foliage, five positions within these three foliage layers must be considered. This leads to evaluations at 15 light intensities for the functioning of sunlit foliage. In addition to absorbed radiation, the model calculates stomatal conductance, foliage temperature, internal CO_2 concentration, i.e. within the stomatal cavity, and the hourly rates of photosynthesis and transpiration. Figure 5 shows an example of the contribution of the different foliage layers to each of these variables. In this example, the stomatal conductance for water declines over the day because there is a large air vapour pressure deficit. Consequently, at midday, foliage temperature exceeds ambient temperature, which depresses photosynthesis The reduction in transpiration is less

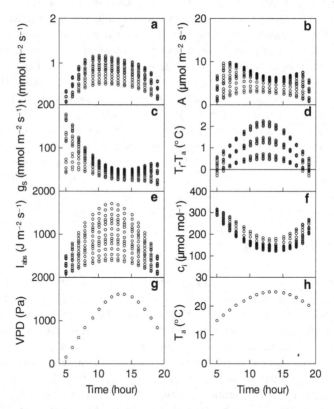

Fig. 5 Instantaneous rates of transpiration and photosynthesis and the variables that determine these rates over 15 foliage layers as modelled with the FORGRO model. Each dot represents the value of a given variable of a layer at a particular moment of the day. (**a**) transpiration, t, (**b**) photosynthesis, A, (**c**) stomatal conductance for water, gs, (**d**) difference between foliage temperature, T_f, and air temperature, T_a, (**e**) absorbed radiation, I_{abs}, (**f**) internal CO_2 concentration, c_i (**g**) air vapour pressure deficit, VPD, (**h**) air temperature, T_a.

pronounced than the reduction of photosynthesis because the foliage temperature is above optimal temperature for photosynthesis. The scaling up in space from the leaf to the canopy scale is then obtained by summing the assimilates and the transpired water over the foliage layers in the canopy.

4.2   Scaling Up in Time

Integration of forest growth over time refers to the daily and annual accumulation of above- and belowground biomass, soil organic matter and amount of water in the soil. This integration is first done by accumulating the instantaneous values for photosynthesis and transpiration (per second and per unit ground area) to daily values (per day and unit ground area). Thus, in the model the spatial integration always precedes the temporal integration. In FORGRO, 24 hourly values are accumulated to achieve a daily value; this is done assuming that the instantaneous values are constant per hour. Figure 6 shows examples of daily output for a 2-year period for a *Pinus pinaster* forest with a *Molinia caerulea*

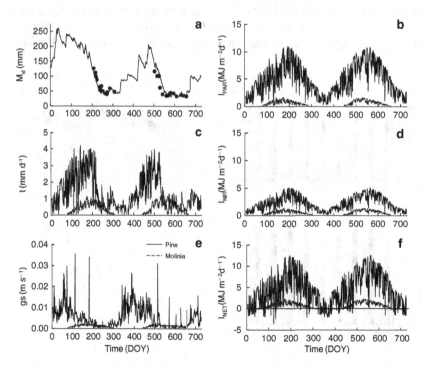

Fig. 6 Daily model output of a *Pinus pinaster* stand with *Molinia caerulea* understory in Bray, France, as calculated by the FORGRO model. (**a**) soil moisture, M_{sl}, data provided by Dr. Denis Loustau, (**b**) absorbed photosynthetically active radiation, I_{PAR}, (**c**) transpiration, t, (**d**) absorbed near-infrared radiation, I_{NIR}, (**e**) stomatal conductance, g_s, (**f**) net radiation, I_{NET}.

grass understory in Bray in the south-east of France. Daily incoming global radiation and LAI distribution determine the absorbed PAR and NIR for both *Pinus* and *Molinia*, and the net radiation is derived by subtracting the outgoing long-wave radiation from the incoming short-wave radiation. In this example, there is feedback between stomatal conductance, soil moisture content and transpiration, because transpiration reduces the soil moisture content, thus increasing the soil moisture deficit which causes the stomata to close thereby reducing transpiration. This instantaneous effect of soil moisture deficit on whole-tree functioning is additional to the long-term effect of water shortage on the development of leaf area index presented in Fig. 4.

Figure 7 shows the results of accumulating daily values to attain an annual value for the Bray site. It shows large variability in annual precipitation. The transpiration by the pine also shows large differences between years, whereas the transpiration by the understory is relatively constant, and is on average responsible for 20% of the total transpiration. Net primary productivity (NPP) is allocated to the different tree components, resulting in a change in the biomass of foliage, branches, stem, coarse roots and fine roots. Periodic thinning reduces the amount of biomass. Tree growth expressed as annual volume increment provides an output that can be tested against independent observations.

Once a process model has been validated at both the daily and the annual scales, its output can be used either to upscale to larger spatial scales, or to evaluate the long-term consequences of climate change scenarios. For example, radiation use

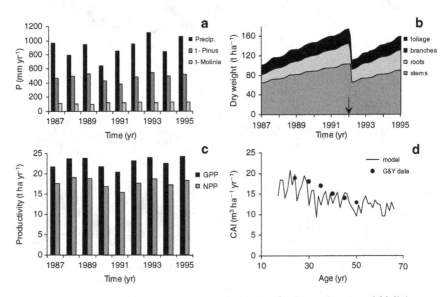

Fig. 7 Annual values of: (**a**) precipitation and transpiration for *Pinus pinaster* and *Molinia caerulea*, (**b**) biomass per tree component of *Pinus pinaster*, (**c**) gross primary productivity, GPP, and net primary productivity, NPP, of *Pinus pinaster*, and (**d**) current annual increment, CAI, of *Pinus pinaster*. The *arrow* in panel (**b**) indicates a thinning event.

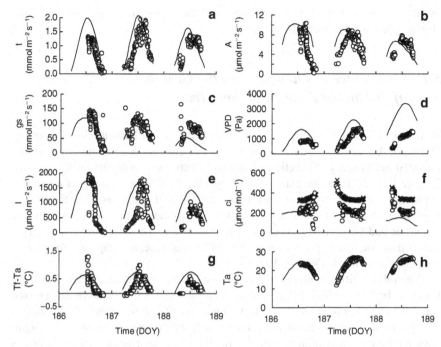

Fig. 8 Validation of model simulations at the leaf level. The simulated values are of the upper layer of the sunlit leaf area (see Fig. 5). The observations are on the sunlit needles of *Pinus* pinaster at Bray, France. Data provided by Dr. Denis Loustau. (**a**) transpiration, *t*, (**b**) photosynthesis, *A*, (**c**) stomatal conductance for water, *gs*, (**d**) vapour pressure deficit of the air, *VPD*, (**e**) irradiance, *I*, (**f**) internal CO_2 concentration, c_i, (**g**) foliage minus air temperature, $T_f - T_a$, (**h**) air temperature.

efficiency provides a simple statistic for evaluating interannual activity, as well as for validations against remote sensing models based on light use efficiency (e.g. Running's and Xiao's models, see Chap. 11 in current volume). Validation of a process-based model entails comparing the model output with independent data at different spatial and temporal scales, including (a) the timing of the onset and cessation of the growing season for phenology (Kramer 1994a, b, 1995a, b), (b) leaf values (Fig. 8) (c) daily values (Figs. 6a, 9b), (Kramer et al. 2002) and (d) long-term growth and yield (Fig. 9d) and ecosystem carbon budgets (Mohren et al. 1999).

5 Effects of Climate Change on Growth of Trees

The combined effects of climate change on the annual development and growth of trees can be evaluated with a model that combines the annual cycle of development of trees and the processes that drive tree growth as described above.

In this section, we present several examples of the importance of phenology for the climate change assessment of tree growth.

5.1   Importance of Phenology on the Growth of Temperate Zone Deciduous Trees

Tree species differ in their response of the timing of bud burst to global warming (Table 1). An earlier bud burst leads to a longer growing season and thus enhancesannual net primary productivity but also affects the probability of spring frost damage (Fig. 2). It can be expected that the climate change response of a species' net primary productivity in a monoculture will not be the same as when the species is growing in a mixed stand of species differing in their temperature response the timing of bud burst. This is because competition for light is asymmetrical: larger individuals obtain a disproportional share of the resource, and suppress the growth of smaller individuals (Weiner 1990).

When the FORGRO model was used to evaluate the impact of changes in the timing of bud burst with increasing temperature on NPP in monoculture and a mixed species stand it was found that the effects of phenological differences between species became more pronounced in the mixed stand, where species compete for light, compared with monocultures (Fig. 9). In this example, climatic warming brings about a steady decline of the NPP of monocultures of *Fagus, Quercus* and *Betula*. The temperature effect on whole-tree respiration thus exceeds the effect of climate change on gross photosynthesis and the duration of the growing season. In mixed species stands, however, *Fagus* benefits more from the higher temperature than the other species and becomes more competitive, at the expense of the other species.

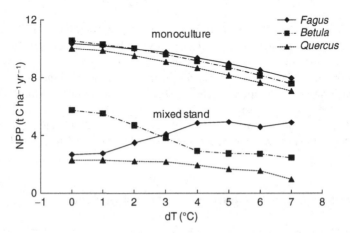

Fig. 9 FORGRO model predictions of net primary production (NPP), for three species with different phenological responses to temperature (see Table 1, Fig. 2). Reproduced with permission from Kramer et al. (1996). © Inter-Research.

5.2   Importance of Phenology on the Growth of Boreal Coniferous Trees

Coniferous tree species in the boreal zone are vulnerable to late spring frosts, which damage the needles (i.e. reduce leaf area) and suppress photosynthetic capacity. Both the development of needle area (19) and (20) and the recovery of photosynthetic capacity (21)–(23) depend on the level of frost hardiness, which itself depends on the state of ontogenetic development.

When the importance of the effect of frost hardiness on needle area and on the recovery of photosynthetic capacity for gross photosynthesis of Scots pine was analysed using FORGRO, it was found that frost hardiness affected the recovery of photosynthetic capacity much more than it affected leaf area development (Fig. 10).

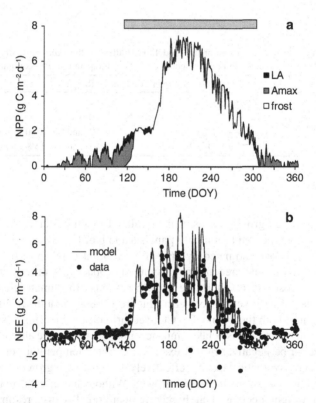

Fig. 10 (a) Effect of frost on net primary productivity (NPP) of Scots pine in Hyytiälä, Finland. *frost* – taking account of effect of frost damage on needle area (LA) and on recovery of photosynthetic capacity (A_{max}) account. *LA* – only taking account of effect of frost on photosynthetic capacity. *Amax* – only taking account of effect of frost on needle area. *Grey bar* indicates the growing season. (b) Comparison of simulated and observed net ecosystem carbon exchange (NEE) at Hyytiälä, Finland. Data provided by Dr. Timo Vessala. Reproduced with permission from (Mohren et al. 1999). © SPB Academic Publishing.

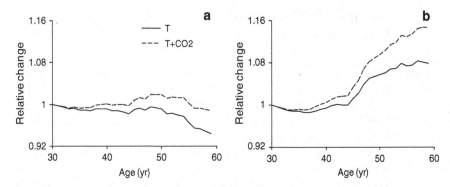

Fig. 11 Change in stem weight of *Pinus sylvestris* at Hyytiälä, Finland, relative to the current climate. (**a**) without and (**b**) with the effects of frost damage.

Table 2 Climate change scenarios for the modelling studies for the *Pinus sylvestris* in Finland (scenarios 1–3) and *P. pinaster* in France (scenarios 1–4). T temperature, P precipitation

Scenario No	Linear increase over 50 years	Range
1	Reference	–
2	T	0–3°C
3	T + CO$_2$	2) and 350–500 µmol mol^{-1}
4	T + CO$_2$ + P	2) and 3) −15% precipitation

The duration of the growing season is simulated much better if the effects of frost are taken into account (compare panels a and b of Fig. 10). In this example, the annual NPP is overestimated by 26% if no account is taken of the effects of frost hardiness on needle area and photosynthetic capacity. Failure to incorporate the effects of frost can result in the growth response to climate change being negative instead of positive (or vice versa). Figure 11 shows that the relative change of stem weight in both the T- and T + CO$_2$-scenario (see Table 2) is positive when the effect of late spring frosts is taken into account, whereas in the absence of frosts the effect would be negative. In this example, elevated temperature enhances the recovery of photosynthetic capacity, effectively extending the growing season (see Fig. 9a) thereby leading to increased growth. Without the effects of late spring frosts, stem weight decreased at higher temperature because respiration was stimulated more than assimilation. This effect is not completely counteracted by an increase in atmospheric CO$_2$ concentration. However if effects of frost are included, then the effect of increasing atmospheric CO$_2$ concentration on growth more than counteracts the effect of temperature on respiration.

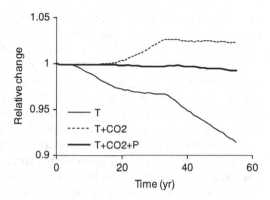

Fig. 12 Change in stem weight of *Pinus pinaster* at the Bray site in southern France under different climate scenarios (Table 2) relative to current climate. The abrupt change in response is caused by a management intervention.

5.3   Importance of Phenology on the Growth of Mediterranean Coniferous Trees

In the Mediterranean zone the impact of climate change works through the effects of water availability rather than of temperature (see Sect. 3.3). Water availability strongly influences the leaf area development of both the pine upper story and the grass understory (Fig. 4). Figure 12 shows the effects of different climate change scenarios (Table 2) on stem weight of *Pinus pinaster* at Bray, France relative to the reference scenario. The results show that elevated temperature plus atmospheric CO_2 concentration stimulate stem weight accumulation, whereas on its own, elevated temperature reduces stem weight accumulation relative to current climate conditions. The smaller leaf area that is the consequence of reduced precipitation counteracts the beneficial effect of elevated CO_2, but not to the same degree as scenario 2.

6   Future Research

The next challenge in the evaluation of climate change impacts on tree growth is to assess whether trees have the potential to adapt to climate change because climate change is likely to exert strong evolutionary pressure on (Parmesan 2006). Some authors have expressed concern that trees are unable to adapt to such changes because the rate of climate change is rapid relative to the longevity of individual trees (Davis and Shaw 2001), trees may not have adequate genetic diversity to adapt to the changing environmental conditions (Davis and Kabinski 1992), and trees may not be able to disperse to newly available habitat fast enough to outstrip the rate of global change, as the landscape they have to cross is highly fragmented (Jump and Penuelas 2005). Other authors, however, point to evident characteristics that uniquely empower trees to withstand environmental changes (Hamrick and Godt 1996):

trees have high phenotypic plasticity that allows them to withstand large environmental fluctuations during their lifetime (Rehfeldt et al. 2002); there are high levels of genetic diversity for allozymes and nuclear markers within – rather than between – populations (Buiteveld et al. 2007; Leonardi and Menozzi 1995); and gene flow – especially of pollen – occurs over large distances, thereby exchanging favourable genetic variants between isolated stands (Petit and Hampe 2006).

Adaptive responses of trees are particularly important at the edges of the geographic distribution of trees because as demographic processes at the leading edge differ from those at the rear edge of a species area (Hampe 2005; Thuiller et al. 2008). At the northern of the distribution of tree species of the northern hemisphere, is phenology – particularly the timing of the onset of the growing season – an important adaptive trait, as it balances the effect of frost and the effective use of the available growing. At the southern edge of the distribution, it is rather water limitation and successful recruitment that determines the rate of adaptive response of traits related to water availability. As Hamrick (2004) points out, the rate of adaptation to environmental change depends largely on successful recruitment events during the lifetime of a tree, and less so on the longevity of individual trees. Indeed, trees have overlapping generations and produce large numbers of seeds at regular intervals. During the first few decades of forest development tree numbers decline from several millions per hectare to a few hundred or less, resulting in a strong genetic selection (Geburek 2005).

To assess the adaptive potential of the phenology of trees, a process-based growth model needs to be combined with a quantitative genetic model. The genetic model then characterizes selected parameters of the process model using a multi-locus/multi-allele genetic system for individual trees. The individual-tree model then includes gene flow by seed and pollen dispersal and tree mortality, in addition to the processes related to growth described above for FORGRO (Kramer et al. 2008).

Figure 13 shows an example of the adaptive response of the critical state of chilling for bud burst of *Fagus sylvatica* to climate change scenarios after running the model

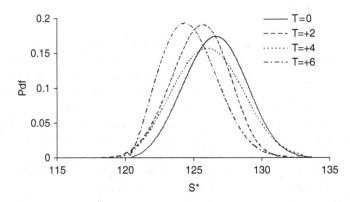

Fig. 13 Adaptive response of critical state of chilling, S*, of a *Fagus sylvatica* to elevated temperature scenarios. Pdf indicates the probability density function of the Weibull distribution fitted through the relative frequency of S* in the simulated beech population.

simulation over 300 years at the northern edge of the geographic distribution of European beech. The model was initialized with a stand of about 2 ha containing 181 trees for which position, height, diameter at breast height, crown dimensions and a number of years with bud burst data were known. Position and crown dimension are important to determine the relative contribution of neighbouring father trees of the seeds of a target mother tree. The genetic model was initialized for the critical state of chilling (S_o^*) of the phenological model. This means that allelic effects were assigned to alleles of a 10-loci and two allele system for each of this parameters. Allelic effects were statistically determined such that (a) the distribution of budburst dates in the stand coincided with that of the simulated beech population, and (b) initial allele frequencies followed a theoretical skewed distribution (Kramer et al. 2008). The latter is important because the initial distribution of allele frequency has a strong impact on the rate of the adaptive response. The assumed effect of frost on plant fitness was that if ambient temperature drops below −2°C within 5 days after budburst, seedlings (<50 cm) are killed and both leaves and flowers of adult trees are dropped. The foliage of adult trees can be restored from a reserve pool, but the trees are assumed not to flower again the same season. The frequency of alleles change during the simulation as seedlings are killed or trees contribute less to the next generation than other trees depending on their genetic make-up. Consequently, the distribution of the values of the critical state of chilling in the population changes.

The first results indicate a selection for a lower critical state of chilling, i.e. for lower chilling requirement, thus adaptation towards earlier bud burst − which is indeed earlier than can be attained by the phenotypic plastic response of bud burst alone. Although this example is only one representation of the selection process on a single parameter, which includes much stochasticity, it does indicate that there is some potential for an adaptive response of phenology to climate change. Tentative results of the adaptive response of stomatal conductance to soil water availability at the southern edge of the distribution of beech indicate that the regeneration may become much more erratic in a future climate, resulting in a lack of adaptive response of this particular trait.

Acknowledgements This research is part of the strategic research program "*Sustainable spatial development of ecosystems, landscapes, seas and regions*" which is funded by the Dutch Ministry of Agriculture, Nature Conservation and Food Quality, and carried out by Wageningen University Research Centre. KK was also supported by the EU-FP6 Network of Excellence *EVOLTREE* (contract no. 016322). Joy Burrough advised on the English.

References

Bergh, J., McMurtrie, R.E. and Linder, S. (1998) Climatic factors controlling the productivity of Norway spruce: A model-bases analysis. For. Ecol. Manag. 110, 127–139.
Buiteveld, J., Vendramin, G.G., Leonardi, S., Kamer, K. and Geburek, T. (2007) Genetic diversity and differentiation in European beech (*Fagus sylvatica L.*) stands varying in management history. For. Ecol. Manag. 247, 98–106.

Cannell, M.G.R. (1990) Modelling the phenology of trees. Silva Carelica 15, 11–27.

Cannell, M.G.R. and Smith, R.I. (1983) Thermal time, chill days and prediction of budburst in Picea sitchensis. J. Appl. Ecol. 20, 951–963.

Cannell, M.G.R. and Thornley, J.H.M. (2000) Modelling the components of plant respiration: Some guiding principles. Ann. Bot. 85, 45–54.

Chuine, I. (1998) Modelisation de la phenologie des arbres de la zone temperee et ses implications en biologie evolutive, Ecole Nationale Superieure Agronomique de Montpellier, Montpellier.

Chuine, I., Cour, P. and Rousseau, D.D. (1998) Fitting models predicting dates of flowering of temperate-zone trees using simulated annealing. Plant Cell Environ. 21, 455–466.

Davis, M.B. and Kabinski, C. (1992). Changes in geographical range from greenhouse warming: effects on biodiversity in forests. In: Peters, R.L. and Lovejoy, T.E. (eds.), Global Warming and Biological Diversity. Yale University Press, Yale, pp. 297–309.

Davis, M.B. and Shaw, R.G. (2001) Range shift and adaptive responses to Quaternary climate change. Science 292, 673–679.

Fuchigami, L.H., Weiser, C.J., Kobayashi, K., Timmis, R. and Gusta, L.V. (1982). A degree growth stage model and cold acclimation in temperate woody plants. In: P.H. Li and A. Sakai (eds.), Plant Cold Hardiness and Freezing Stress. Plant Proceedings of an International Seminar on Plant Hardiness Held at the Sapporo Educational and Cultural Hall. Sappora, Japan, August 11–14, 1981.

Geburek, T. (2005). Sexual reproduction in forest trees. In: Geburek, T. and Turok, J. (eds.), Conservation and Management of Forest Genetic Resoucres in Europe. Arbora Publishers, Zvolen

Goudriaan, J. and Van Laar, H.H. (1994) Modelling Potential Crop Growth Processes. Kluwer, The Netherlands.

Grote, R. (1998) Integrating dynamic morphological properties into forest growth modelling: II Allocation and mortality. Forest Ecol. Manag. 111, 193–210.

Hampe, A. (2005) Conserving biodiversity under climate change: the rear edge matters. Ecol. Lett. 8, 461–467.

Hamrick, J.L. (2004) Response of forest trees to global environmental changes. Forest Ecol. Manag. 197, 323–335.

Hamrick, J.L. and Godt, M.J.W. (1996) Effects of life history traits on genetic diversity in plant species. Phil. T. Roy. Soc. B 351, 1291–1298.

Hänninen, H. (1990) Modelling the annual growth rhythm of trees: conceptual, experimental, and applied aspects. Silva Carelica 15, 35–45.

Hänninen, H. (1995) Effects of climatic change on trees from cool and temperate regions: an ecophysiological approach to modelling of budburst phenology. Can. J. Bot. 73, 183–199.

Hänninen, H. and Kramer, K. (2007) A framework for modelling the annual cycle of trees in boreal and temperate regions. Silva Fenn. 41, 167–205.

Hänninen, H. and Lundell, R. (2007). Dynamic models in plant ecophysiology. In: Taulavuori, E. and Taulavuori, K. (eds.), Physiology of Northern Plants Under Changing Environment. Research Signpost, Kerala, India, pp. 157–175.

Hari, A.J. (1972) Physiological state of development in biological models of growth and maturation. Ann. Bot. Fenn. 9, 107–115.

Hari, P.L. and Rasanen, P. (1970) A dynamic model of the daily heigth increment of plants. Ann. Bot. Fenn. 7, 375–378.

Jump, A. and Penuelas, J. (2005) Running to stand still: adaptation and the response of plants to rapid climate change. Ecol. Lett. 8, 1010–1020.

Kellomäki, S., Hänninen, H. and Kolström, M. (1995) Computations on frost damage to Scots pine under climatic warming in boreal conditions. Ecol. Appl. 5, 42–52.

Kellomäki, S., Vaisanen, H., Hänninen, H., Kolström, T., Lauhanen, R., Mattila, U. and Pajari, B. (1992) A simulation model for the succession of the boreal forest ecosystem. Silva Fenn. 26, 1–18.

Koski, V. and Sievänen, R. (1985). Timing growth cessation in relation to the variations in the growing season. In: P.M.A. Tigerstedt, P. Puttonen & V. Koski (eds). Crop Physiology of Forest Trees. Proceedings of an International Conference on Managing Forest Trees as Cultivated Plants held in Finland, July 23–28, 1985.

Kramer, K. (1994a) A modelling analysis of the effects of climatic warming on the probability of spring frost damage to tree species in The Netherlands and Germany. Plant Cell Environ. 17, 367–377.

Kramer, K. (1994b) Selecting a model to predict the onset of growth of Fagus sylvatica. J. Appl. Ecol. 31, 172–181.

Kramer, K. (1995a) Modelling comparison to evaluate the importance of phenology for the effects of climate change in growth of temperate-zone deciduous trees. Climate Res. 5, 119–130.

Kramer, K. (1995b) Phenotypic plasticity of the phenology of seven European tree species in relation to climatic warming. Plant Cell Environ. 18, 93–104.

Kramer, K., Buiteveld, J., Forstreuter, M., Geburek, T., Leonardi, S., Menozzi, P., Povillon, F., Schelhaas, M.J., du Cros, E.T., Vendramin, G.G. and van der Werf, D.C. (2008) Bridging the gap between ecophysiological and genetic knowledge to assess the adaptive potential of European Beech. Ecol. Modell., 333–353.

Kramer, K., Friend, A., and Leinonen, I. (1996) Modelling comparison to evaluate the importance of phenology and spring frost damage for the effects of climate change on growth of mixed temperate-zone deciduous forests. Climate Res. 7, 31–41.

Kramer, K., Leinonen, I., Bartelink, H.H., Berbigier, P., Borghetti, M. Bernhofer, Ch., Cienciala, E., Dolman, A.J., Froer, O. Gracia, C.A., Granier, A., Grünwald, T. Hari, P. Jans, W., Kellomäki, S., Loustau, D., Magnani, F., Markkanen, T., Matteucci, G., Mohren, G.M.J., Moors, E., Nissinen, A., Peltola, H., Sabaté, S., Sanchez, A., Sontag, M., Valentini R. and Vesala, T. (2002) Evaluation of 6 process-based forest growth models using eddy-covariance measurements of CO_2 and H_2O fluxes at 6 forest sites in Europe. Global Change Biol. 2002, 1–18.

Kramer, K., Leinonen, I. and Loustau, D. (2000) The importance of phenology for the evaluation of impacts of climate change on growth of boreal, temperate and Mediterranean forests ecosystems: an overview. Int. J. Biometeorol. 44, 67–75.

Kropff, M. (1993). Mechanisms of the competition for water. In: Kropff, M. and H., V.L.H. (eds.), Modelling Crop-Weed Interactions. CAB International, Wallingford, UK, pp. 63–76.

Leinonen, I. (1996a) Dependence of dormancy release on temperature in different origins of pinus sylvestris and betula pendula seedlings. Scand. J. Forest Res. 11, 122–128.

Leinonen, I. (1996b) A simulation model for the annual frost hardiness and freeze damage of Scots Pine. Ann. Bot. 78, 687–693.

Leinonen, I., Repo, T., Hänninen, H. and Burr, K.E. (1995) A second-order dynamic model for the frost hardiness of trees. Ann. Bot. 76, 89–95.

Leonardi, S. and Menozzi, P. (1995) Genetic variability of Fagus sylvatica L. (beech) in Italy: the role of postglacial recolonization. Heredity 75, 35–44.

Leuning, R. (1995) A critical appraisal of a combined stomatal- photosynthesis model for C-3 plants. Plant Cell Environ. 18, 339–355.

Leuning, R., Kelliher, F.M., Depury, D.G.G. and Schulze, E.D. (1995) Leaf nitrogen, photosynthesis, conductance and transpiration: scaling from leaves to canopies. Plant Cell Environ. 18, 1183–1200.

Loustau, D., Berbigier, P. and Granier, A. (1992) Interception loss, throughfall and stemflow in a maritime pine stand. II. An application of Gash's analytical model of interception. J. Hydrol. 138, 469–485.

Loustau, D., Berbigier, P., Roumagnac, P., Arruda-Pacheco, C., David, J.S., Ferreira, M.I., Pereira, J.S. and Tavares, R. (1996) Transpiration of a 64-year-old maritime pine stand in Portugal 1. Seasonal course of water flux through maritime pine. Oecologia 107, 33–42.

Mäkelä, A. and Hari, P. (1986) Stand growth models based on carbon uptake and allocation in individual trees. Ecol. Modell. 33, 205–229.

Mäkelä, A., Hari, P., Berninger, F., Hänninen, H. and Nikinmaa, E. (2004) Acclimation of photosynthetic capacity in Scots pine to the annual cycle of temperature. Tree Physiol. 24, 369–376.

Mohren, G.M.J., Bartelink, H.H., Kramer, K., Magnani, F., Sabaté, S. and Loustau, D. (1999) Modelling long-term effects of CO_2 increase and climate change on European forests, with emphasis on ecosystem carbon budgets. In: Ceulemans, R.J.M., Veroustraete, F., Gond, V. and van Rensbergen, J.B.H.F. (eds.), Forest Ecosystem Modelling, Upscaling and Remote Sensing. SPB Academic Publishing bv, The Netherlands, pp. 179–192.

Parmesan, C. (2006) Ecological and evolutionairy responses to recent climate change. Annu. Rev. Ecol. Syst. 37, 912–929.

Pelkonen, P. (1980) The uptake of carbon dioxide in Scots pine during spring. Flora 169, 386–397.

Pelkonen, P. and Hari, P. (1980) The independence of the springtime recovery of CO_2 uptake in Scots pine on temperature and internal factors. Flora 169, 398–404.

Penning de Vries, F.W., Brunsting, A.H. and van Laar, H.H. (1974) Products, requirements and efficiency of biosynthesis: a quantitative approach. J. Theor. Biol. 45, 339–377.

Petit, R.J. and Hampe, A. (2006) Some Evolutionary Consequences of Being a Tree. Annu. Rev. Ecol. Evol. Syst. 37, 187–214.

Reaumur, R.A.F.D. (1735) Observations du thermomètre, faites à Paris pendant l'année 1735, comparées avec celles qui ont été faites sous la ligne, à l'isle de France, à Alger et quelques unes de nos isles de l'Amérique. Mem. Paris Acad. Sci. 1735 545 ff.

Rehfeldt, G.E., Tchebakova, N.M., Parfenova, Y.I., Wykoff, W.R., Kuzmina, N.A. and Milyutin, L.I. (2002) Intraspecific responses to climate in Pinus sylvestris. Global Change Biol. 8, 912–929.

Repo, T. (1992) Seasonal changes of frost hardiness in Picea abies and Pinus sylvestris in Finland. Can. J. For. Res. 22, 1949–1957.

Repo, T. (1993) Impedance spectroscopy and temperature acclimation of forest treesUniversity of Joensuu, Faculty of Forestry, Joensuu, pp.53p.

Repo, T. and Lappi, J. (1989) Estimation of standard error of impedance-estimated frost resistance. Scand. J. Forest Res. 4, 67–74.

Repo, T., Mäkelä, A. and Hänninen, H. (1990a) Modelling frost resistance of trees. Silva Carelica 15, 61–74.

Repo, T., Tuovinen, T. and Savolainen, T. (1990b) Estimation of an electrical model of plant tissue using the impedance locus. Silva Carelica 15, 51–59.

Ryan, M.G. (1995) Foliar maintenance respiration of subalpine and boreal trees and shrubs in relation to nitrogen content. Plant Cell Environ. 18, 765–772.

Ryan, M.G., Hubbard, R.M., Pongracic, S., Raison, R.J. and Mcmurtrie, R.E. (1996) Foliage, fine-root, woody-tissue and stand respiration in Pinus radiata in relation to nitrogen status. Tree Physiol. 16, 333–343.

Sarvas, R. (1972) Investigations on the annual cycle of development on forest trees active period. Commun. Inst. For. Fenn. 76, 110.

Sarvas, R. (1974) Investigations on the annual cycle of development of forest trees. Autumn dormancy and winter dormancy. Commun. Inst. For. Fenn. 84, 1–101.

Spitters, C.J.T., Toussaint, H.A.J.M. and Goudriaan, J. (1986) Separating the diffuse and direct component of global radiation and its implications for modellingf canopy photosynthesis. Agric. For. Meteorol. 38, 217–229.

Thuiller, W., Albert, C., Araujo, M.B., Berry, P.M., Cabeza, M., Guisan, A., Hickler, T., Midgley, G.F., Paterson, J., Schurr, F.M., Sykes, M.T. and Zimmermann, N.E. (2008) Predicting global change impacts on plant species' distributions: Future challenges. Perspect. Plant Ecol. Evol. Syst. 9, 137–152.

Valentine, H.T. (1988) A carbon-balance model of stand growth: a derivation employing the Pipe-model theory and the self-thinning rule. Ann. Bot. 62, 389–396.

Weiner, J. (1990) Asymmetric competition in plant populations. Trends Ecol. Evol. 5, 360–364.

Appendix

The annual cycle of development for boreal and temperate zone trees. See Fig. 1 for a schematic representation of the four phenological phases and processes.

Potential Rates

Phase 1: active growth

$$R_a(t) = \begin{cases} 0, T(t) < T_b \\ T(t) - T_b, T(t) \geq T_b \end{cases}$$ (Eqn. A1)

$$R_o(t) = \begin{cases} 0, T(t) < T_{o,\min} \\ \dfrac{1}{1 + e^{(a_o \cdot (T(t) + b_o))}}, T(t) \geq T_{o,\min} \end{cases}$$ (Eqn. A2)

Phase 2: lignification

$$R_1(t) = \begin{cases} 0, T(t) < T_b \\ T(t) - T_b, T(t) \geq T_b \end{cases}$$ (Eqn. A3)

Phase 3: rest

$$Rr(t) \begin{cases} 0, T(t) T_{r,\min} \\ \dfrac{T(t) - T_{r,\min}}{T_{r,opt} - T_{r,\min}}, T_{r,\min} \leq T(t) \leq T_{r,opt} \\ \dfrac{T(t) - T_{r,\max}}{T_{r,opt} - T_{r,\max}}, T_{r,opt} \langle T(t) \leq T_{r,\max} \\ 0, T(t) \rangle T_{r,\max} \end{cases}$$ (Eqn. A4)

Phase 4: quiescence

$$R_o(t) = \begin{cases} 0, T(t) < T_{o,\min} \\ \dfrac{1}{1 + e^{(a_o \cdot (T(t) + b_o))}}, T(t) \geq T_{o,\min} \end{cases}$$ (Eqn. A5)

Frost hardiness

$$R_h(t) = \frac{1}{\tau_h} \cdot (\hat{S}_h(t) - S_h(t))$$ (Eqn. A6)

$$\Delta \hat{S}_{hT}(t) = \begin{cases} \Delta \hat{S}_{hT,\min}, T(t) \rangle T_{h,1} \\ \Delta \hat{S}_{hT,\max} \cdot \left(1 - \dfrac{T_{\min}(t) - T_{h,2}}{T_{h,1} - T_{h,2}}\right), T_{h,2} \leq T(t) T_{h,1} \\ \Delta \hat{S}_{hT,\max}, T(t) \langle T_{h,2} \end{cases}$$ (Eqn. A7)

$$\Delta \hat{S}_{hP}(t) = \begin{cases} \Delta \hat{S}_{hP,\min}, NL(t) \langle NL_{h,1} \\[2mm] \dfrac{\Delta \hat{S}_{hP,\max}}{NL_{h,2} - NL_{h,1}} \cdot (NL(t) - NL_{h,1}), NL_{h,1} \le NL(t) \le NL_{h,2} \\[2mm] \Delta \hat{S}_{hP,\max}, NL(t) \rangle NL_{h,2} \\[2mm] 0, Phase = 1 \wedge C_h = 0 \end{cases} \qquad \text{(Eqn. A8)}$$

Recovery of photosynthetic capacity

$$R_p = \frac{1}{\tau_p} \cdot \left(\frac{1}{1 + c_p \cdot a_p^{-(\hat{S}_p - S_p)}} - \frac{1}{1 + c_p \cdot a_p^{(\hat{S}_p - S_p)}} \right) \qquad \text{(Eqn. A9)}$$

States

Phase 1: active growth

$$S_a(t) = \int_{t_{a,i}}^{t} C_{aT} \cdot R_a(t) \cdot dt + C_{ap} \cdot NL(t) \qquad \text{(Eqn. A10)}$$

$$S_o(t) = \int_{t_{a,i}}^{t} C_o(t) \cdot R_o(t) \cdot dt \qquad \text{(Eqn. A11)}$$

end of active growth if $S_a(t) \rangle S_a^*$ then:
• Phase = 2
• $S_l(t) = 0$
• $t_{l,i} = 0$

Phase 2: lignification

$$S_l(t) = \int_{t_{l,i}}^{t} R_l(t) \cdot dt \qquad \text{(Eqn. A12)}$$

end of lignification if $S_l(t) \rangle S_l^*$ then:
• Phase = 3
• $S_o(t) = 0$
• $S_r(t) = 0$
• $t_{o,i} = 0$
• $t_{r,i} = 0$

Phase 3: rest

$$S_r(t) = \int_{t_{r,j}}^{t} R_r(t) \cdot dt \qquad \text{(Eqn. A13)}$$

end of rest (= rest completion) if $S_r(t) \rangle S_r^*$ then:
• Phase = 4

Phase 4: quiescence

$$S_o(t) = \int_{t_{o,i}}^{t} C_o(t) \cdot R_o(t) \cdot a \qquad \text{(Eqn. A14)}$$

$$C_o(t) = \begin{cases} 0, S_r(t) < S_r^* \\ 1, S_r(t) \ge S_r^* \end{cases} \qquad \text{(Eqn. A15)}$$

end of quiescence (= bud burst) if $S_o(t) \rangle S_o^*$ then:
• Phase = 1
• $S_a(t) = 0$
• $t_{a,i} = 0$

Frost hardiness

$$S_h(t) = \int_0^t R_h(t) \cdot dt \qquad \text{(Eqn. A16)}$$

$$\hat{S}_h(t) = \hat{S}_{h,\min} + C_h(t) \cdot \left(\Delta\hat{S}_{hT}(t) + \Delta\hat{S}_{hP}(t) \right) \qquad \text{(Eqn. A17)}$$

$$C_h(t) = \begin{cases} MAX\left(0, 1 - \dfrac{S_o(t)}{S_h^*}\right), Phase = 1 \\[3mm] \dfrac{S_l(t)}{S_l^*}, Phase = 2 \\[3mm] 1, Phase = 3 \\[3mm] MAX\left(0, 1 - \dfrac{S_o(t)}{S_{o,h}^*}\right), Phase = 4 \end{cases} \qquad \text{(Eqn. A18)}$$

Fraction of needle area damaged by frost

$$d_f = a_f + b_f \cdot e^{c_f \cdot S_h(t)} \qquad \text{(Eqn. A19)}$$

$$D(t) = \frac{1}{1 + e^{d_f \cdot (S_h(t) - T\min(t))}} \qquad \text{(Eqn. A20)}$$

Recovery of photosynthetic capacity

$$S_p(t) = \int_0^t R_p(t) \cdot dt \qquad \text{(Eqn. A21)}$$

$$\hat{S}_p(t) = b_p \cdot T(t) \qquad \text{(Eqn. A22)}$$

$$K_p(t) = \frac{S_p(t)}{S_p^*}$$ (Eqn. A23)

Table A1 Variables for models of the annual cycle of trees

Symbol	Description	Units
C_h	Capacity for stationary state of frost hardiness to respond to temperature or photoperiod	
C_o	Capacity for ontogenetic development	
D	Proportion of foliage damaged by frost	
$\Delta\hat{S}_{hT}$	Change of stationary state of frost hardiness due to change in temperature	
$\Delta\hat{S}_{hP}$	Change of stationary state of frost hardiness due to change in photoperiod	
K_p	efficiency of the actual photosynthetic capacity relative to the annual maximum at similar environmental conditions	
NL	Duration of night	h d^{-1}
R_a	Rate of change of active growth	°Cd d^{-1}
R_h	Rate of change of frost hardiness	°C d^{-1}
R_l	Rate of change of lignification	°Cd d^{-1}
R_o	Rate of ontogenetic development	FU d^{-1}
R_p	Rate of change of recovery of photosynthetic capacity	D^{-1}
R_r	Rate of change of rest	CU d^{-1}
$\hat{S}_h$	Stationary state of frost hardiness	°C
$\hat{S}_p$	Stationary state of recovery of photosynthetic capacity	°C
S_a	State of active growth	°Cd
S_h	State of frost hardiness	°C
S_l	State of lignification	°Cd
S_o	State of ontogenetic development	FU
S_p	State of recovery of photosynthetic capacity	
S_r	State of rest	CU
t	Time	d
$t_{a,i}$	Time of initiation of active growth	day of year
$t_{o,i}$	Time of initiation of ontogenetic development	day of year
$t_{r,i}$	Time of initiation of rest	day of year
T	Average daily temperature	°C
T_{min}	Minimum daily temperature	°C

Table A2 Parameters for models of the annual cycle of trees

Symbol	Description	Value	Source
a_f	Coefficient for effect of frost hardiness damage on amount of foliage	−0.1435	(1)
a_o	Coefficient for ontogenetic development	−0.11	(2)
a_p	Coefficient for rate of recovery of photosynthetic capacity	2	
b_f	Coefficient for effect of frost hardiness damage on amount of foliage	−1.4995	(1)

(continued)

Table A2 (continued)

Symbol	Description	Value	Source
b_0	Coefficient for ontogenetic development	−37.6	(2)
b_p	Coefficient for stationary state of recovery of photosynthetic capacity	600	
c_f	Coefficient for effect of frost hardiness damage on amount of foliage	0.1071	(1)
c_p	Coefficient for rate of recovery of photosynthetic capacity	100	(3)
CaT	Capacity of state of active growth to respond to temperature	1	
CaP	Capacity of state of active growth to respond to photoperiod	0	
$\Delta\hat{S}_{hP,\,min}$	Minimum change of stationary state of frost hardiness induced by a change in photoperiod	0	
$\Delta\hat{S}_{hP,\,max}$	Maximum change of stationary state of frost hardiness induced by a change in photoperiod	−18.5	
$\Delta\hat{S}_{hT,\,min}$	Minimum change of stationary state of frost hardiness induced by a change in temperature	0	
$\Delta\hat{S}_{hT,\,max}$	Maximum change of stationary state of frost hardiness induced by a change in photoperiod	−47	
$NL_{h,\,1}$	Lower limit of range of photoperiod within which frost hardiness changes	10	
$NL_{h,\,2}$	Upper limit of effective range of photoperiod within which frost hardiness changes	16	
$\hat{S}_{h,\,min}$	Minimal state of frost hardiness	−4.5	(1)
$S^*_{o,\,h}$	Critical state of ontogenetic development where capacity of stationary level of frost hardiness to respond to change in temperature or photoperiod is zero	4.5	(1)
S^*_o	Critical state of ontogenetic development to end quiescence	2.4	(2)
S^*_p	Critical state of recovery of photosynthetic capacity so that full capacity to recover is attained	2.708 (=6,500/ (24 × 100)	(1)
S^*_r	Critical state of ontogenetic development to end rest	28	
S^*_a	Critical state of active growth to end active growth phase	510	
S^*_l	Critical state of lignification to end lignification phase	300	
τ_h	Time constant for rate of change of frost hardiness	5	
τ_p	Time constant for rate of recovery of photosynthetic capacity	4.167E−4	
T_b	Base temperature for accumulation of thermal time	5	
$T_{h,\,1}$	Upper limit of range of temperature within which frost hardiness changes	10	
$T_{h,\,2}$	Lower limit of range of temperature within which frost hardiness changes	−16	
$T_{o,\,min}$	Minimum temperature for rate of change of ontogenetic development	0	(2)
$T_{r,\,max}$	Maximum temperature for rate of change of rest	16.5	(2)
$T_{r,\,min}$	Minimum temperature for rate of change of rest	−19.42	(2)
$T_{r,\,opt}$	Optimum temperature for rate of change of rest	−1.2	(2)

(1) Leinonen et al. (1995), (2) Kramer (1994a), (3) Pelkonen and Hari (1980)

Part III
Upscaling and Global View

Remote Sensing Phenology: Status and the Way Forward

Bradley C. Reed, Mark D. Schwartz, and Xiangming Xiao

Abstract A number of approaches using a variety of satellite remote sensing products have been used to derive metrics related to the timing of biological events (or land surface phenology, LSP). The advantages of utilizing remote sensing for phenology applications are the ability to capture the continuous expression of phenology patterns across the landscape and the ability to retrospectively observe phenology from archived satellite data sets (e.g. Landsat and Advanced Very High Resolution Radiometer). However, LSP databases have not yet been satisfactorily validated due to the difficulty in obtaining sufficiently extensive ground observations throughout the growing season. A multi-level validation approach that uses ground observations, dedicated web cameras, and high, medium, and coarse spatial resolution satellite data is needed to give scientists an improved level of confidence in utilizing the data. Many of these shortcomings are being addressed by phenology networks across the globe such as the U.S. National Phenology Network. Even without extensive validation, a number of applications areas have employed LSP data successfully, including studies on ecosystems analysis, disasters, land use, and climate change. Land surface phenology promises to continue contributing to these types of applications, and will also likely serve as an important early indicator of environmental effects of climate change.

B.C. Reed (✉)
U.S. Geological Survey, Geographic Analysis and Monitoring, Reston, VA, USA
e-mail: reed@usgs.gov

M.D. Schwartz
Department of Geography, University of Wisconsin-Milwaukee, Milwaukee, WI
e-mail: mds@uwm.edu

X. Xiao
Center for Spatial Analysis, University of Oklahoma, Norman, OK, USA
e-mail: xiangming.xiao@ou.edu

A. Noormets (ed.), *Phenology of Ecosystem Processes*,
DOI 10.1007/978-1-4419-0026-5_10, © Springer Science+Business Media, LLC 2009

1   Introduction

The study of vegetation phenology using remote sensing – also referred to as land surface phenology (LSP) - has experienced considerable progress over the past two decades, both in terms of generating the basic satellite data sets that are required for documenting phenology over large areas, and in terms of developing methodologies for creatively working with the data sets to derive metrics that describe the seasonality of vegetation. While progress has certainly been made, many challenges remain in this field; primarily in understanding the ecological meaning of the LSP estimates and in validating the results with in situ observations, systematic photography, or climatology models.

The term "land surface phenology" refers to the seasonal pattern of variation in vegetated land surfaces observed from remote sensing (Friedl et al. 2006; Henebry et al. 2005). This is distinct from observations of individual plants or species, as space-based observations aggregate information on the timing of heterogeneous vegetation development over pixel-sized areas (usually ranging from 250 m to 8 km in size). This aggregation often disassociates the response signal of the landscape from that of the individual species, yet is important for representing landscape-scale processes (e.g. water, energy, and carbon fluxes) in biosphere-atmosphere interaction and other models (Friedl et al. 2006). The United States National Phenology Network (USA-NPN) states that remote sensing provides the potential to move from plant specific observations to complete, continuous expressions of phenological patterns on the landscape (Betancourt et al. 2005), which are important for characterizing large area phenology and for better parameterizing ecological and climate models (i.e., specifying when to change values for seasonally dependent parameters such as albedo, surface roughness, transpiration, etc).

The development of the study of land surface phenology has, to date, been largely producer- rather than user-driven. The pathway of LSP from research to validation to operations, modeling, and forecasting is only just underway, but a body of literature has been produced to demonstrate the importance and potential of land surface phenology. This chapter outlines the development of remote sensing phenology research, from the sensors that have been used for characterizing phenology, commonly employed spectral transforms, to validation efforts and applications.

2   Satellite Remote Sensing

A number of satellite-based sensors with a variety of spectral, spatial, and temporal characteristics have been utilized for LSP studies, each providing certain advantages and disadvantages (Table 1). The Landsat series of satellites offers two primary advantages; a 30 m spatial resolution that is appropriate for landscape characterization and a data archive that extends back to the 1970s. However, Landsat's 16-day repeat cycle does not readily support the frequent observations that are necessary during rapidly changing phenological stages. But Landsat availability is undergoing

Table 1 Satellite sensors and data sets utilized for land surface phenology studies

Satellite	Sensor	Operation	Resolution	Frequency
Landsat	MSS	1973–1985	79 m	18 days
Landsat	TM	1984–present	30 m	16 days
Landsat	ETM+	1999–present	30 m	16 days
SPOT	Vegetation	1999–present	1 km	1–2 days
NOAA	AVHRR	1982–present	8 km	twice monthly
NOAA	AVHRR	1989–present	1 km	biweekly
Terra	MODIS	2000–present	250 m, 500 m, 1 km	1–2 days
Aqua	MODIS	2002–present	250 m, 500 m, 1 km	1–2 days
Envisat	MERIS	2002–present	300 m	1–3 days

an important change that promises to rapidly advance, if not revolutionize, multitemporal satellite analysis studies (Woodcock et al. 2008). The availability of web enabled, free-of-charge data through the U.S. Geological Survey (USGS) will make Landsat imagery accessible to all researchers, who will then have the opportunity to develop new methodologies that can potentially overcome the shortcomings of this satellite series.

The Systéme Probatoire pour l'Observation de la Terre (SPOT) Vegetation sensor has been collecting near-daily surface reflectance data since 1999. One widely used SPOT Vegetation product is their 1 km resolution, 10-day normalized difference vegetation index (NDVI) composites (10-day synthesis or S10 product). This ongoing data set provides a well georeferenced collection that is widely available for LSP studies (e.g. Delbart et al. 2006).

The advanced very high resolution radiometer (AVHRR) has been providing data for land surface studies since the 1980s. A variety of AVHRR collections are available for LSP studies. For example, global, twice monthly data at 8 km resolution are available for the post-1982 period (Global Inventory Modeling and Mapping Studies – GIMMS; Tucker et al. 2004); conterminous United States coverage is available at biweekly intervals at 1 km resolution since 1989 (Eidensink 1992). One of the primary advantages of AVHRR is its relatively long-term continuity; there are data since the 1980s and current plans call for this sensor family to continue into the 2020s with the National Oceanic and Atmospheric Administration (NOAA) and the European Organisation for the Exploitation of Meteorological Satellites (EUMETSAT) sharing operations.

The newer generation of sensors that are well suited for phenological studies includes the Moderate Resolution Imaging Spectroradiometer (MODIS) and Medium Resolution Imaging Spectrometer (MERIS) sensors. They each have additional spectral bands that are utilized for a variety of applications, and importantly for phenological studies, maintain bands in the visible and near infrared wavelengths. There are currently MODIS sensors aboard two satellites, Terra and Aqua, collecting data in 36 spectral bands at 250 m, 500 m, or 1,000 m resolution. MODIS data are processed into a suite of data products, including surface reflectance, vegetation indices, land surface temperature, and others (Justice et al. 2002). The MODIS Land Cover Dynamics product (MOD12Q2) is a yearly summary of several phenology metrics at

1 km spatial resolution (Zhang et al. 2003). MERIS, aboard the Envisat satellite, has global coverage every 3 days and collects data in 15 spectral bands at 300 m resolution, which is well suited for phenology studies (e.g. Verstraete et al. 2008).

Alternative approaches to LSP studies may involve non-optical sensors, such as Advanced Microwave Scanning Radiometer (AMSR-E) or other radar sensors, which can be used to detect moisture conditions of the land surface (including vegetation and soil; Doubková and Henebry 2006).

An issue of concern to land surface phenology studies is the long-term continuity of sensors that are best suited for such studies. Even though the AVHRR series of sensors has been operating since the 1980s, there are issues with continuity between sensors as they have differing overpass times (morning vs. afternoon) with varying illumination conditions, and different spectral band response characteristics. The current generation of satellites (MODIS and MERIS) have spatial resolutions that are better suited for LSP studies, but the value of the data would be significantly increased if a connection to past sensors, such as the AVHRR, and to future sensors, such as the Visible Infrared Imager/Radiometer Suite (VIIRS) can be made (Murphy et al. 2001). The development of transition algorithms between sensors is essential for LSP-related global change studies (e.g. Gallo et al. 2004, 2005).

3 Vegetation Indices

A number of spectral transformations are commonly utilized to prepare sensor data for LSP applications. The transformations are designed to enhance spectral reflect-ance and emissive characteristics of vegetation or environmental conditions that are related to phenological development. These conditions include changing leaf area, soil moisture, and vegetation water content. The set of transforms collectively referred to as vegetation indices utilize the reflective characteristics of vegetation in the red and near infrared wavelengths. Plants generally absorb energy in the 0.6–0.7 μm wavelengths of the electromagnetic spectrum (chlorophyll absorption of red energy) and reflect very strongly in the near infrared (NIR). Satellite sensors that are designed for land surface analysis applications usually collect data in discrete bands that approximate the red and near infrared wavelengths in order to differentiate and highlight varying land surface conditions. Vegetation indices utilize the reflective characteristics of vegetation often by some kind of ratio and/or differencing of the red and NIR reflectance bands or utilize other reflective wavelengths for reducing the effects of atmospheric aerosols or soil background (blue reflectance) and for highlighting leaf water content (shortwave infrared).

The normalized difference vegetation index (1) is a commonly used index for vegetation studies (Goward et al. 1985; Tucker and Sellers 1986; Malingreau et al. 1989; Loveland et al. 1991; Townshend et al. 1994). The NDVI is strongly coupled to red reflectance, which is related to the photosynthetic capacity of vegetation and biophysical variables such as the fraction of photosynthetically active radiation (fPAR) and fractional green cover (Huete et al. 1997). Other indices have been developed to reduce canopy background effects and atmospheric contamination, including the soil adjusted vegetation index (SAVI, Huete 1988) and the soil and atmospherically resistant vegetation index (SARVI, Kaufman and Tanré 1992). The SAVI and SARVI are more tightly linked to near infrared reflectance and to structural parameters such as leaf area index (LAI) and biomass. The enhanced vegetation index (EVI; (2)) is an extension of the progress made in reducing soil and atmospheric effects (Huete et al. 2002) designed to improve sensitivity in high biomass regions and to reduce the canopy background signal and atmospheric influences. The EVI includes the blue reflectance band to correct for the influence of atmospheric aerosols on red reflectance.

$$NDVI = \frac{\rho_{NIR} - \rho_{Red}}{\rho_{NIR} + \rho_{Red}} \tag{1}$$

$$EVI = 2.5 \frac{\rho_{NIR} - \rho_{Red}}{\rho_{NIR} + 6\rho_{Red} - 7\rho_{Blue} + 1} \tag{2}$$

3.1 Linearity of Indices

One issue that must be understood when working with NDVI is its nonlinearity with respect to ground biomass or leaf area and its saturation at high greenness levels. Huete et al. (2002) found it to saturate asymptotically in high biomass regions, such as in the Amazon. The nonlinearity and saturation issues are addressed by the EVI which is more sensitive to canopy variations in high biomass regions (Huete et al. 2002), but it requires noise removal parameters and reduces sensitivity to modest land surface change under low to moderate greenness conditions. Figure 1 illustrates the saturation issue as the NDVI value varies only about 0.06 between 25 May and 30 September, while EVI varies by nearly 0.30, thus detecting differences in canopy conditions, even while in full leaf. The global environment monitoring index (GEMI) and wide dynamic range vegetation index (WDRVI) are alternative vegetation indices that do not enjoy widespread usage, but warrant additional investigation for LSP studies. GEMI is a non-linear index that seeks to reduce atmospheric perturbations without external information on atmospheric composition (Pinty and Verstraete 1992). The WDRVI is a modification of NDVI that increases the correlation with vegetation fraction by adding a weighting coefficient, a, (suggested value of 0.1–0.2) to the numerator and denominator of the NDVI equation (Gitelson 2004; Eqn. 3).

$$WDRVI = \frac{a\rho_{NIR} - \rho_{Red}}{a\rho_{NIR} + \rho_{Red}} \tag{3}$$

A number of studies have explored the shortwave infrared (SWIR) spectral bands for investigating vegetation water content (Hunt and Rock 1989; Gao 1996; Ceccato et al. 2001, 2002; Xiao et al. 2002). Results from these studies (Ceccato et al. 2002) suggest that a combination of NIR and SWIR bands have the potential for retrieving leaf and canopy water content. A few water-oriented vegetation indices were developed for characterization of leaf and canopy water content, e.g. Moisture Stress Index (MSI; Hunt and Rock 1989), Normalized Difference Water Index (NDWI; Gao 1996), and Land Surface Water Index (LSWI; Xiao et al. 2002). LSWI (Eqn. 4) is calculated as the normalized ratio between NIR and SWIR bands (Xiao et al. 2002).

$$LSWI = \frac{\rho_{NIR} - \rho_{SWIR}}{\rho_{NIR} + \rho_{SWIR}} \tag{4}$$

The performance of three of these indices (NDVI, EVI, and LSWI) is discussed in more detail in the following chapter by Xiao et al. in the current volume. From a graphical standpoint, Fig. 1 shows that each of the indices calculated for a single pixel in Harvard Forest, Massachusetts, USA, is sensitive to vegetation phenology. Each index exhibits a rapid increase around the first of May and a sustained decrease beginning around the first of October. However, the NDVI and especially the LSWI appear to be sensitive to non-vegetative phenomena as well, exhibiting a winter increase in value that is likely related to snow cover and/or high soil moisture. The EVI shows less variance during the non-growing season.

Several researchers (e.g. Delbart et al. 2006; Kathuroju et al. 2007; Studer et al. 2007) suggest that improved integration of snow cover into land surface phenology derivations could potentially help the definition of phenological stages. However, this is an additional complication as often there may be an herbaceous layer underneath the snow that begins immediately photosynthesizing when the snow cover melts. Obviously, temperature conditions that drive snow melt are suitable for photosynthesis and there may, indeed, be points in time where greenup (photosynthesis) has begun and there is still partial snow cover.

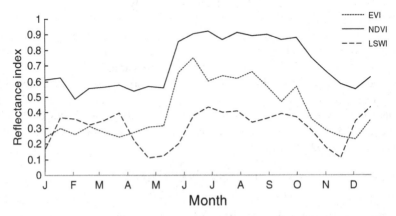

Fig. 1 Enhanced vegetation index, normalized difference vegetation index, and land surface water index plotted for one 500 m MODIS pixel in Harvard Forest, Massachusetts, USA throughout 2005.

4 Publicly Available Land Surface Phenology Products

There are currently three primary publicly available remote sensing phenology data sets; one derived from AVHRR and two derived from MODIS. The AVHRR data set, produced by the US Geological Survey (http://edc2.usgs.gov/phenological), covers 1989-present and is distributed at 1,000 m spatial resolution (Reed et al. 1994). One of the MODIS products (http://accweb.nascom.nasa.gov/data) was produced by NASA for the North America Carbon Program (NACP), for 2001–2005 and is available at a variety of spatial resolutions (250 m, 500 m, and 1,000 m), using different MODIS products as source data; EVI, NDVI, gross primary production (GPP), leaf area index (LAI) and fraction of photosynthetically active radiation (FPAR) (Gao et al. 2008). The other MODIS data set was produced by Boston University as part of the NASA Earth Observing System Data and Information System and uses the MODIS Nadir Bidirectional Reflectance Distribution Function (BRDF)-Adjusted Reflectance EVI product as source data (https://lpdaac.usgs.gov/lpdaac/products/modis_product_table/land_cover/dynamics_yearly_l3_global_1km/v5/terra). The data set currently includes 2001–2004 at 1,000 m (Zhang et al. 2003).

5 Validation Efforts

Many studies have employed various forms of climate data to validate satellite-derived measures of phenology related to canopy duration. These approaches include use of station temperature and precipitation values (e.g. Bogaert et al. 2002; Wang et al. 2003; Deng et al. 2007), growing degree-days and other temperature summations (e.g. Zhang et al. 2004a; de Beurs and Henebry 2005; Fisher and Mustard 2007; Fisher et al. 2007), or gridded temperature and precipitation (e.g. Zhou et al. 2003; Tateishi and Ebata 2004; Goetz et al. 2005). Other studies have used phenology models driven by climate data (Schwartz and Reed 1999; Schwartz et al. 2002; Kathuroju et al. 2007). A smaller number of studies have compared satellite-derived measures directly to plant phenology, either from multiple locations in a network (Delbart et al. 2005; Studer et al. 2007) or at a few individual sites (Kang et al. 2003; Chen et al. 2005; Zhang et al. 2006; Fisher and Mustard 2007).

Other validations strategies are less common. They include general comparisons to land cover types in different locations (Hoare and Frost 2004; Bradley et al. 2007), and use of contiguous uniform (pixel-level) land cover sites (White et al. 1997). Many early or methodological studies essentially used no validation at all (Lloyd 1990; Reed et al. 1994; Slayback et al. 2003; Cao et al. 2004). One recent study made comparisons to various surface radiation-derived measures, such as Photosynthetically Active Radiation (PAR) and Plant Area Index (PAI; Ahl et al. 2006). Two others employed ground-level imagery, either along transects (Fisher et al. 2006) or from a fixed location atop a flux tower (Richardson et al. 2007).

While most studies have used some form of climate information to validate satellite-derived measures of phenology and a smaller number used phenological models driven by surface climate data, direct validation of remote sensing phenology measures with surface phenological information has been limited by a lack of widely distributed surface data sources. Thus, the vast majority of studies have incorporated few such direct comparisons except for a few plant species or at selected locations. The only solution to this problem is to expand use of data from existing phenology networks (such as those in Germany and some other European countries, Menzel 2003), and develop and expand new ground-based phenology observation networks, such as the USA National Phenology Network (Betancourt et al. 2007).

5.1   Validation and Scale

Even as more direct observations of phenology become available, a serious impediment to comparing traditional surface phenological measures with satellite-derived measures is scale. While high temporal resolution satellite data record information for areas no smaller than 250 × 250 m, surface phenological data are recorded for a small number of individual plants. Another concern is that conventional phenological data are typically recorded as the date of occurrence of a limited number of events (first leaf, first bloom, etc.), which is challenging to compare to continuous vegetation indices. All these issues are so serious that some have suggested that comparison of individual pixels to surface data is not possible, and that comparisons must be based on probabilistic assessment of the response from large numbers of pixels (White and Nemani 2006). Schwartz and Liang (in preparation) counter that surface phenological data have never been collected with a technique sufficient to adequately compare them to satellite measures. In an ongoing effort to obtain ground data directly comparable to remotely sensed reflectance, they have tagged almost 900 native trees in two 600 × 600 m study areas located in a mixed boreo-temperate forest in northern Wisconsin. By monitoring phenological development of these trees every second day over the approximately 30-day period from bud-burst to full leaf expansion, using an observation protocol geared toward compatibility with satellite measurements (stage and percent of canopy at that stage), understanding of the variation of individual species responses can be better scaled-up into community and ultimately landscape-level phenology. In addition, below-canopy light levels are being monitored to provide additional assistance in translating conventional phenological measurements into a continuous signal.

The link between phenology estimated from remote sensing and ground observations is often weaker than one would hope. Badeck et al. (2004) hypothesized that this is caused by qualitative differences between the observations, as well as the heterogeneity of pixel composition in satellite scenes. They state that the modest correlations between ground and satellite observations should be expected and that a key to resolving these problems lies with improved spatial interpolation of ground observation data. The two scales of observation should be thought of as complementary, with scaling studies a rich area of prospective research. A promising development in validating LSP is the use of inexpensive downward- or outward-looking cameras to record phenological development around eddy covariance monitoring sites where the instrument towers afford an above-canopy view (Richardson et al. 2007). Widespread adoption of these devices will greatly increase the data available to improve understanding of the general relationships among surface phenology and various satellite-derived measures.

5.2 Climate-Based Validation

As long as national-level and global surface phenological data are lacking, there will be an acute need for selected applications employing phenological models. For example, the Spring Index (SI) models (developed from cloned lilac and honeysuckle data) have a demonstrated potential to represent seasonally integrated changes in temperature for selected plant species (Schwartz et al. 2006). These models are most applicable to temperature responsive forest trees and shrubs and agricultural crops planted in temperate regions with adequate rainfall. The SI models have been thoroughly tested at over 300 sites across the Northern Hemisphere, and show a consistent response across varied regions (Schwartz et al. 2006). The greatest advantages of using SI models for evaluating basic phenological trends in the context of global change research are (1) SI models can be generated at any location that has a daily maximum-minimum temperature time series, so they can be produced and evaluated over much larger geographic areas than any currently available conventional phenological data; and (2) SI model output is consistent over all areas, which may not be true for conventional data due to different species and event definitions, as well as inconsistencies in human observations.

The SI models show valid output in all areas that the lilac and honeysuckle plants would receive sufficient chilling in the cold season and sufficient heat in the warm season to theoretically survive (survival is theoretical because the models do not address potential summer water stress or winter cold mortality). While SI model outputs do not reproduce all the details of multi-species phenology data at any site, or the specific phenology of some types of plants, these models process weather data into indices directly related to growth and development of many plants. As such, they provide a baseline assessment of each location's general phenological response (with the above noted limitations) over a standard period, supplying a needed context for evaluating and comparing regional or local-scale studies. Further, SI model output is expected to be robust under future conditions where climate departs significantly from the historical mean, as the models are optimized for continental-scale applications and included input data from a study area that extended southward from the U.S. Northeast to North Carolina, and then westward to Oklahoma and North Dakota (approximately 35 to 49°N and −104 to −68°W; Schwartz 1997; Hayhoe et al. 2007).

6 Applications of Land Surface Phenology

As land surface phenology studies move from producer-driven to user-driven, a variety of applications demonstrate the utility of satellite derived phenology estimates. This section describes studies that contribute to our understanding of ecosystems (e.g. carbon cycling, invasive species), environmental disasters (e.g. drought, fire), land use change, and climate change, and there is new research directed toward phenological forecasting.

6.1 Ecosystem Studies

To tie LSP to phenological processes, Fisher et al. (2006) state that there are two challenges; scaling from field observations to satellite imagery and deriving comparable phenological metrics. The field-to-satellite scaling step requires a very large number of ground observations over a relatively large area. The challenge of comparable phenological metrics refers to the issue of interpreting the biophysical meaning (e.g. albedo, photosynthesis, surface roughness changes) of LSP estimates – a real difficulty with heterogeneous land cover.

In another study illustrating the difficulties connecting LSP to ground observations, Ahl et al. (2006) used MODIS data in 2002 over a deciduous forest in northern Wisconsin to observe that the satellite-derived products tended to predict onset of greenness and maturity between 1 and 21 days earlier than field observations. The authors state that more research on the influence of phenology on carbon and water exchange is needed, especially for the transition period between budbreak and full photosynthetic capacity.

Specifically, more needs to be known about growing season length as a limiting factor of productivity and carbon flux, and better parameterizing models for changing surface conditions, such as albedo or surface roughness is necessary. Increased knowledge in these areas would lead to improvement in models that change the value of certain parameters at pre-determined times of the year. Improving our understanding of the issues underlying these questions can be achieved by analysis of multiple, simultaneously collected data at intensive phenology observation sites (Betancourt et al. 2005). These can be instrumented research sites, such as Ameriflux or Long Term Ecological Research stations where observations of weather, biogeochemistry, satellite imagery, and plant phenology can be collected. While some of these efforts are underway (Richardson et al. 2007), the USA-NPN plans to facilitate more such studies.

6.2 Disasters

Westerling et al. (2006) relate early spring (defined by hydrological estimates) to increased forest wildfire activity in the western US. Increased spring and summer temperatures, along with earlier snowmelt produce a relatively large increase in cumulative moisture deficit by midsummer and subsequent increased fire danger. Incorporating ground and satellite-derived phenology information into further study of fire severity and possible early warning of fire is a logical next step for such studies.

Brown et al. (2008) include start of season anomalies (derived from AVHRR) as a key input to a vegetation drought response index (VegDRI). The VegDRI incorporates climate-based drought indicators with satellite-derived vegetation metrics and other biophysical data to identify drought conditions in near-real time. The phenology component of their model helps to distinguish a delayed growing season from one that is experiencing stressful growing conditions.

6.3   Land Use

Changes in land surface phenology have been used as an indicator of land use change. de Beurs and Henebry (2004) used 8 km AVHRR data to identify agricultural land cover change in Kazakhstan. They state that institutional change was manifested as land cover change, principally as an increase in fallow land (and pioneering weedy species) that was reflected in changing land surface phenology. Similarly, Reed (2006) identified agricultural practices in Saskatchewan, Canada as a source of changing land surface phenology using a similar 8 km AVHRR data set.

White et al. (2002) used satellite based start of season estimates in a study of the urban heat island effect on seasonality of deciduous broadleaf forest. Urbanization was associated with an expansion of the growing season by 7.6 days during the 1990s. Most of this effect was caused by an earlier start of the growing season. In a similar study using MODIS data from 2001, Zhang et al. (2004b) identified an increase in the growing season of about 15 days in urban areas relative to adjacent unaffected rural areas. They saw an urbanization influence on the length of the growing season up to 10 km beyond the edge of the urban areas.

6.4   Phenology Response to Climate Change

Zhang et al. (2007), utilize AVHRR global vegetation index data (GVI) to characterize a diverse response of vegetation phenology to a warming climate. They identify a latitudinal transition zone where the landscape changes from an earlier to a later trend in the time of greenup from north to south due to differential fulfillment of chilling requirements. This latitudinal transition zone appears to be shifting northward at a rate of 0.1° of latitude per year.

Bunn and Goetz (2006) used a 22-year record (1982–2003) of satellite observations from AVHRR to identify trends in circumpolar photosynthetic activity. They report disparate seasonality trends over this period between tundra areas (greening) and boreal forests (browning). They speculate that the boreal forest may be responding to climate change in unexpected ways, thus necessitating an expanded observation network and revisiting ecosystem process models. Reed (2006) identified similar patterns in Alaska, but suggests that the boreal forest "browning" may, in part, be due to insect and/or fire disturbance. Regardless of the cause, it is apparent that continued observation and investigation in this northern biome is warranted.

6.5 Forecasting

Real time monitoring and short-term forecasting of land surface phenology can contribute significantly to land management, human health, and other applications (White and Nemani 2006). Phenology forecasts, based on remote sensing data, coupled with uncertainty estimates could be used for estimating future crop conditions, fire danger, and potentially contribute to early warning of drought conditions.

Kathuroju et al. (2007) used LSP derived from AVHRR to develop prognostic phenology models and tested these models against climatological phenology values from AVHRR. The prognostic model did not perform any better than using a mean date model for most of their model runs. The authors state that other, more advanced sensors or the addition of snow and atmospheric information may produce better results. They also suggest that species-specific models coupled with LSP may be more appropriate approach.

7   Conclusions

The study of land surface phenology both contributes to and supports basic scientific inquiry. The public availability of LSP data sets, provides an objective (though not thoroughly validated) indication of plant development. The ongoing applications that use these data will necessarily lead to incremental improvements in each of the data sets and, in turn, lead to more extensive usage. LSP not only contributes to applications, but in and of itself provides an indication of the environmental response to a changing climate, especially at the tundra/boreal boundary. Studies such as this will be expanded to cover other ecosystems to assess their variable responses.

Land surface phenology has been approached using a variety of satellite sensors varying from optical sensors such as Landsat, AVHRR, MODIS, and MERIS to microwave sensors such as AMSR-E. The majority of studies have utilized high temporal observation satellite sensors (near-daily) that have moderate spatial resolution (from 250 m to 8 km). These characteristics provide an opportunity to gauge phenology with the needed precision, but present challenges in terms of accuracy, especially in regard to specific plant types or communities. A number of spectral transforms, usually vegetation indices, have been utilized to derive phenology metrics, such as start and end of growing season, but there is still a need to further evaluate options to determine the optimal approach for different vegetation types.

Recent developments in web-enabled, free-of-charge Landsat data may well open the door to rapid progress in land surface phenology studies. The 30 m Landsat spatial resolution offers an appropriate scale for connecting ground-based observations to the family of high temporal frequency satellite sensors. Once techniques are developed for effectively utilizing Landsat for phenology studies, the phenology record can be extended back in time to the 1970s, thus providing valuable, multi-decadal information that can be used for climate change studies.

There is a strong need to couple field observations, ecological and weather data, and satellite collections to better understand the annual dynamics of the vegetated landscape. Land surface phenology is moving toward multiple characterizations of what was formerly characterized as a single definition of the start of the growing season. We need to characterize successive stages of a continuous process – such as budbreak and full photosynthetic capacity – and understand the influence of environmental factors such as climate, weather, and nutrients on these stages. To better address these complexities, more ground observations specifically designed for comparison to LSP are needed. This could involve a multi-scale field approach that includes individual plant observations, ground-based environmental sensor instrumentation, digital web-cam images, high-resolution satellite data (or aerial photography), Landsat, and high temporal resolution imagery (such as MODIS). A thorough analysis of what information is detectable at what scale of observation would contribute substantially to our ability to monitor ecological effects of a changing climate.

References

Ahl, D.E., Gower, S.T., Burrows, S.N., Shabanov, N.V., Myneni, R.B. and Knyazikhin, Y. (2006) Monitoring spring canopy phenology of a deciduous broadleaf forest using MODIS. Remote Sens. Environ. 104, 88–95.

Badeck, F-W., Bondeau, A., Bottcher, K., Doktor, D., Lucht, W., Schaber, J. and Sitch, S. (2004) Responses of spring phenology to climate change. New Phytol. 162, 295–309.

Betancourt, J.L., Schwartz, M.D., Breshears, D.D., Brewer, C.A., Frazer, G., Gross, J.E., Mazer, S.J., Reed, B.C., Wilson, B.E. (2007) Evolving plans for the USA National Phenology Network. EOS Trans. AGU 88, 211.

Betancourt, J.L, Schwartz, M.D., Breshears, D.D., Cayan, D.R., Dettinger, M.D., Inouye, D.W. Post, E. and Reed, B. (2005) Implementing a U.S. National Phenology Network. EOS Trans. AGU 86, 539–541.

Bogaert, J., Zhou, L., Tucker, C.J., Myneni, R.B. and Ceulemans, R. (2002) Evidence for a persistent and extensive greening trend in Eurasia inferred from satellite vegetation index data. J. Geophys. Res. 107, (D11) 10.1029/2001JD001075.

Bradley, B.A., Jacob, R.W., Hermance, J.F. and Mustard, J.F. (2007) A curve fitting procedure to derive inter-annual phenologies from time series of noisy satellite NDVI data. Remote Sens. Environ. 106, 137–145.

Brown, J.F., Wardlow, B.D., Tadesse, T., Hayes, M.J. and Reed, B.C. (2008) The Vegetation Drought Response Index (VegDRI), A new integrated approach for monitoring drought stress in vegetation, GIScience Remote Sens. 45, 16–46.

Bunn, A.G. and Goetz, S.J. (2006) Trends in satellite-observed circumpolar photosynthetic activity from 1982 to 2003: The influence of seasonality, cover type, and vegetation density. Earth Interact. 10, 1–19.

Cao, M., Prince, S.D., Small, J. and Goetz, S.J. (2004) Remotely sensed interannual variations and trends in terrestrial net primary productivity 1981–2000. Ecosystems 7, 233–242.

Ceccato, P., Flasse, S., Tarantola, S., Jacquemoud, S. and Gregoire, J.M. (2001) Detecting vegetation leaf water content using reflectance in the optical domain. Remote Sens. Environ. 77, 22–33.

Ceccato, P., Gobron, N., Flasse, S., Pinty, B. and Tarantola, S. (2002) Designing a spectral index to estimate vegetation water content from remote sensing data, Part 1 - Theoretical approach. Remote Sens. Environ. 82, 188–197.

Chen, X., Hu, B., and Yu, R. (2005) Spatial and temporal variation of phenological growing season and climate change impacts in temperature eastern China. Global Change Biol. 11, 1118–1130.

Delbart, N., Picard, G., Toan, T.L., Kegoat, L. and Quegan, S. (2005) Spring phenology in Siberia in 1982–2004, observations by remote sensing, modeling, and impact on the terrestrial carbon budget. Geophys. Res. Abstr. 7, 01283.

Delbart, N., Toan, T.O., Kergoat, L. and Fedotova, V. (2006) Remote sensing of spring phenology in boreal regions; A free of snow-effect method using NOAA-AVHRR and SPOT-VGT data (1982–2004). Remote Sens. Environ. 101, 52–62.

de Beurs, K.M. and Henebry, G.M. (2005) Land surface phenology and temperature variation in the International Geosphere-Biosphere Program high-latitude transects. Global Change Biol. 11, 779–790.

de Beurs, K.M. and Henebry, G.M. (2004) Land surface phenology, climatic variation, and institutional change, Analyzing agricultural land cover change in Kazakhstan. Remote Sens. Environ. 89, 497–509.

Deng, F., Su, G. and Liu, C. (2007) Seasonal variation of MODIS Vegetation Indexes and their statistical relationship with climate over the subtropic evergreen forest in Zhejiang, China. IEEE Geosci. Remote Sens. Lett. 4, 236–240.

Doubková, M. and Henebry, G.M. (2006) Synergistic use of AMSR-E and MODIS data for understanding grassland land surface phenology. IGARSS 2006, Denver, Colorado, July 30-August 4, 2006.

Eidenshink, J.C. (1992) The 1990 Conterminous U.S. AVHRR Data Set. Photogramm. Eng. Rem. Sens. 58, 809–813.

Fisher, J.I. and Mustard, J.F. (2007) Cross-scalar satellite phenology from ground, Landsat, and MODIS data. Remote Sens. Environ. 109, 261–273.

Fisher, J.I., Mustard, J.F. and Vadeboncoeur, M.A. (2006) Green leaf phenology at Landsat resolution, Scaling from the field to the satellite. Remote Sens. Environ. 100, 265–279.

Fisher, J. I., Richardson, A.D. and Mustard, J. F. (2007) Phenology model from surface meteorology does not capture satellite-based green-up estimations. Global Change Biol. 13, 707–721.

Friedl, M., Henebry, G.M., Reed, B.C., Huete, A., White, M.A., Morisette, J.T., Nemani, R., Zhang, X. and Myneni, R. (2006) Land Surface Phenology, A Community White Paper requested by NASA, ftp,//zeus.geog.umd.edu/Land_ESDR/Phenology_Friedl_white-paper.pdf, Apr 2006.

Gallo, K., Ji, L., Reed, B.C., Eidenshink, J. and Dwyer, J. (2005) Multi-platform comparisons of MODIS and AVHRR normalized difference vegetation index data. Remote Sens. Environ. 99, 221–231.

Gallo, K., Ji, L., Reed, B.C., Dwyer, J. and Eidenshink, J. (2004) Comparison of MODIS and AVHRR 16-day normalized difference vegetation index composite data, Geophys. Res. Lett. 31 (L07502), doi,10.1029/2003GL019385.

Gao, B.C. (1996) NDWI - A normalized difference water index for remote sensing of vegetation liquid water from space. Remote Sens. Environ. 58, 257–266.

Gao, F., Morisette, J.T., Wolfe, R.E., Ederer, G., Pedelty, J., Masuoka, E., Myneni, R., Bin, T. and Nightingale, J. (2008) An algorithm to produce temporally and spatially continuous MODIS-LAI time series, IEEE Geosci. Remote Sens. Lett. 5, 60–64.

Gitelson, A.A. (2004) Wide dynamic range vegetation index for remote quantification of biophysical characteristics of vegetation. J. Plant Physiol. 161, 165–173.

Goetz, S.J., Bunn, A.G., Fiske, G.J. and Houghton, R.A. (2005) Satellite-observed photosynthetic trends across boreal North America associated with climate and fire disturbance. Proc. Natl. Acad. Sci. USA 102, 13521–13525.

Goward, S.N., Tucker, C.J. and Dye, D.G. (1985) North American vegetation patterns observed with the NOAA-7 advanced very high resolution radiometer. Vegetation 64, 3–14.

Hayhoe, K., Wake, C., Huntington, T.G., Luo, L., Schwartz, M.D., Sheffield, J., Wood, E., Anderson, B., Bradbury, J., DeGaetano, A., Troy, T.J. and Wolfe, D. (2007) Past and future changes in climate and hydrological indicators in the U.S. Northeast. Clim. Dyn. 28, 381–407.

Henebry, G.M., de Beurs, K.M. and Gitelson, A.A. (2005) Land surface phenologies of Uzbekistan and Turkmenistan between 1982 and 1999. Arid Ecosyst. 11, 25–32.

Hoare, D. and Frost, P. (2004) Phenological description of natural vegetation in southern Africa using remotely-sensed vegetation data. Appl. Veg. Sci. 7, 19–28.

Huete, A., Didan, K., Miura, T., Rodriguez, E.P., Gao, X. and Ferreira, L.G. (2002) Overview of the radiometric and biophysical performance of the MODIS vegetation indices. Remote Sens. Enviorn. 83, 195–213.

Huete, A.R., Liu, H.Q., Batchily, K. and van Leeuwen, W. (1997) A comparison of vegetation indices over a global set of TM images for EOS-MODIS. Remote Sens. Environ. 59, 440–451.

Huete, A.R. (1988) A soil adjusted vegetation index (SAVI). Remote Sens. Environ. 25, 295–309.

Hunt, E.R. and Rock, B.N. (1989) Detection of changes in leaf water-content using near-infrared and middle-infrared reflectances. Remote Sens. Environ. 30, 43–54.

Justice, C.O., Townshend, J.R.G., Vermote, E.F., Masuoka, E., Wolfe, R.E., Saleous, N., Roy, D.P. and Morisette, J.T. (2002) An overview of MODIS Land data processing and product status. Remote Sens. Environ. 83, 3–15.

Kang, S., Running, S.W., Lim, J-H., Zhao, M., Park, C.R. and Loehman, R. (2003) A regional phenology model from detecting onset of greenness in temperate mixed forests, Korea. An application of MODIS leaf area index. Remote Sens. Environ. 86, 232–242.

Kathuroju, N., White, M.A., Symanzik, J., Schwartz, M.D., Powell, J.A. and Nemani, R. (2007) On the use of the Advanced Very High Resolution Radiometer for development of prognostic land surface phenology models. Ecol. Modell. 201, 144–156.

Kaufman, Y.J. and Tanre, D. (1992) Atmospherically resistant vegetation index (ARVI) for EOS-MODIS. IEEE Trans. Geosci. Remote Sens. 30, 261–270.

Lloyd, D. (1990) A phenological classification of terrestrial vegetation cover using shortwave vegetation index imagery. Int. J. Remote Sens. 12, 2269–2279.

Loveland, T.R., Merchant, J.W., Ohlen, D.O. and Brown, J.F. (1991) Development of a land-cover characteristics database for the conterminous U.S. Photogramm. Eng. Rem. Sens. 57, 1453–1463.

Malingreau, J.P., Tucker, C.J. and Laporte, N. (1989) AVHRR for monitoring global tropical deforestation. Int. J. Remote Sens. 10, 855–867.

Menzel, A. (2003) Europe. In: Schwartz, M.D. (Ed.) Phenology, An Integrative Environmental Science. Kluwer Academic Publishers, Dordrecht, Netherlands, pp. 45–56.

Murphy, R.E., Barnes, W.L., Lyapustin, A.I., Privette, J., Welsch, C., DeLuccia, F., Swenson, H., Schueler, C.F., Ardanuy, P.E. and Kealy, P.S.M. (2001) Using VIIRS to provide data continuity with MODIS. IEEE Geoscience and Remote Sensing Symposium 03, 1212–1214.

Pinty, B. and Verstraete, M.M. (1992) GEMI, a non-linear index to monitor global vegetation from satellites. Vegetatio 101, 5–20.

Reed, B.C. (2006) Trend analysis of time-series phenology of North America derived from satellite data. GIScience Remote Sens. 43, 24–38.

Reed, B.C., Brown, J.F., VanderZee, D., Loveland, T.R., Merchant, J.W. and Ohlen, D.O. (1994) Measuring phenological variability from satellite imagery. J. Veg. Sci. 5, 703–714.

Richardson, A.D., Jenkins, J.P., Braswell, B.H., Hollinger, D.Y., Ollinger, S.V. and Smith, M.L. (2007) Use of digital webcam images to track spring green-up in a deciduous forest. Oecologia 152, 323–334.

Schwartz, M.D. (1997) Spring index models: an approach to connecting satellite and surface phenology. In: Lieth, H., Schwartz, M.D. (Eds.) Phenology in Seasonal Climates. Backhuys Publishers, Leiden, Netherlands, pp. 23–38.

Schwartz, M.D. and Reed, B.C. (1999) Surface phenology and satellite sensor-derived onset of greenness: an initial comparison. Int. J. Remote Sens. 20, 3451–3457.

Schwartz, M.D., Reed, B.C. and White, M.A. (2002) Assessing satellite-derived start-of-season measures in the conterminous USA. Int. J. Climatol. 22, 1793–1805.

Schwartz, M.D., Ahas, R. and Aasa, A. (2006) Onset of spring starting earlier across the Northern Hemisphere. Global Change Biol. 12, 343–351.

Slayback, D.A., Pinzon, J.E., Los, S.O. and Tucker, C.J. (2003) Northern hemispheric photosynthetic trends 1982–99. Global Change Biol. 9, 1–15.

Studer, S., Stöckli, R., Appenzeller, C. and Vidale, P.L. (2007) A comparative study of satellite and ground-based phenology. Int. J. Biometeorol. 51, 405–414.

Tateishi, R. and Ebata, M. (2004) Analysis of phenological change patterns using 1982–2000 Advanced Very High Resolution Radiometer (AVHRR) data. Int. J. Remote Sens. 25, 2287–2300.

Townshend, J.R.G., Justice, C.O. and Skole, D. (1994) The 1 km resolution global data set, needs of the International Geosphere Biosphere Programme. Int. J. Remote Sens. 15, 3417–3441.

Tucker, C.J., Pinzon, J.E. and Brown, M.E. (2004) Global Inventory Modeling and Mapping Studies, NA94apr15b.n11-VIg, 2.0, Global Land Cover Facility, University of Maryland, College Park, Maryland, 04/15/1994.

Tucker, C.J., and Sellers, P.J. (1986) Satellite remote sensing of primary productivity, Int. J. Remote Sens. 7, 1395–1416.

Verstraete, M.M., Gabron, N., Aussedat, O., Robustelli, M., Pinty, B., Widlowski, J.L., Lavergne, T. and Taberner, M. (2008) An automatic procedure to identify key vegetation phenology events using the JRC-FAPAR products. Adv. Space Res. 41, 1773–1783.

Wang, J., Rich, P.M. and Price, K.P. (2003) Temporal responses of NDVI to precipitation and temperature in the central Great Plains, USA. Int. J. Remote Sens. 24, 2345–2364.

Westerling, A.L., Hidalgo, H.G., Cayan, D.R. and Swetnam, T.W. (2006) Warming and earlier spring increase Western U.S. forest wildfire activity. Science 18, 940–943.

White M.A. and Nemani, R. (2006) Real-time monitoring and short-term forecasting of land surface phenology. Remote Sens. Environ. 104, 43–49.

White, M.A., Nemani, R.R., Thornton, P.E. and Running, S.W. (2002) Satellite evidence of phenological differences between urbanized and rural areas of the eastern United States deciduous broadleaf forest. Ecosystems 5, 260–273.

White, M.A., Thornton, P.E. and Running, S.W. (1997) A continental phenology model for monitoring vegetation responses to interannual climatic variability. Global Biogeochem. Cycles 11, 217–234.

Woodcock, C.E., Allen, R., Anderson, M., Belward, A., Bindschadler, R., Cohen, W., Gao, F., Goward, S.N., Helder, D., Helmer, E., Nemani, R., Oreopoulos, L., Schott, J., Thenkabail, P.S., Vermote, E.F., Vogelmann, J., Wulder, M.A. and Wynne, R. (2008) Free access to Landsat imagery. Science 320, 1011.

Xiao, X., Boles, S., Liu, J.Y., Zhuang, D.F. and Liu, M.L. (2002) Characterization of forest types in Northeastern China, using multi-temporal SPOT-4 VEGETATION sensor data. Remote Sens. Environ. 82, 335–348

Zhang, X., Tarpley, D. and Sullivan, J.T. (2007) Diverse responses of vegetation phenology to a warming climate, Geophys. Res. Lett. 34, L19405.

Zhang, X., Friedl, M.A. and Schaaf, C.B. (2006) Global vegetation phenology from Moderate Resolution Imaging Spectroradiometer (MODIS). Evaluation of global patterns and comparison with in situ measurements. J. Geophys. Res. 111, G04017.

Zhang, X., Friedl, M.A., Schaaf, C.B. and Strahler, A.H. (2004a) Climate controls on vegetation phenological patterns in northern mid- and high latitudes inferred from MODIS data. Global Change Biol. 10, 1133–1145.

Zhang, X., Friedl, M.A., Schaaf, C.B. and Strahler, A.H. (2004b) The footprint of urban climates on vegetation phenology. Geophys. Res. Lett. 31, L12209.

Zhang, X., Friedl, M.A., Schaaf, C.B., Strahler, A.H., Hodges, J.C.F., Gao, F., Reed, B.C., Huete, A. (2003) Monitoring vegetation phenology using MODIS. Remote Sens. Environ. 84, 471–475.

Zhou, L., Kaufmann, R. K., Tian, Y., Myneni, R.B. and Tucker, C. J. (2003) Relation between interannual variations in satellite measures of northern greenness and climate between 1982 and 1999. J. Geophys. Res. 108 (D1), 4004, doi:10.1029/2002JD002510.

Land Surface Phenology: Convergence of Satellite and CO_2 Eddy Flux Observations

Xiangming Xiao, Junhui Zhang, Huimin Yan, Weixing Wu, and Chandrashekhar Biradar

Abstract Land surface phenology (LSP) is a key indicator of ecosystem dynamics under a changing environment. Over the last few decades, numerous studies have used the time series data of vegetation indices derived from land surface reflectance acquired by satellite-based optical sensors to delineate land surface phenology. Recent progress and data accumulation from CO_2 eddy flux towers offers a new perspective for delineating land surface phenology through either net ecosystem exchange of CO_2 (NEE) or gross primary production (GPP). In this chapter, we discussed the potential convergence of satellite observation approach and CO_2 eddy flux observation approach. We evaluated three vegetation indices (Normalized Difference Vegetation Index, Enhanced Vegetation Index, and Land Surface Water Index) in relation to NEE and GPP data from five CO_2 eddy flux tower sites, representing five vegetation types (deciduous broadleaf forests, evergreen needle-leaf forest, temperate grassland, cropland, and tropical moist evergreen broadleaf forest). This chapter highlights the need for the community to combine satellite observation approach and CO_2 eddy flux observation approach, in order to develop better understanding of land surface phenology.

X. Xiao (✉) and C.M. Biradar
Department of Botany and Microbiology, and Center for Spatial Analysis,
University of Oklahoma, OK, USA
e-mail: xiangming.xiao@ou.edu; chandra.biradar@ou.edu

J. Zhang
Institute of Ecology, Chinese Academy of Sciences, Beijing, China
e-mail: junhui.zhang@126.com

H. Yan and W. Wu
Institute of Geographic Science and Natural Resources Research,
Chinese Academy of Sciences, Beijing, China
e-mail: hanhm@igsnrr.ac.cn; wuweixing2003@163.com

A. Noormets (ed.), *Phenology of Ecosystem Processes*,
DOI 10.1007/978-1-4419-0026-5_11, © Springer Science+Business Media, LLC 2009

1   Introduction

Phenology is the study of periodic biological events in the animals and plants (Lieth 1974; Schwartz 2003). Plant phenology is often studied in a hierarchical structure, ranging from plant organs (first leaf, first flower), individual plants, (% leaf expansion, % flowering), community (% of vegetation) to landscape. Phenology of animals and plants are sensitive to changes in weather and climate. Changes in phenology of plants will affect the carbon cycle, water cycle and energy fluxes through photosynthesis and evapotranspiration, which are closely related to food security, water resources availability and climate.

Numerous in situ observations from researchers and volunteers (e.g. gardeners) have documented various phenological phases of plants (e.g. leaf-on, fall leaf coloring and leaf-off, date of the first flowering) over years. For instance, in situ plant phenology data have been collected at the Hubbard Brook and Harvard Forest Long-Term Ecological Research (LTER) sites for several decades and these data clearly document a marked trend towards an earlier onset of spring over the years (Richardson et al. 2006).

The satellite-based Earth observation systems in 1980s opened a new frontier in the field of phenology from the landscape perspective. Satellite-based optical remote sensing platforms (e.g. NOAA AVHRR sensors) provided daily observations of the land surface for the entire Earth, and the land surface reflectance as recorded in the optical sensors are associated with the biophysical and biochemical properties of vegetation and soils. Vegetation indices, a mathematical transformation as calculated from surface reflectance of different spectral bands (e.g. red and near infrared), have been widely used to track vegetation dynamics in the land surface. The Normalized Difference Vegetation Index (NDVI), calculated from images of NOAA AVHRR sensors, is now the longest time series data for phenology study, and the results from the analysis of AVHRR-based NDVI revealed significant changes in spring phenology of vegetation in 1980s–1990s (Myneni et al. 1997; White et al. 2005). Recently, phenology data products were generated by the Science Team of the Moderate Resolution Imaging Spectroradiometer (MODIS), as part of the NASA Earth Observing System (EOS) program (Zhang et al. 2006, 2003).

The CO_2 eddy covariance method and instrument system in 1990s also opened a new frontier in the field of phenology from the ecosystem and landscape perspective. Continuous CO_2 flux measurements at CO_2 eddy tower sites over a year offer unprecedented opportunity to quantify phenological phases at the ecosystem- and landscape levels from the eco-physiological perspective. It is thought that even modest changes in the length or magnitude of the plant growing season could result in large changes in annual gross primary production in deciduous broadleaf forests (Goulden et al. 1996). An analysis of net ecosystem exchange of CO_2 (NEE) between forest ecosystems and the atmosphere during 1991–2000 in Harvard Forest also suggested that weather and seasonal climate (e.g. light, temperature, and moisture) regulated seasonal and interannual fluctuations of carbon uptake in a temperate deciduous broadleaf forest (Barford et al. 2001). Nowadays, there are more than 600 eddy flux tower sites in various biomes of the world; and these CO_2 flux sites formed several networks (e.g. AmeriFlux, ChinaFlux, EuroFlux and AsiaFlux). The networks of CO_2 eddy flux tower sties play an increasing important role in determining whether individual ecosystems are carbon sink or carbon source (Baldocchi et al. 2001).

Availability of large-volume and valuable datasets from both CO_2 eddy flux towers and satellite remote sensing offer unprecedented opportunity to cross-validating the phenological observations at ecosystem and landscape levels, as observed from the satellite approach and flux tower. It makes possible to address the following scientific questions: To what degree will the photosynthetically active period as delineated by GPP be consistent with the phenology as delineated by satellite observations? Which vegetation index is better to delineate land surface phenology (land surface dynamics in spring and fall)? Can several vegetation indices be used together to improve delineation of land surface phenology? Will phenology delineated from eco-physiological approach and satellite observation approach be consistent with the phenology delineated from the bio-climatic approach and *in situ* observations (scaling-up from in situ individual plants to landscapes)?

In this chapter, we present the data from CO_2 eddy flux towers and satellite images over several terrestrial ecosystems. Our objective is to illustrate the (1) usefulness of CO_2 eddy flux data for delineating phenology of terrestrial ecosystems from the perspective of eco-physiology; and (2) sensitivity of three vegetation indices derived from satellite images for delineating land surface phenology from the perspective of satellite-based Earth observation. These case studies may shed some light on the potential convergence between satellite observation approach and tower-based eco-physiological approach.

2 Methods

In this chapter we focused on land surface phenology (LSP) and vegetation phenology: Vegetation phenology (plant community level or ecosystem level) describes plant phenology for all species in a plant community, or vegetation. In situ observations of individual ecosystem types are often carried out at a specific location, and document differentiated dynamics of individual species of a plant community in responses to climate. Land surface phenology (landscape level) describes temporal dynamics of landscape that is often a mix of soil, vegetation, water bodies, etc. Phenology at the landscape scale poses challenge to observers, because of its complexity; and it often generates confusions among observers because observers may use different approaches. Research approaches for land surface phenology could be roughly grouped into three categories by their observation platforms: meteorological observation; satellite-based reflectance; and CO_2 eddy flux tower.

Meteorological Approach. The first major approach is based on meteorological measurements at weather stations (Lieth 1974; White et al. 1997). For example, frost-free period is a good indicator for plant growing season length; cumulated temperature over 0°C, 5°C is often used to delineate the starting date of the plant growing season. The meteorological approach has the longest history in phenology, and is often incorporated into vegetation models to predict future change of phenology in response to climate change in twenty-first century, or to re-construct phenology in the past.

CO_2 *Eddy Flux Tower Approach.* The second major approach is the eco-physiological approach, based on eddy flux tower platforms (Churkina et al. 2005). The eddy covariance technique measures net exchange of CO_2 and water between terrestrial ecosystems and the atmosphere, and energy fluxes at very short time interval (10 Hz aggregated to 30 min) over years, which together provide precise measurements of ecosystem metabolisms over time (Wofsy et al. 1993; Baldocchi et al. 2001; Falge et al. 2002). Since the early 1990s, hundreds of eddy flux tower sites have been established and cover all major biome types in the world. The footprint size of an eddy flux tower site varies, dependent upon many factors such as the height of the tower, wind speed, and topography, and it ranges from hundreds of meters to a few kilometers. It is important to note that the footprint sizes of CO_2 eddy flux tower are comparable to the spatial resolution of several major satellite observation platforms (e.g. MODIS, VGT). Therefore, several studies have compared the dynamics of satellite-derived vegetation indices with CO_2 fluxes from the flux towers, in an effort to establish the linkage between ecosystem metabolism (CO_2 flux) and satellite-based observation of vegetation dynamics (Xiao et al. 2004b). In this chapter, our discussion focuses on both satellite-based approach and CO_2 flux tower approach.

Satellite-Based Surface Reflectance Approach: The third major approach is based on space-borne satellite observations. Satellite observations provide information on changes in biophysical (e.g. leaf area index) and biochemical (e.g. chlorophyll content, water content) parameters in the land surface (Zhang et al. 2005). Time series of NDVI data derived from AVHRR data since 1980 highlight its potential for monitoring land surface phenology (Stockli and Vidale 2004; White et al. 2005; Philippon et al. 2007). NDVI-derived land surface phenology was often evaluated with time series data of ground-based plant phenology, and the results from a recent study in Switzerland over 1982–2001 indicated that satellite-derived phenology is very susceptible to snow cover (Studer et al. 2007).

2.1 CO_2 *Eddy Flux Tower Approach for Land Surface Phenology*

Application of eddy covariance technique to measures net ecosystem exchange (NEE) of CO_2 between terrestrial ecosystems and the atmosphere dates back to 1974 (Shaw et al. 1974). In 1990, the first year-long continuous CO_2 flux measurements by eddy covariance technique were conducted at Harvard Forest site, Massachusetts (Wofsy et al. 1993). CO_2 flux tower sites provide integrated CO_2 flux measurements over footprints with sizes and shapes (linear dimensions typically ranging from hundreds of meters to several kilometers) that vary with the tower height, canopy physical characteristics and wind velocity (Baldocchi et al. 1996).

The NEE of CO_2 between the terrestrial ecosystem and the atmosphere, as measured at half-hourly interval throughout a year, is the difference between gross primary production (GPP) and ecosystem respiration (Reco):

$$NEE = GPP - Reco \tag{1}$$

An analysis of NEE from 1991 to 2000 at Harvard Forest suggested that weather and climate (e.g. light, temperature and moisture) regulated seasonal and interannual fluctuations of carbon uptake in a temperate deciduous broadleaf forest (Barford et al. 2001). A number of other studies have demonstrated a major role for plant growing season length (GSL) in the terrestrial carbon cycle (Myneni et al. 1997; White et al. 1999). An earlier study used NEE data from 28 flux tower sites to delineate a carbon uptake period (CUP), defined as the number of days with negative NEE values (net CO_2 uptake by terrestrial ecosystems), and then used CUP as an approximate estimate of GSL (Churkina et al. 2005). There is a strong correlation between annual NEE and CUP for temperate broadleaf forests (Baldocchi and Wilson 2001; White and Nemani 2003).

Partitioning of NEE into ecosystem respiration and GPP is an active research field (Falge et al. 2002), and the resultant GPP data offers a way to delineate the plant photosynthetically active period (PAP), defined as the number of consecutive days with GPP values of greater than zero (or a threshold value, for example, 1 g C m^{-2} day^{-1}, if given some estimates of uncertainty in NEE measurements and estimation of ecosystem respiration). The PAP can be used as an approximate estimate of GSL.

Both CUP and PAP are definitions from the perspective of eco-physiological approach, and theoretically PAP may have stronger correlation with GSL and vegetation indices than does CUP, because the latter is also affected by ecosystem respiration. In this chapter we examine the relationships between vegetation indices with both NEE and GPP.

2.2 Satellite-Based Approach for Land Surface Phenology

2.2.1 Spectral, Spatial and Temporal Characteristics of Optical Sensors

The time series data of from the following three optical sensors have been widely used for phenological studies.

The Advanced Very High Resolution Radiometer (AVHRSSR) Sensors. The AVHRR sensor was originally designed for weather and climate study, and aims to provide radiance data for investigation of clouds, snow and ice extent, temperature of radiating surface and sea surface temperature. It has two spectral bands: red band (580–680 nm) and near infrared (725–1,100 nm), and acquires daily images for the globe at 1-km to 4-km spatial resolution. Numerous studies have used these two spectral bands to study vegetation condition (Myneni et al. 1997). However AVHRR data has some limitation for vegetation studies such as lack of calibration, poor geometry and high level of noise due to large pixel size and limited cloud screening (Goward et al. 1991).

The Vegetation (VGT) Sensors: The VGT sensor onboard the SPOT-4 satellite was launched in March 24, 1998, as the first space-borne moderate resolution sensor designed for vegetation study. The VGT sensor onboard the SPOT-4 satellite has four spectral bands: blue (430–470 nm), red (610–680 nm), near infrared (NIR, 780–890 nm) and shortwave infrared (SWIR, 1,580–1,750 nm). The sensor has a spatial resolution of 1,165 m × 1,165 m and a swath of 2,250 km. VGT sensor onboard the SPOT-5 satellite was launched on April 5, 2002, and is still operational. The VEGETATION program is co-financed by the European Union and conducted under the supervision of CNES (National Centre for Space Studies, France). The VGT program provides daily and 10-day synthesis (composite) products at 1-km spatial resolution. The standard 10-day composite data (VGT-S10) are freely available to the public (http://free.vgt.vito.be). The temporal compositing method for generating standard 10-day synthetic products (VGT-S10) is to select an observation with the maximum NDVI value within a ten-day period. There are three 10-day composites within a month: day 1–10, day 11–20, and day 21 to the end of the month.

The MODerate resolution Imaging Spectroradiometer (MODIS) Sensors: The MODIS sensor onboard the NASA Terra satellite was launched in December 1999. The MODIS sensor has 36 spectral bands, seven of which are designed for the study of vegetation and land surfaces: blue (459–479 nm), green (545–565 nm), red (620–670 nm), near infrared (NIR1: 841–875 nm; NIR2: 1,230–1,250 nm), and shortwave infrared (SWIR1: 1,628–1,652 nm, SWIR2: 2,105–2,155 nm). Daily global imagery is provided at spatial resolutions of 250-m (red and NIR1) and 500-m (blue, green, NIR2, SWIR1, SWIR2). The MODIS Land Science Team provides a suite of standard MODIS data products to the users, including the 8-day composite MODIS Surface Reflectance Product (MOD09A1). Each 8-day composite (MOD09A1) includes estimates of surface spectral reflectance for the seven spectral bands at 500-m spatial resolution. In the production of MOD09A1, atmospheric corrections for gases, thin cirrus clouds and aerosols are implemented (Vermote and Vermeulen 1999). MOD09A1 composites are generated in a multi-step process that first eliminates pixels with a low observational coverage, and then selects an observation with the minimum blue-band value during an 8-day period (http://modis-land.gsfc.nasa.gov/MOD09/MOD09ProductInfo/MOD09_L3_8-day.htm).

2.2.2 Vegetation Indices

A number of vegetation indices have been developed for broad-waveband optical sensors over the last three decades, and the following three vegetation indices have been used for the study of land surface phenology: Normalized Difference Vegetation Index (NDVI; Tucker 1979), Enhanced Vegetation Index (EVI; Huete et al. 1997) and Land Surface Water Index (LSWI; Xiao et al. 2002b).

$$NDVI = \frac{\rho_{nir} - \rho_{red}}{\rho_{nir} + \rho_{red}} \qquad (2)$$

$$EVI = 2.5 \times \frac{\rho_{nir} - \rho_{red}}{\rho_{nir} + 6 \times \rho_{red} - 7.5 \times \rho_{blue} + 1} \tag{3}$$

$$LSWI = \frac{\rho_{nir} - \rho_{swir}}{\rho_{nir} + \rho_{swir}} \tag{4}$$

For MODIS data, surface reflectance values from blue, red, NIR (841–875 nm), SWIR1$_1$ (1,628–1,652 nm) are used in calculation of NDVI, EVI and LSWI.

NDVI is calculated as the normalized ratio between NIR and red bands (Tucker 1979). Numerous studies have shown that NDVI is closely correlated with leaf area index (LAI), a biophysical parameter of the vegetation canopy (Gao et al. 2000). Fraction of photosynthetically active radiation absorbed by vegetation canopy (FAPARcanopy) is also assumed to be a linear or non-linear function of NDVI (Potter et al. 1993; Myneni and Williams 1994; Ruimy et al. 1994; Prince and Goward 1995; Justice et al. 1998). The LAI–NDVI relationship and NDVI–FAPAR relationship (Knyazikhin et al. 1998; Myneni et al. 2002) were developed largely from analysis of images from the AVHRR sensor onboard NOAA meteorological satellites, and are also used in the standard LAI/FAPAR product (MOD15A2; Myneni et al. 2002) from the MODIS sensor onboard the Terra satellite. The empirical relationships among LAI–NDVI–FAPAR are the dominant paradigm and the foundation for a number of satellite-based Production Efficiency model (PEM) that estimate gross primary production (GPP) or net primary production (NPP) of terrestrial ecosystems at the global scale (Potter et al. 1993; Ruimy et al. 1994; Prince and Goward 1995; Justice et al. 1998).

EVI is calculated from red, NIR and blue bands. An earlier study that used airborne multispectral data has shown that EVI is linearly correlated with the green leaf area index (LAIgreen) in crop fields (Boegh et al. 2002). Evaluation of radiometric and biophysical performance of EVI calculated from MODIS data indicated that EVI remain sensitive to canopy variation (Huete et al. 2002). In another study that compared NDVI and EVI data derived VGT images for Northern Asia over the period of 1998–2001, the results indicated EVI is less sensitive to residual atmospheric contamination due to aerosols from extensive fire in 1998 (Xiao et al. 2003).

LSWI is calculated as the normalized ratio between NIR and SWIR bands (Xiao et al. 2002b). As SWIR band is sensitive to leaf water content, a number of studies have explored the SWIR spectral bands (e.g. 1.6 and 2.1 mm) for vegetation water content (Hunt et al. 1987; Hunt and Rock 1989; Gao 1996; Serrano et al. 2000; Ceccato et al. 2001, 2002a, b; Xiao et al. 2002a, b; Roberts et al. 2003), and the results from these studies (Ceccato et al. 2002a, b) suggested that a combination of NIR and SWIR bands have the potential for retrieving leaf and canopy water content (equivalent water thickness, EWT, g cm^{-2}). A few water-oriented vegetation indices were developed for characterization of leaf and canopy water content, e.g. Moisture Stress Index (MSI; Hunt and Rock 1989), Normalized Difference Water Index (NDWI; Gao 1996), and Land Surface Water Index (LSWI; Xiao et al. 2002a, b). LSWI was previously called as the Normalized Difference Infrared Index (NDII; Hunt and Rock 1989), Very limited numbers of field studies on leaf water content have been carried out (Xiao et al. 2004a). Analysis of time series LSWI data have shown that LSWI is useful for improving land cover classification and phenology (Xiao et al. 2002a, b; Boles et al. 2004; Delbart et al. 2005; Sakamoto et al. 2007).

3 Land Surface Phenology of Forests, Grasslands and Cropland: Five Case Studies

In this section, we presented data from five CO_2 eddy flux tower sites, representing four vegetation types (deciduous broadleaf forest, evergreen needleleaf forest, temperate grassland, cropland, and evergreen tropical broadleaf forest). Time series data of vegetation indices (NDVI, EVI and LSWI) were compared with NEE and GPP data from the tower sites.

3.1 Land Surface Phenology of Temperate Deciduous Broadleaf Forests

Here we present CO_2 eddy flux data and satellite images for two eddy flux tower sites in USA and China as case studies that use CO_2 flux data for interpreting phenology (spring and fall) of multi-temporal satellite images at moderate spatial resolution.

3.1.1 Harvard Forest (USA)

The eddy flux tower site (42.54°N, 72.18°W, 340 m elevation) is located within Harvard Forest, Petersham, Massachusetts, USA. Vegetation at the site is primarily a 60–80-year-old deciduous broadleaf forest, and dominant species composition includes red oak (*Quercus rubra*), red maple (*Acer rubrum*), black birch (*Betula lenta*), white pine (*Pinus strobes*) and hemlock (*Tsuga Canadensis*). Annual mean temperature is about 7.9°C and annual precipitation is about 1,066 mm. On the average, plant growing season lasts for 161 days (Waring et al. 1995).

Eddy flux measurements of CO_2, H_2O and energy at Harvard Forest have been collected since 1991 (Wofsy et al. 1993; Goulden et al. 1996; Barford et al. 2001). Daily data of maximum and minimum temperature (°C), precipitation (mm) and photosynthetically active radiation (PAR, mol m^{-2} d^{-1}) in 2000 were obtained from the website of Harvard Forest (http://www-as.harvard.edu/data/nigec-data.html). Daily measured NEE flux data and derived GPP and ecosystem respiration (Reco) at Harvard Forest in 2000 were provided by researchers at Harvard Forest. Daily climate and CO_2 flux data were aggregated to the 10-day interval as defined by the 10-day composite VGT images (days 1–10, 11–20, and 21 to the end of the month).

We acquired the VGT-S10 data (http://free.vgt.vito.be) over the period of January 1–10, 2000 to December 21–31, 2000 (a time-series data of 36 VGT-S10 images) for the study site. Three vegetation indices (NDVI, EVI and LSWI) were calculated for all the 10-day composite images (VGT-S10). A detailed description of the pre-processing and calculation of vegetation indices from VGT-S10 data can/ be found in (Xiao et al. 2003).

Figure 1 shows the seasonal cycle of NEE and GPP in 2000 at Harvard Forest. The CUP lasts from early May to early October, whereas PAP is from late March to late October (Fig. 1). GPP values were near zero in winter season (December, January, and February), because the deciduous dominated canopy is bare and low air temperature and frozen soil inhibit photosynthetic activities of conifer trees. GPP began to increase

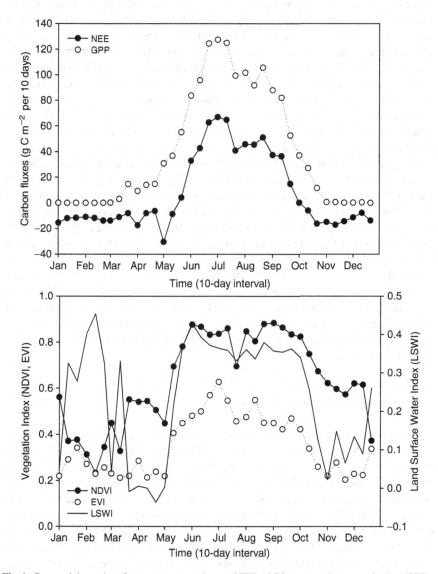

Fig. 1 Seasonal dynamics of net ecosystem exchange (NEE) of CO_2, gross primary production (GPP), Enhanced Vegetation Index (EVI), Normalized Difference Vegetation Index (NDVI), and Land Surface Water Index (LSWI) in 2000 at the eddy flux tower in Harvard Forest, Massachusetts, USA. Here we used 10-day composites of VEGETASTION land surface reflectance data (VGT-S10).

in late March (mostly attributed to leaves of understory plants) and rose rapidly in late April to early Mary (mostly attributed to leaf emergence and expansion of upper canopy plants). GPP declined rapidly after its peak in late June to early July, despite that the fact that LAI changed little over the period between July and September.

The time series of LSWI data have a distinct seasonal cycle with a spring trough and a fall trough (Fig. 1). The high LSWI values in late fall, winter and early spring are attributed to snow cover (above or below the canopy). The green-up period (from bud burst to full expansion of leaves) is defined as the period from the date that had the minimum LSWI in spring to the date that had the maximum LSWI in early summer (early June at Harvard Forest site). Therefore, the plant growing season can be simply delineated as the period from the spring trough to the fall trough (Fig. 1). This LSWI-based delineation of plant growing season was first proposed in an early study of forests in Northern China (Xiao et al. 2002b), and late used in the study of forests in Russia (Delbart et al. 2005).

The time series of EVI data also have a distinct seasonal cycle (Fig. 1). The rapid increase of EVI in late April to early May also makes it relatively easy to define the starting dates of plant growing season.

The seasonal dynamic of NDVI (Fig. 1), particularly in the spring, is not as distinct as EVI and LSWI time series data. Note that NDVI has been widely used to delineate plant growing season; and a number of studies used a NDVI threshold value (Myneni et al. 1998; Jenkins et al. 2002). It is a reasonable choice and approach when NDVI data from AVHRR sensors were analyzed, as AVHRR sensors have only red and NIR bands. The NDVI-based threshold approach for delineating plant growing season (starting and ending dates of plant growing season) is clearly constrained by what threshold NDVI values to be used (White and Nemani 2006), as it might vary depending upon NDVI datasets generated by different research groups, vegetation types and vegetation conditions. For example, a threshold of 0.25 NDVI value from a AVHRR dataset was used to define plant growing season in the northern latitudes (Myneni et al. 1998), but a threshold of 0.45 NDVI value was used for eastern USA (Jenkins et al. 2002).

3.1.2 Changbai Mountain (China)

The CO_2 eddy flux tower site (42.40°N, 128.10°E) is located within No. 1 Plot of the Changbai Mountains Forest Ecosystem Research Station (CBM-FERS), Jilin Province, China. The CBM-FERS is one of Chinese Ecosystem Research Network (CERN) stations, Chinese Academy of Sciences. The climate belongs to the temperate continental climate influenced by monsoon, with four distinct seasons: windy spring, hot and rainy summer, cool autumn and cold winter. Annual mean temperature of the flux site ranges from 0.9 to 4.0°C and mean annual precipitation is 695 mm y^{-1} (1982–2004). The site has a flat topography, and the soil is classified as dark brown forest soil originating from volcanic ashes. The site is covered by on average 200-year-old, multi-layered, uneven-aged, multi-species mixed forest consisting of *Pinus koraiensis*, *Tilia amurensis*, *Acer mono*, *Fraxinus mandshurica*, *Quercus mongolica* and 135 other species. The mean canopy height is 26 m. A dense understory, consisting of broad-leaved shrubs, has a height of 0.5–2 m. The peak LAI is ~6.1 m^2 m^{-2}. CO$_2$ flux measurements at the site started in August 2002.

Figure 2 shows the seasonality of NEE and GPP at the Changbai site in 2003. The CUP ranges from early May to mid September, whereas PAP ranges from early April to late October.

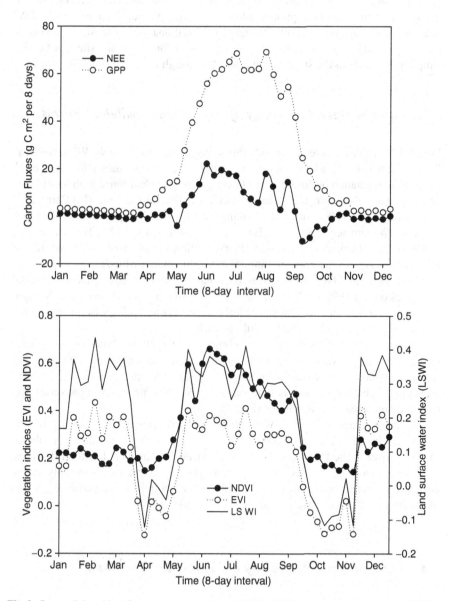

Fig. 2 Seasonal dynamics of net ecosystem exchange (NEE) of CO$_2$, gross primary production (GPP), Enhanced Vegetation Index (EVI), Normalized Difference Vegetation Index (NDVI), and Land Surface Water Index (LSWI) in 2003 at the eddy flux tower in Changbain Mountain, Jilin province, China. Here we used 8-day composites of MODIS land surface reflectance product (MOD09A1).

Both LSWI and EVI time series data have distinct seasonal cycles with a spring trough and a fall trough (Fig. 2). LSWI values remained high during January to late February due to snow cover in the ground and canopy, but started to decline rapidly in March. After snowmelt in late March, EVI and LSWI values started to rise rapidly, corresponding well with the increases of GPP in early April (Fig. 2). In the fall season, both LSWI and EVI reached the lowest values by late October, corresponding well with the end of the photosynthetically active period. In comparison, NDVI reaches its lowest value in early November. Visual analysis of these three vegetation indices suggests that phenology of deciduous forests at this site can be also simply derived from the spring trough and fall trough of LSWI (Fig. 2).

3.2   Land Surface Phenology of Evergreen Needleleaf Forest

We used field data collected at an eddy flux tower site (45.20°N, 68.74°W) in Howland Forest, Maine, USA, where evergreen coniferous trees dominate (Hollinger et al. 1999). The vegetation of this 90-year-old evergreen needleleaf forest is about 41% red spruce (*Pinus rubens* Sarg), 25% eastern hemlock (*Tsuga canadensis* (L.) Carr.), 23% other conifers and 11% hardwoods (Hollinger et al. 1999). The peak LAI of the forest stand is approximately 5.3 m^2 m^{-2}. Eddy flux measurements of CO_2, H_2O and energy at the site have being conducted since 1996 (Hollinger et al. 1999, 2004) and the site is part of the AmeriFlux network (http://public.ornl.gov/ameriflux/).

Both NEE and GPP time series data have distinct and consistent seasonal dynamics during 1998–2001 (Fig. 3). The flux tower site is carbon source throughout the winter season (November to March). GPP started to increase in late March and reached its peak values in late July to early August.

The low NDVI values in January–February are largely attributed to snow cover in the site (Fig. 3). NDVI time series data over a year does not show clear signal of seasonal dynamics in a year, which is to large degree attributed to saturation of NDVI in a forest with relatively high value of LAI. In contrast, EVI time series data have a distinct seasonal dynamics (Fig. 3).

The seasonal dynamics of LSWI is unique and characterized by a "spring trough" and a "fall trough." The high LSWI values in winter and early spring are attributed to snow cover in the forest stands. As snow melted in late March, LSWI declined. The "spring trough" corresponds to the beginning of photosynthetically active period of evergreen needleleaf forest, and the "fall trough" corresponds to the end of photosynthetically active period (Xiao et al. 2004a).

3.3   Land Surface Phenology of Temperate Grassland

The CO_2 eddy flux tower site (43.55°N, 116.68°E) of grassland is located within the Inner Mongolia Grassland Ecosystem Research Station (IMGERS) in Xilingol

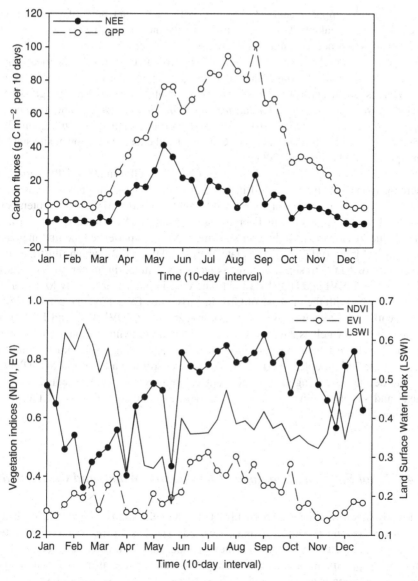

Fig. 3 Seasonal dynamics of net ecosystem exchange (NEE) of CO_2, gross primary production (GPP) of CO_2, Enhanced Vegetation Index (EVI), Normalized Difference Vegetation Index (NDVI), and Land Surface Water Index (LSWI) in 2000 at the eddy flux tower in Howland, Maine, USA. Here we used 10-day composites of VEGETASTION land surface reflectance data (VGT-S10).

League, Inner Mongolia, China (Fu et al. 2006a, b). The IMGERS is one of the Chinese Ecosystem Research Network (CERN), Chinese Academy of Sciences. The study area has flat topography with an elevation of 1,000 m. Annual mean precipitation is approximately 350.9 mm, with more than 80% of precipitation occurring in the peak plant growing season of July and August. The growing season is generally short because much of the region is snow-covered from late October to early April. The mean annual temperature is approximately −0.4°C; the mean temperature of the coldest month (January) is −19.5°C; and the mean temperature of the warmest month (July) is 20.8°C. The dominant species of Xilingol grassland include *Leymus Chinensis, Achnatherum sibiricum, Stipa gigantean* and *Agropyron michnoi* (Xiao et al. 1995, 1996). Major soil type is chestnut soil with about 3% organic matter. Flux measurements of CO_2, H_2O and energy at the tower site started on April 23, 2003 (Fu et al. 2006a).

Figure 4 shows the seasonal dynamics of NEE and GPP in 2004. GPP started to increase in early May, and ended in late September. The plant growing season in 2004, defined as the carbon uptake period, was from early May to late September.

LSWI values were high in February and March, and declined rapidly in late March due to snowmelt. Right after snowmelt, NDVI experienced the first increase from early March (NDVI = 0.1) to late March (NDVI = 0.2), which illustrates the sensitivity of NDVI to soil moisture, as wet soils resulted in higher NDVI values. Note that both EVI and NDVI started to increase in late April to early May, corresponding well with the increase of GEE in early May. Evidently, one can use those dates with consistent increases of vegetation indices (NDVI, EVI and LSWI) in spring after snowmelt as the starting date of the plant growing season. LSWI values reached its lowest value in late September, corresponding well with the ending dates of the carbon uptake period. While one can still use a threshold of NDVI or EVI to define the ending date of the plant growing season, LSWI seems to offer a clean and simple alternative approach to delineate the ending dates of the plant growing season.

3.4 Land Surface Phenology of Croplands: Wheat and Corn Fields

The eddy flux tower site (36.95°N, 116.60°E, 28 m elevation) is located in Yucheng County, Shandong Province, China. The study site has a crop rotation of winter wheat and maize in a year. Annual mean temperature at this site is about 13.1°C and annual precipitation is approximately 528 mm. On the average, mean annual sunshine duration is about 2,640 h and frost-free period is about 200 days.

Eddy flux measurement system were located in the center of a crop field within a large and homogeneous cropland area, and CO_2, H_2O and energy fluxes have been simultaneously measured since 2003 (Li et al. 2006; Zhao et al. 2007). Daily data of maximum and minimum temperature (°C), precipitation (mm) and photosynthetically active radiation (PAR, mol m^{-2} d^{-1}) were also available from this site for this study.

Daily flux data of NEE, GPP and ecosystem respiration (Reco) at the flux sites were generated from the half-hourly flux data. Half-hourly values were calculated from the covariance of the fluctuations in vertical wind speed and CO_2 concentration

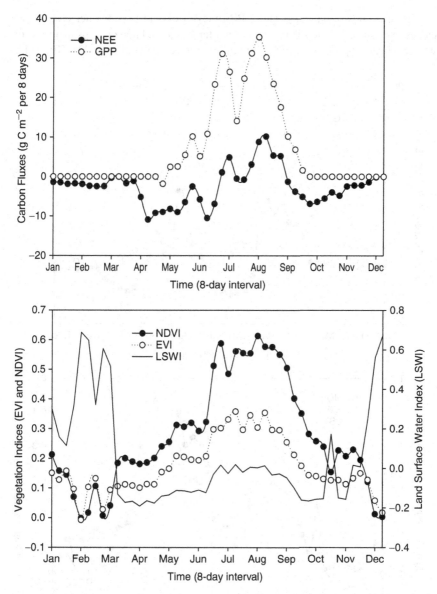

Fig. 4 Seasonal dynamics of net ecosystem exchange (NEE) of CO_2, gross primary production (GPP), Enhanced Vegetation Index (EVI), Normalized Difference Vegetation Index (NDVI), and Land Surface Water Index (LSWI) in 2004 at the eddy flux tower of grassland ecosystem in Xilingol, Inner Mongolia, China. Here we used 8-day composites of MODIS land surface reflectance product (MOD09A1).

measured at 5 Hz. We calculated the 8-day sums of GPP and NEE from the daily GPP and NEE data, in order to be consistent with the 8-day composite satellite images we used.

Figure 5 shows that NEE and GPP time series had two distinct crop growth cycles, corresponding to the rotation of winter wheat and maize crops in a year. For winter wheat crop, which were planted in previous year (fall of 2002), GPP values

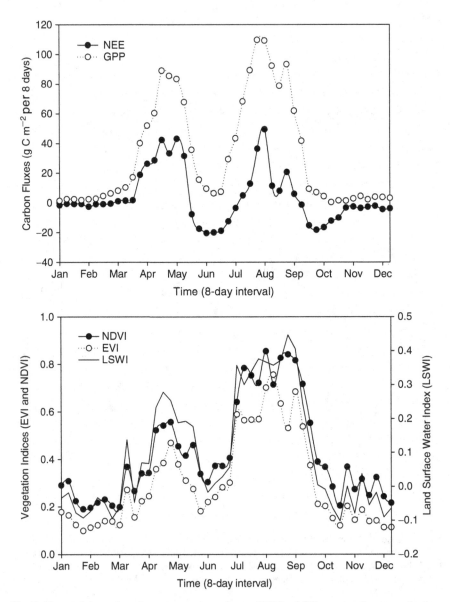

Fig. 5 Seasonal dynamics of net ecosystem exchange (NEE) of CO_2, gross primary production (GPP), Enhanced Vegetation Index (EVI), Normalized Difference Vegetation Index (NDVI), and Land Surface Water Index (LSWI) in 2004 at the eddy flux tower of croplands in Yuchen, Hebei province, China. Here we used 8-day composites of MODIS land surface reflectance product (MOD09A1).

were close to zero during January to February, and started to increase by March, as air temperature increased in spring. Winter wheat crop reached its maximum GPP value (~89 g C m^{-2} 8 d^{-1}) by late April and then declined gradually. By early June winter wheat was harvested. Within 2-weeks after the harvest of winter wheat, summer maize crops were planted. For maize crops, GPP values increased rapidly and reached its maximum values (~103 g C m^{-2} 8 d^{-1}) by early August. Maize crops were harvested in early October. For this cropland site, the CUP has little difference from the PAP (Fig. 5).

NDVI, EVI and LSWI time series data all have bimodal temporal curves, corresponding to the rotation of winter wheat – maize crops. For the first cycle, both NDVI and EVI values started to increase by early March as winter wheat plants turn green, and reached a plateau by late April. After the harvest of winter wheat in early June, both NDVI and EVI values dropped. The timing of NDVI and EVI dynamics for winter wheat is consistent with the phenological or seasonal cycle of winter wheat. For the second cycle, both NDVI and EVI started to substantially increase by early July, corresponding to the planting of summer maize crop. After the harvest of maize in late September, both NDVI and EVI values dropped substantially.

3.5 Land Surface Phenology of Seasonally Moist Tropical Evergreen Broadleaf Forest

The CO$_2$ eddy flux tower site (2.85°S, 54.97°W) is located in the Tapajos National Forest, south of Santarem, Para, Brazil. It is an old-growth, seasonally wet tropical evergreen forest. Annual precipitation is approximately 1,920 mm with distinct wet and dry seasons. The 7-month wet season is usually from December through June, and the dry season is from July to November. Continuous measurement of CO$_2$, H$_2$O and energy fluxes at the site has been conducted since April 2001.

The analysis of CO$_2$ flux data at the site reported that this site maintained high gross primary production during the dry season (Saleska et al. 2003). The observed H$_2$O flux data at the flux tower site show that the evapotranspiration at the site was higher in dry season than in wet season (Fig. 6). It was proposed that seasonally moist tropical evergreen forest have evolved two adaptive mechanisms in an environment with strong seasonal variation of light and water (1) deep root system to access water from deep soils, and (2) leaf phenology for maximum utilization of light (Xiao et al. 2005).

EVI values increased gradually from July to November (Fig. 6), representing a canopy dynamic process within the dry season. This EVI dynamics is temporally consistent with two processes that occur in the canopy during the dry season (1) fall of old leaves in the canopy at the early dry season and (2) emergence of new leaves in the late dry season (Xiao et al. 2005). In comparison, NDVI time series data from the MODIS sensor have low values in wet season (from January to June), mostly due to atmospheric condition (e.g. thin clouds), but remain similar high values in

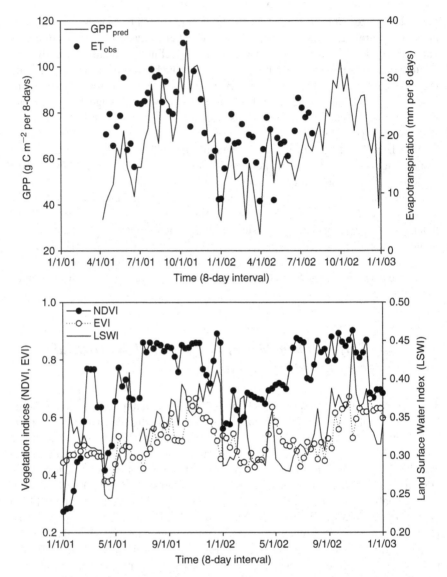

Fig. 6 Seasonal dynamics of observed evapotranspiration, predicted gross primary production from a model (Xiao et al. 2005), Enhanced Vegetation Index (EVI), Normalized Difference Vegetation Index (NDVI), and Land Surface Water Index (LSWI) in 2001–2002 at the eddy flux tower of evergreen tropical forest in Tapajos, Brazil. Here we used 8-day composites of MODIS land surface reflectance product (MOD09A1).

dry season from June to November, which shows no temporal dynamics in dry season (Fig. 6). Mature evergreen tropical forests usually have high LAI (>4 m^2 m^{-2}), which results in saturation of NDVI. The high LAI also presents challenges for phenological studies. Consequently, there are a limited number of field-based phenological studies for evergreen forests in tropical South America (Van Schaik et al. 1993; Wright and van Schaik 1994).

The long-held and conventional wisdom has been that mature evergreen tropical forests have a monotonous growing season, in other words, there is no seasonal dynamics of leaf phenology in evergreen tropical forests. This conventional wisdom was recently challenged by eddy flux data (Saleska et al. 2003) and analysis of EVI and LSWI data from Vegetation and MODIS sensors (Xiao et al. 2005) at a seasonally moist evergreen tropical forest site. Recent analyses of EVI data over the Amazon basin further suggest that there is large seasonal dynamics of leaf phenology in mature forests in the Amazon basin (Huete et al. 2006; Xiao et al. 2006).

4 Summary

Phenology of plants has been studied from several perspectives, including in-situ observations of leaf and flowering, bio-climate, eco-physiology and satellite observation. The eco-physiological approach focuses on continuous data from CO_2 eddy flux tower sites. Year-long continuous CO_2 flux data from the eddy flux tower sites enable the scientific community to delineate the carbon uptake period (CUP), which is defined by net ecosystem exchange (NEE) of CO_2 between the terrestrial ecosystems and the atmosphere, and the photosynthetically active period (PAP), which is defined by gross primary production (GPP). To what degree will the photosynthetically active period (PAP) as delineated by GPP be consistent with the phenology as delineated by satellite observations? Which vegetation index is better to delineate land surface phenology (land surface dynamics in spring and fall)? Can several vegetation indices be used together to improve delineation of land surface phenology? Will phenology delineated from eco-physiological approach and satellite observation approach be consistent with the phenology delineated from the bio-climatic approach and in-situ observations (scaling-up from in-situ individual plants to landscapes)? The case studies presented in this chapter provide exploratory data analyses that compare CO_2 flux data (NEE and GPP) with vegetation indices, which shed new lights on the questions mentioned above.

Among the three vegetation indices (NDVI, EVI and LSWI) we discussed in this chapter, NDVI has the longest time series data (from early 1980s to present), and is most widely used and well documented in the literature. Time series of NDVI data is useful for delineating land surface phenology, Careful comparisons between NDVI and CO_2 flux data, as reported in this chapter, highlights the limitation and problems of NDVI for land surface phenology, in particular for forests with large values of leaf area index. One can still use a threshold of NDVI or EVI to define the beginning and ending date of the plant growing season, however LSWI seems to offer a clean and simple alternative approach to delineate the ending dates of the plant growing season.

LSWI and EVI time series data together have shown distinct seasonal cycles with a spring trough and a fall trough. The increases of the EVI and LSWI values have shown strong relationship with the increases of GPP. Both EVI and LSWI provide new and complementary data to delineate land surface phenology, and they are clearly more consistent with the CO_2 flux data than does NDVI in forest ecosystems.

Accuracy assessment and validation of land surface phenology is a long-term effort that requires coordination from both the remote sensing community and the CO_2 eddy flux community. At present, there are more than 600 CO_2 eddy flux tower sites across various biomes in the world, including different land use, management, disturbance stages. The eddy flux community needs to partition half-hourly NEE data into GPP and ecosystem respiration in a consistent method, and provide GPP data to users in a timely fashion. The remote sensing community needs to provide time series data of vegetation indices for the eddy flux tower sites in a timely fashion, including daily satellite images. In addition, the scientific community also needs to facilitate integration of other in situ field observations (e.g. leaf emergence and leaf fall, canopy dynamics as observed from web cameras) with CO_2 fluxes and remote sensing data. The advance of internet, computer and web technology makes it possible for the scientific community to develop a citizen-based network of plant phenology through participation of individual citizens and scientists, which would offer the capacity of near-real time monitoring of land surface phenology.

Acknowledgements This study was supported by NASA Land Cover and Land Use Change Program (the Northern Eurasia Earth Science Partnership Initiative (NEESPI); NN-H-04-Z-YS-005-N, and NNG05GH80G), and NASA Interdisciplinary Science program (NAG5-11160, NAG5-10135), and National Key Research and Development Program of China (2002CG412501) and International Partnership Project of Chinese Academy of Sciences (CXTD-Z2005-1).

References

Baldocchi, D., Falge, E., Gu, L.H., Olson, R., Hollinger, D., Running, S., Anthoni, P., Bern-hofer, C., Davis, K., Evans, R., Fuentes, J., Goldstein, A., Katul, G., Law, B., Lee, X.H., Malhi, Y., Meyers, T., Munger, W., Oechel, W., U, K.T.P., Pilegaard, K., Schmid, H.P., Valentini, R., Verma, S., Vesala, T., Wilson, K. and Wofsy, S. (2001) FLUXNET: A new tool to study the temporal and spatial variability of ecosystem-scale carbon dioxide, water vapor, and energy flux densities. Bull. Am. Meteorol. Soc. 82, 2415–2434.

Baldocchi, D., Valentini, R., Running, S., Oechel, W. and Dahlman, R. (1996) Strategies for measuring and modelling carbon dioxide and water vapour fluxes over terrestrial ecosys-tems. Global Change Biol. 2, 159–168.

Baldocchi, D. and Wilson, K. (2001) Modeling CO_2 and water vapor exchange of a temperate broadleaved forest across hourly to decadal time scales. Ecol. Modell. 142, 155–184.

Barford, C.C., Wofsy, S.C., Goulden, M.L., Munger, J.W., Pyle, E.H., Urbanski, S.P., Hutyra, L., Saleska, S.R., Fitzjarrald, D. and Moore, K. (2001) Factors controlling long- and short-term sequestration of atmospheric CO_2 in a mid-latitude forest. Science 294, 1688–1691.

Boegh, E., Soegaard, H., Broge, N., Hasager, C.B., Jensen, N.O., Schelde, K. and Thomsen, A. (2002) Airborne multispectral data for quantifying leaf area index, nitrogen concentra-tion, and photosynthetic efficiency in agriculture. Remote Sens. Environ. 81, 179–193.

Boles, S.H., Xiao, X.M., Liu, J.Y., Zhang, Q.Y., Munkhtuya, S., Chen, S.Q. and Ojima, D. (2004) Land cover characterization of Temperate East Asia using multi-temporal VEGETATION sensor data. Remote Sens. Environ. 90, 477–489.

Ceccato, P., Flasse, S. and Gregoire, J.M. (2002a) Designing a spectral index to estimate vegetation water content from remote sensing data – Part 2. Validation and applications. Remote Sens. Environ. 82, 198–207.

Ceccato, P., Flasse, S., Tarantola, S., Jacquemoud, S. and Gregoire, J.M. (2001) Detecting vegetation leaf water content using reflectance in the optical domain. Remote Sens. Envi-ron. 77, 22–33.

Ceccato, P., Gobron, N., Flasse, S., Pinty, B. and Tarantola, S. (2002b) Designing a spectral index to estimate vegetation water content from remote sensing data: Part 1 – Theoretical approach. Remote Sens. Environ. 82, 188–197.

Churkina, G., Schimel, D., Braswell, B.H. and Xiao, X.M. (2005) Spatial analysis of growing season length control over net ecosystem exchange. Global Change Biol. 11, 1777–1787.

Delbart, N., Kergoat, L., Le Toan, T., Lhermitte, J. and Picard, G. (2005) Determination of phenological dates in boreal regions using normalized difference water index. Remote Sens. Environ. 97, 26–38.

Falge, E., Baldocchi, D., Tenhunen, J., Aubinet, M., Bakwin, P., Berbigier, P., Bernhofer, C., Burba, G., Clement, R., Davis, K.J., Elbers, J.A., Goldstein, A.H., Grelle, A., Granier, A., Guomundsson, J., Hollinger, D., Kowalski, A.S., Katul, G., Law, B.E., Malhi, Y., Meyers, T., Monson, R.K., Munger, J.W., Oechel, W., Paw, K.T., Pilegaard, K., Rannik, U., Reb-mann, C., Suyker, A., Valentini, R., Wilson, K. and Wofsy, S. (2002) Seasonality of eco-system respiration and gross primary production as derived from FLUXNET measure-ments. Agric. For. Meteorol. 113, 53–74.

Fu, Y.L., Yu, G.R., Sun, X.M., Li, Y.N., Wen, X.F., Zhang, L.M., Li, Z.Q., Zhao, L. and Hao, Y.B. (2006a) Depression of net ecosystem CO_2 exchange in semi-arid Leymus chinensis steppe and alpine shrub. Agric. For. Meteorol. 137, 234–244.

Fu, Y.L., Yu, G.R., Wang, Y.F., Li, Z.Q. and Hao, Y.B. (2006b) Effect of water stress on ecosystem photosynthesis and respiration of a *Leymus chinensis* steppe in Inner Mongolia. Sci. China D 49, 196–206.

Gao, B.C. (1996) NDWI – A normalized difference water index for remote sensing of vegeta-tion liquid water from space. Remote Sens. Environ. 58, 257–266.

Gao, X., Huete, A.R., Ni, W.G. and Miura, T. (2000) Optical-biophysical relationships of vegetation spectra without background contamination. Remote Sens. Environ. 74, 609–620.

Goulden, M.L., Munger, J.W., Fan, S.M., Daube, B.C. and Wofsy, S.C. (1996) Exchange of carbon dioxide by a deciduous forest: Response to interannual climate variability. Science 271, 1576–1578.

Goward, S.N., Markham, B., Dye, D.G., Dulaney, W. and Yang, J.L. (1991) Normalized Difference Vegetation Index measurements from the Advanced Very High-Resolution Ra-diometer. Remote Sens. Environ. 35, 257–277.

Hollinger, D., Aber, J., Dail, B., Davidson, E.A., Goltz, S.M., Hughes, H., Leclerc, M.Y., Lee, J.T., Richardson, A.D., Rodrigues, C., Scott, N.A., Achuatavarier, D. and Walsh, J. (2004) Spatial and temporal variability in forest-atmosphere CO_2 exchange. Global Change Biol. 10, 1689–1706.

Hollinger, D.Y., Goltz, S.M., Davidson, E.A., Lee, J.T., Tu, K. and Valentine, H.T. (1999) Seasonal patterns and environmental control of carbon dioxide and water vapour exchange in an ecotonal boreal forest. Global Change Biol. 5, 891–902.

Huete, A., Didan, K., Miura, T., Rodriguez, E.P., Gao, X. and Ferreira, L.G. (2002) Overview of the radiometric and biophysical performance of the MODIS vegetation indices. Remote Sens. Environ. 83, 195–213.

Huete, A.R., Didan, K., Shimabukuro, Y.E., Ratana, P., Saleska, S.R., Hutyra, L.R., Yang, W.Z., Nemani, R.R. and Myneni, R. (2006) Amazon rainforests green-up with sunlight in dry season. Geophys. Res. Lett. 33, L06405, doi:10.1029/2005GL025583.

Huete, A.R., Liu, H.Q., Batchily, K. and vanLeeuwen, W. (1997) A comparison of vegetation indices over a global set of TM images for EOS-MODIS. Remote Sens. Environ. 59, 440–451.

Hunt, E.R. and Rock, B.N. (1989) Detection of changes in leaf water-content using near-infrared and middle-infrared reflectances. Remote Sens. Environ. 30, 43–54.

Hunt, E.R., Rock, B.N. and Nobel, P.S. (1987) Measurement of leaf relative water-content by infrared reflectance. Remote Sens. Environ. 22, 429–435.

Jenkins, J.P., Braswell, B.H., Frolking, S.E. and Aber, J.D. (2002) Detecting and predicting spatial and interannual patterns of temperate forest springtime phenology in the eastern US. Geophys. Res. Lett. 29, 2201, doi:10.1029/2001GL014008.

Justice, C.O., Vermote, E., Townshend, J.R.G., Defries, R., Roy, D.P., Hall, D.K., Salomon-son, V.V., Privette, J.L., Riggs, G., Strahler, A., Lucht, W., Myneni, R.B., Knyazikhin, Y., Running, S.W., Nemani, R.R., Wan, Z.M., Huete, A.R., van Leeuwen, W., Wolfe, R.E., Giglio, L., Muller, J.P., Lewis, P. and Barnsley, M.J. (1998) The Moderate Resolution Im-aging Spectroradiometer (MODIS): Land remote sensing for global change research. IEEE Trans. Geosci. Rem. Sens. 36, 1228–1249.

Knyazikhin, Y., Martonchik, J.V., Myneni, R.B., Diner, D.J. and Running, S.W. (1998) Syn-ergistic algorithm for estimating vegetation canopy leaf area index and fraction of ab-sorbed photosynthetically active radiation from MODIS and MISR data. J. Geophys. Res. D103, 32257–32275.

Li, J., Yu, Q., Sun, X., Tong, X., Ren, C., Wang, J., Liu, E., Zhu, Z. and Yu, G. (2006) Carbon dioxide exchange and the mechanism of environmental control in a farmland ecosystem in North China Plain. Sci. China D 46, 226–240.

Lieth, H. (Ed.) (1974) Phenology and Seasonality Modeling. Springer, New York, pp. 444.

Myneni, R.B., Hoffman, S., Knyazikhin, Y., Privette, J.L., Glassy, J., Tian, Y., Wang, Y., Song, X., Zhang, Y., Smith, G.R., Lotsch, A., Friedl, M., Morisette, J.T., Votava, P., Ne-mani, R.R. and Running, S.W. (2002) Global products of vegetation leaf area and fraction absorbed PAR from year one of MODIS data. Remote Sens. Environ. 83, 214–231.

Myneni, R.B., Keeling, C.D., Tucker, C.J., Asrar, G. and Nemani, R.R. (1997) Increased plant growth in the northern high latitudes from 1981 to 1991. Nature 386, 698–702.

Myneni, R.B., Tucker, C.J., Asrar, G. and Keeling, C.D. (1998) Interannual variations in satellite-sensed vegetation index data from 1981 to 1991. J. Geophys. Res. D103, 6145–6160.

Myneni, R.B. and Williams, D.L. (1994) On the relationship between fAPAR and NDVI. Remote Sens. Environ. 49, 200–211.

Philippon, N., Jarlan, L., Martiny, N., Camberlin, P. and Mougin, E. (2007) Characterization of the interannual and intraseasonal variability of West African vegetation between 1982 and 2002 by means of NOAA AVHRR NDVI data. J. Clim. 20, 1202–1218.

Potter, C.S., Randerson, J.T., Field, C.B., Matson, P.A., Vitousek, P.M., Mooney, H.A. and Klooster, S.A. (1993) Terrestrial ecosystem production – a process model-based on global satellite and surface data. Global Biogeochem. Cycles 7, 811–841.

Prince, S.D. and Goward, S.N. (1995) Global primary production: A remote sensing approach. J. Biogeogr. 22, 815–835.

Richardson, A.D., Bailey, A.S., Denny, E.G., Martin, C.W. and O'Keefe, J. (2006) Phenology of a northern hardwood forest canopy. Global Change Biol. 12, 1174–1188.

Roberts, D.A., Dennison, P.E., Gardner, M.E., Hetzel, Y., Ustin, S.L. and Lee, C.T. (2003) Evaluation of the potential of Hyperion for fire danger assessment by comparison to the Airborne Visible/Infrared Imaging Spectrometer. IEEE Trans. Geosci. Rem. Sens. 41, 1297–1310.

Ruimy, A., Saugier, B. and Dedieu, G. (1994) Methodology for the estimation of terrestrial net primary production from remotely sensed data. J. Geophys. Res. D99, 5263–5283.

Sakamoto, T., Van Nguyen, N., Kotera, A., Ohno, H., Ishitsuka, N. and Yokozawa, M. (2007) Detecting temporal changes in the extent of annual flooding within the Cambodia and the Vietnamese Mekong delta from MODIS time-series imagery. Remote Sens. Environ. 109, 295–313.

Saleska, S.R., Miller, S.D., Matross, D.M., Goulden, M.L., Wofsy, S.C., da Rocha, H.R., de Camargo, P.B., Crill, P., Daube, B.C., de Freitas, H.C., Hutyra, L., Keller, M., Kirchhoff, V., Menton, M., Munger, J.W., Pyle, E.H., Rice, A.H. and Silva, H. (2003) Carbon in amazon forests: Unexpected seasonal fluxes and disturbance-induced losses. Science 302, 1554–1557.

Schwartz, M.D. (Ed.) (2003) *Phenology: An Integrative Environmental Science.* Kluwer, Dordrecht, The Netherlands, pp. 592.

Serrano, L., Ustin, S.L., Roberts, D.A., Gamon, J.A. and Penuelas, J. (2000) Deriving water content of chaparral vegetation from AVIRIS data. Remote Sens. Environ. 74, 570–581.

Shaw, R.H., Silversides, R.H. and Thurtell, G.W. (1974) Some observations of turbulence and turbulent transport within and above plant canopies. Bound.-Lay. Meteorol. 5, 429–449.

Stockli, R. and Vidale, P.L. (2004) European plant phenology and climate as seen in a 20-year AVHRR land-surface parameter dataset. Int. J. Remote Sens. 25, 3303–3330.

Studer, S., Stockli, R., Appenzeller, C. and Vidale, P.L. (2007) A comparative study of satel-lite and ground-based phenology. Int. J. Biometeorol. 51, 405–414.

Tucker, C.J. (1979) Red and photographic infrared linear combinations for monitoring vegetation. Remote Sens. Environ. 8, 127–150.

Van Schaik, C.P., Terborgh, J.W. and Wright, S.J. (1993) The phenology of tropical forests – adaptive significance and consequences for primary consumers. Ann. Rev. Ecol. Syst. 24, 353–377.

Vermote, E.F. and Vermeulen, A. (1999) Atmospheric correction algorithm: Spectral reflec-tance (MOD09), MODIS Algorithm Technical Background Document, version 4.0University of Maryland, Department of Geography, pp.107.

Waring, R.H., Law, B.E., Goulden, M.L., Bassow, S.L., Mccreight, R.W., Wofsy, S.C. and Bazzaz, F.A. (1995) Scaling gross ecosystem production at Harvard Forest with remote-sensing – a comparison of estimates from a constrained quantum-use efficiency model and eddy-correlation. Plant Cell Environ. 18, 1201–1213.

White, M.A., Hoffman, F., Hargrove, W.W. and Nemani, R.R. (2005) A global framework for monitoring phenological responses to climate change. Geophys. Res. Lett. 32, L04705.

White, M.A. and Nemani, A.R. (2003) Canopy duration has little influence on annual carbon storage in the deciduous broad leaf forest. Global Change Biol. 9, 967–972.

White, M.A. and Nemani, R.R. (2006) Real-time monitoring and short-term forecasting of land surface phenology. Remote Sens. Environ. 104, 43–49.

White, M.A., Running, S.W. and Thornton, P.E. (1999) The impact of growing-season length variability on carbon assimilation and evapotranspiration over 88 years in the eastern US deciduous forest. Int. J. Biometeorol. 42, 139–145.

White, M.A., Thornton, P.E. and Running, S.W. (1997) A continental phenology model for monitoring vegetation responses to interannual climatic variability. Global Biogeochem. Cycles 11, 217–234.

Wofsy, S.C., Goulden, M.L., Munger, J.W., Fan, S.M., Bakwin, P.S., Daube, B.C., Bassow, S.L. and Bazzaz, F.A. (1993) Net Exchange of CO_2 in a mid-latitude Forest. Science 260, 1314–1317.

Wright, S.J. and van Schaik, C.P. (1994) Light and the phenology of tropical trees. The Am. Nat. 143, 192–199.

Xiao, X., Boles, S., Frolking, S., Salas, W., Moore, B., Li, C., He, L. and Zhao, R. (2002a) Observation of flooding and rice transplanting of paddy rice fields at the site to landscape scales in China using VEGETATION sensor data. Int. J. Remote Sens. 23, 3009–3022.

Xiao, X., Boles, S., Liu, J.Y., Zhuang, D.F. and Liu, M.L. (2002b) Characterization of forest types in Northeastern China, using multi-temporal SPOT-4 VEGETATION sensor data. Remote Sens. Environ. 82, 335–348.

Xiao, X., Shu, J., Wang, Y., Ojima, D. and Bonham, C. (1996) Temporal variation in above-ground biomass of Leymus chinense steppe from species to community levels in the Xilin River basin, Inner Mongolia, China. Vegetation 123, 1–12.

Xiao, X.M., Braswell, B., Zhang, Q.Y., Boles, S., Frolking, S. and Moore, B. (2003) Sensitiv-ity of vegetation indices to atmospheric aerosols: continental-scale observations in North-ern Asia. Remote Sens. Environ. 84, 385–392.

Xiao, X.M., Hagen, S., Zhang, Q.Y., Keller, M. and Moore, B. (2006) Detecting leaf phenol-ogy of seasonally moist tropical forests in South America with multi-temporal MODIS images. Remote Sens. Environ. 103, 465–473.

Xiao, X.M., Hollinger, D., Aber, J., Goltz, M., Davidson, E.A., Zhang, Q.Y. and Moore, B. (2004a) Satellite-based modeling of gross primary production in an evergreen needleleaf for-est. Remote Sens. Environ. 89, 519–534.

Xiao, X.M., Wang, Y.F., Jiang, S., Ojima, D.S. and Bonham, C.D. (1995) Interannual Varia-tion in the Climate and Aboveground Biomass of Leymus-Chinense Steppe and Stipa-Grandis Steppe in the Xilin River Basin, Inner-Mongolia, China. J. Arid Environ. 31, 283–299.

Xiao, X.M., Zhang, Q.Y., Braswell, B., Urbanski, S., Boles, S., Wofsy, S., Moore, B. and Ojima, D. (2004b) Modeling gross primary production of temperate deciduous broadleaf forest using satellite images and climate data. Remote Sens. Environ. 91, 256–270.

Xiao, X.M., Zhang, Q.Y., Saleska, S., Hutyra, L., De Camargo, P., Wofsy, S., Frolking, S., Boles, S., Keller, M. and Moore, B. (2005) Satellite-based modeling of gross primary pro-duction in a seasonally moist tropical evergreen forest. Remote Sens. Environ. 94, 105–22.

Zhang, Q.Y., Xiao, X.M., Braswell, B., Linder, E., Baret, F. and Moore, B. (2005) Estimating light absorption by chlorophyll, leaf and canopy in a deciduous broadleaf forest using MODIS data and a radiative transfer model. Remote Sens. Environ. 99, 357–371.

Zhang, X.Y., Friedl, M.A. and Schaaf, C.B. (2006) Global vegetation phenology from Moder-ate Resolution Imaging Spectroradiometer (MODIS): Evaluation of global patterns and compari-son with in situ measurements. J. Geophys. Res. 111, G04017.

Zhang, X.Y., Friedl, M.A., Schaaf, C.B., Strahler, A.H., Hodges, J.C.F., Gao, F., Reed, B.C. and Huete, A. (2003) Monitoring vegetation phenology using MODIS. Remote Sens. En-viron. 84, 471–475.

Zhao, F.H., Yu, G.R., Li, S.G., Ren, C.Y., Sun, X.M., Mi, N., Li, J. and Ouyang, Z. (2007) Canopy water use efficiency of winter wheat in the North China Plain. Agric. Water Manage. 93, 99–108.

Index

Printed in the United States
By Bookmasters